Phenological Synchrony and Bird Migration

CRC Press
Taylor & Francis Group
Boca Raton London New York

CRC Press is an imprint of the
Taylor & Francis Group, an **informa** business

STUDIES IN AVIAN BIOLOGY

A Publication of The Cooper Ornithological Society

www.crcpress.com/browse/series/crcstdavibio

Studies in Avian Biology is a series of works published by The Cooper Ornithological Society since 1978. Volumes in the series address current topics in ornithology and can be organized as monographs or multi-authored collections of chapters. Authors are invited to contact the series editor to discuss project proposals and guidelines for preparation of manuscripts.

See complete series list on page [227].

Volume 47
Studies in Avian Biology
Cooper Ornithological Society

Phenological Synchrony and Bird Migration

Changing Climate and Seasonal Resources in North America

EDITED BY

Eric M. Wood
University of Wisconsin, Madison

Jherime L. Kellermann
Oregon Institute of Technology

CRC Press
Taylor & Francis Group
Boca Raton London New York

CRC Press is an imprint of the
Taylor & Francis Group, an **informa** business

Cover photo by Mike McDowell: Magnolia Warbler (*Setophaga magnolia*). Pheasant Branch, Wisconsin, May 17, 2014 (spring migration).

CRC Press
Taylor & Francis Group
6000 Broken Sound Parkway NW, Suite 300
Boca Raton, FL 33487-2742

First issued in paperback 2017

© 2015 by The Cooper Ornithological Society
CRC Press is an imprint of Taylor & Francis Group, an Informa business

No claim to original U.S. Government works

ISBN-13: 978-1-4822-4030-6 (hbk)
ISBN-13: 978-1-138-57578-3 (pbk)

Library of Congress Cataloging-in-Publication Data

Phenological synchrony and bird migration: changing climate and seasonal resources in North America /
    editors, Eric M. Wood and Jherime L. Kellermann.
        pages cm. -- (Studies in avian biology ; 47)
    "A CRC title."
    Includes bibliographical references and index.
    ISBN 978-1-4822-4030-6 (alk. paper)
        1. Birds--Migration--Climatic factors--North America. 2. Birds--Adaptation--North America. I. Wood,
Eric M., editor. II. Kellermann, Jherime L., editor.

QL698.9.P54 2015
598.156'80973--dc23   2014028997

Visit the Taylor & Francis Web site at
http://www.taylorandfrancis.com

and the CRC Press Web site at
http://www.crcpress.com

CONTENTS

Part IV • Fall Migration

CONTRIBUTORS

EVAN M. ADAMS
Biodiversity Research Institute
652 Main Street
Gorham, ME 04038
evan.adams@briloon.org

LORIANNE BARNETT
USA National Phenology Network
1955 E. 6th St.
Tucson, AZ 85721
lorianne@usanpn.org

DAVID N. BONTER
Cornell Lab of Ornithology
159 Sapsucker Woods Road
Ithaca, NY 14850
dnb23@cornell.edu

JAY D. CARLISLE
Intermountain Bird Observatory
Department of Biological Sciences
Boise State University
1910 University Drive
Boise, ID 83725
jaycarlisle@boisestate.edu

EMILY B. COHEN
Migratory Bird Center
Smithsonian Conservation Biology Institute
National Zoological Park
P.O. Box 37012-MRC 5503
Washington, DC 20013-7012
cohene@si.edu

RENÉE L. CORMIER
Point Blue Conservation Science
Petaluma, CA 94954
rcormier@pointblue.org

ROBERT DIEHL
US Geological Survey
Northern Rocky Mountain Science Center
Bozeman, MT 59715
rhdiehl@usgs.gov

ELIZABETH R. ELLWOOD
Department of Biological Science
Florida State University
Tallahassee, FL 32306
eellwood@fsu.bio.edu

CAROLYN A. F. ENQUIST
USA National Phenology Network
1955 E. 6th St.
Tucson, AZ 85721
carolyn@usanpn.org

DAVID N. EWERT
The Nature Conservancy
101 East Grand River
Lansing, MI 48906
dewert@tnc.org

JOSEPH J. FONTAINE
U.S. Geological Survey
Nebraska Cooperative Fish and Wildlife Research
    Unit
The School of Natural Resources
The University of Nebraska
Lincoln, NE 68583
jfontaine2@unl.edu

AMANDA GALLINAT
Department of Biology
Boston University
5 Cummington Mall
Boston, MA 02215
gallinat@bu.edu

LEONARD Z. GANNES
Department of Biology
Cornell College
Mount Vernon, IA 52314
lgannes@cornellcollege.edu

DAVID P. GRUNZEL
School of Biology & Ecology
University of Maine
5722 Deering Hall
Orono, ME 04469
david.grunzel@maine.edu

KIMBERLY R. HALL
The Nature Conservancy
101 East Grand River
Lansing, MI 48906
kimberly_hall@tnc.org

JULIE A. HEATH
Intermountain Bird Observatory
Department of Biological Sciences
Boise State University
1910 University Drive
Boise, ID 83725
julieheath@boisestate.edu

DIANA L. HUMPLE
Point Blue Conservation Science
Petaluma, CA 94954
dhumple@pointblue.org

GREGORY S. KALTENECKER
Intermountain Bird Observatory
Department of Biological Sciences
Boise State University
1910 University Drive
Boise, ID 83725
gregorykaltenecker@boisestate.edu

JHERIME L. KELLERMANN
School of Natural Resources and Environment
University of Arizona
Tucson, AZ 85721
USA National Phenology Network
1955 E. 6th St.
Tucson, AZ 85721
(Current address) Natural Sciences Department
Oregon Institute of Technology
3201 Campus Drive
Klamath Falls, OR 97601
jherime.kellermann@oit.edu

ADRIENNE LEPPOLD
School of Biology & Ecology
University of Maine
5722 Deering Hall
Orono, ME 04469
adrienne.leppold@umit.maine.edu

TREVOR L. LLOYD-EVANS
Manomet Center for Conservation Sciences
125 Manomet Point Road
Plymouth, MA 02360
tlloyd-evans@manomet.org

JENNIFER D. MCCABE
School of Biology & Ecology
University of Maine
5722 Deering Hall
Orono, ME 04469
jennifer.mccabe@umit.maine.edu

ROBERT A. MILLER
Intermountain Bird Observatory
Department of Biological Sciences
Boise State University
1910 University Drive
Boise, ID 83725
robertmiller7@u.boisestate.edu

FRANK R. MOORE
University of Southern Mississippi
Department of Biological Sciences
Hattiesburg, MS 36406
frank.moore@usm.edu

ZOLTÁN NÉMETH
University of California
Department of Neurobiology, Physiology and
    Behavior
Davis, CA 95616
znemeth05@gmail.com

BRIAN J. OLSEN
School of Biology & Ecology
University of Maine
5722 Deering Hall
Orono, ME 04469
brian.olsen@maine.edu

NEIL PAPROCKI
Hawkwatch International
2240 South 900 East
Salt Lake City, UT 84106
npaprock@wildlensinc.org

KRISTINA L. PAXTON
University of Southern Mississippi
Department of Biological Sciences
Hattiesburg, MS 36406
kristina.paxton@gmail.com

EBEN H. PAXTON
U.S. Geological Survey
Pacific Island Ecosystems Research Center
Hawaii Volcanoes National Park, HI 96718
Eben_Paxton@usgs.gov

ANNA M. PIDGEON
Department of Forest and Wildlife Ecology
University of Wisconsin-Madison
1630 Linden Drive
Madison, WI 53706
apidgeon@wisc.edu

RICHARD B. PRIMACK
Department of Biology
Boston University
5 Cummington Mall
Boston, MA 02215
primack@bu.edu

KARINE PRINCÉ
Department of Forest and Wildlife Ecology
University of Wisconsin-Madison
1630 Linden Drive
Madison, WI 53706
prince3@wisc.edu

PAUL G. RODEWALD
Cornell Lab of Ornithology
159 Sapsucker Woods Road
Ithaca, NY 14850
pgr35@cornell.edu

ALYSSA ROSEMARTIN
USA National Phenology Network
1955 E. 6th St.
Tucson, AZ 85721
alyssa@usanpn.org

ERIC J. ROSS
Department of Forest and Wildlife Ecology
University of Wisconsin-Madison
1630 Linden Drive
Madison, WI 53706
ejross3@wisc.edu

NATHANIEL E. SEAVY
Point Blue Conservation Science
Petaluma, CA 94954
nseavy@pointblue.org

ROBERT J. SMITH
Department of Biology
University of Scranton
Scranton, PA18510
robert.smith@scranton.edu

PAUL K. STRODE
Department of Natural Resources and
    Environmental Sciences
University of Illinois
Urbana-ChampaignW-503 Turner Hall
1102 South Goodwin Ave
Urbana, IL 61801
(Current address) 1565 Kendall Drive
Boulder, CO 80305
paul.strode@bvsd.org

RYAN J. STUTZMAN
Nebraska Cooperative Fish and Wildlife Research
    Unit
The School of Natural Resources
The University of Nebraska
Lincoln, NE 68583
ryanstutzman@hotmail.com

CHARLES VAN RIPER III
U.S. Geological Survey
Southwest Biological Science Center
Sonoran Desert Research Station
University of Arizona
Tucson, Arizona 85721
School of Natural Resources and Environment
University of Arizona
Tucson, AZ 85721
charles_van_riper@usgs.gov

ERIC M. WOOD
Department of Forest and Wildlife Ecology
University of Wisconsin-Madison
1630 Linden Drive
Madison, WI 53706
(Current address) Cornell Lab of Ornithology
159 Sapsucker Woods Road
Ithaca, NY 14850
ericmwood@cornell.edu

THEODORE J. ZENZAL JR.
University of Southern Mississippi
Department of Biological Sciences
Hattiesburg, MS 36406
tjzenzal@gmail.com

BENJAMIN ZUCKERBERG
Department of Forest and Wildlife Ecology
University of Wisconsin-Madison
1630 Linden Drive
Madison, WI 53706
bzuckerberg@wisc.edu

EDITORS

Dr. Eric M. Wood is currently a Postdoctoral Quantitative Ecologist in the Conservation Sciences Program at the Cornell Lab of Ornithology. His research investigates the impacts of land use and climate change on species' interactions with their environment. Dr. Wood has over 15 years experience in avian ecology, exploring questions related to spring and fall migration and the breeding period to better understand species–habitat interactions, community and landscape ecology, ecosystem services, and phenology. Dr. Wood began thinking about the relationships between habitat phenology and bird migration as a Master's degree student at Humboldt State University, California. During fieldwork in the northern Sierra Nevada foothills, he noticed apparent patterns of phenological synchrony of flowering blue oak trees (*Quercus douglasii*) and migratory foliage-gleaning birds, and began wondering about the importance of bird migratory arrival at stopover habitats in conjunction with peaks in ephemeral food resources. During Ph.D. work in Wisconsin, Dr. Wood continued developing these ideas and studied a similar phenomenon in Midwestern oak savanna and forested habitats. His projects have since been expanded to explore the effects of climate change and extreme weather on the phenological relationships of migratory birds and their seasonal resources. Dr. Wood is currently continuing with his phenology research and is expanding from his plot-level research to exploring questions at broader spatiotemporal scales.

Dr. Jherime L. Kellermann is currently an Assistant Professor in the Natural Sciences Department at the Oregon Institute of Technology and Science Coordinator at Crater Lake National Park Science and Learning Center. He has nearly 20 years of experience in research and conservation of birds and their habitats in America's Pacific Northwest, Southwest, and Hawaii, as well as the Caribbean and Latin America. After attaining his Master's degree at Humboldt State University, California, and Ph.D. in both wildlife conservation and management at the University of Arizona, Dr. Kellermann worked with the USA National Phenology Network, a national observation program to monitor phenology as a signal of climate change. His research interests include seasonal dynamics of animal migration and habitat ecology, the synchrony or mismatch of interacting species' phenology associated with climate variation and change, and how changing phenology may impact ecosystem services.

PREFACE

Phenology, or the study of plant and animal life-cycle episodes, is a well-studied and oft-admired science that has been an integral part of human culture for millennia. Polynesians used the reoccurrence of migrating Pacific Golden-Plovers (*Pluvialis fulva*) to discover remote islands, naturalists such as Aldo Leopold remarked at the phenology of plants and animals as the "arteries of the land," and Henry David Thoreau collected budburst and flowering data at Walden Pond that is aiding our understanding of historic and current phenological events. Though phenological observations have been conducted for centuries due to their ease of execution, recently there has been a renewed interest in phenology as an integrative science for monitoring the complex biological response to climate change and extreme weather. Nearly all phenological events are influenced by climate and weather, and it has been suggested that phenological monitoring provides a "yardstick" for documenting environmental change due to these factors (Visser and Both 2005). Yet, while there is strong evidence that timing of flowering and other phenological events are indeed shifting, there remains a limited understanding of how these changes impact the intricate relationships of migratory animals with ephemeral resources throughout the annual cycle. Additionally, it is unclear whether there are individual, population, or ecological consequences if these relationships become decoupled due to climate change. Thus, researchers and practitioners are challenged with understanding phenological events and relationships, under a variety of climatic conditions, in order to prioritize management and conservation.

Bird migration is arguably one of the most celebrated and well-studied phenological events, and there is evidence (mainly from Europe) that birds are shifting their migration behaviors due to climate change. However, while there has been a surge in phenology research over the past decade, there remains a paucity of studies for avian species and communities throughout North America. Here, we present this volume of Studies in Avian Biology to explore the critical linkages between migratory birds, their seasonal resources, and shifts in climate change and weather events. We crafted this volume to bring together recent research focused on bird migration and resource phenology in the United States, with a broader focus on birds that breed and winter throughout North America. While this volume in Studies in Avian Biology only scratches the surface of knowledge necessary to understand the effects of climate change on migratory birds properly, we intend for it to act as a compendium for the current state of bird migration and phenology research in North America.

The initial discussions for this project began in 2009. At the time, Jherime and I were Ph.D. students studying bird migration and the phenology of habitat resources in Arizona and Wisconsin, respectively. Upon realizing that we were conducting similar work, we began communicating regularly about similarities and differences we were observing in our research and that of others across dramatically different ecosystems of the United States. Most of our discussions about the current state of phenological monitoring and research acknowledged the richness of papers studying sub-Saharan migratory birds breeding in Europe. We quickly realized that there was a major gap with bird migration phenology research in North America. One of our colleagues and a fellow phenology researcher, Dr. Joseph J. Fontaine, shared a similar assessment, and he challenged us to fill a small portion of this gap in North America. Based on conversations in 2010 and 2011, we developed a road map of two related tasks to achieve this. The first was a special symposia at the fifth meeting of the North American Ornithological Conference in Vancouver, Canada, during August of 2012, titled "Tracking Migratory Stopover Phenology: Climate Change and the Phenological Synchrony of North American Bird Migration with Seasonal Resources." Here, we brought together 11 speakers, 10 based in North America and the other in Europe, presenting on topics ranging from asynchronous changes in phenology of migrating Broad-tailed Hummingbirds and their early-season nectar resources in Colorado and Arizona (speaker David W. Inouye) to historical migration phenology in Thoreau's Concord over 150 years (speaker Elizabeth R. Ellwood).

From this symposium, we organized this volume of Studies in Avian Biology, which was the second task of our road map. With our book, we have brought together over 40 researchers contributing 12 chapters from projects conducted during either spring or fall migration throughout North America. We organized the volume to begin with conservation and management considerations for migratory birds and climate change in both the western and midwestern portions of the United States. We also highlight an exciting new citizen science project—the USA–National Phenology Network Nature's Notebook program—and ways in which these data can be incorporated into conservation research. We then highlight two chapters relating climate on the wintering grounds to spring migration of both short- and long-distance migratory birds. These chapters fall under the theme of migratory connectivity, which aims to understand the carryover effects of events affecting individuals or populations during one distinct segment of the annual cycle to another. Next, we highlight linkages of birds, their seasonal resources, and climate and weather effects at stopover habitats in both spring and fall migration. The last two sections of this volume, on spring and fall migration, are generally focused on understanding synchronous relationships of migratory birds and their seasonal resources, as well as the nature of these relationships in the face of climate change or extreme weather events. At present, it remains unclear whether the decoupling of these synchronous relationships, often termed "phenological mismatch," has broader population level consequences or other deleterious ecological effects, such as disruption of ecosystem services; these chapters present baseline information for potentially exploring such questions in the future.

Our project has benefited from the help of many, including the following referees who offered valuable comments and suggestions for manuscript improvement. We warmly acknowledge F. Beaudry, R. Churchwell, J. Deppe, A. Donnelly, M. Johnson, L. Pomara, C. Rittenhouse, J. Ruth, S. Skagen, and C. Tonra. Furthermore, authors of each chapter of this volume also provided blind comments, and we thank them for their assistance and careful comments. We are grateful to C. Crumly, K. Gallo, and H. Ruggieri of CRC Press, who helped with editorial and publication guidelines. B. K. Sandercock was invaluable with editorial and practical advice and also played an integral role in shifting Studies in Avian Biology between publishers. We also extend a special thanks to him for inviting a proposal from us for this book project. This volume would not have come to fruition if it were not for his helpful and courteous guidance, which he kindly offered during all phases of manuscript preparation.

LITERATURE CITED

Visser, M. E., and C. Both. 2005. Shifts in phenology due to global climate change: the need for a yardstick. Proceedings of the Royal Society of London B 272:2561–2569.

PART ONE

Conservation and Management

CHAPTER ONE

Leaps, Chains, and Climate Change
for Western Migratory Songbirds*

Joseph J. Fontaine, Ryan J. Stutzman, and Leonard Z. Gannes

Abstract. Climate change has increased worldwide temperatures, affected seasonal patterns, and altered important sources of natural selection. To manage wildlife populations successfully, we must understand how patterns and processes of climate change alter trade-offs between sources of selection to predict how individuals may respond, populations may evolve, and management actions may ameliorate the costs of changing climates. Here we discuss how the migratory patterns of leapfrog and chain migration facilitate or constrain responses by migratory songbirds to spatial and temporal variation in climate change across western North America. Based on 52 years of climate data, we show that changes in average minimum monthly temperature differ significantly between the spring migration zone in the desert Southwest and breeding locations throughout western North America, and that these differences are most extreme for populations breeding at low latitudes (37°–49°) and exacerbated for species exhibiting leapfrog migration. Given the importance of climate in the evolution of migratory behaviors, such extreme alterations in the geographical patterns of climate may ultimately threaten the long-term population viability of species dependent on low latitudes for breeding or exhibiting leapfrog migration.

Key Words: chain migration, climate change, leapfrog migration, phenology, stopover habitat.

In response to recent changes in global climate, recognizing the degree to which species react to changes in seasonality is an area of increasing conservation concern, as species that are unable to respond are presumed to be at increased risk of extinction (IPCC 2001, Sæther et al. 2004, Rosenzweig et al. 2008). Some species are responding to changes in seasonality by altering phenology to ensure that annual life cycles coincide with optimal ecological conditions (Walther et al. 2002, Parmesan and Yohe 2003, Root et al. 2003). However, while many plant and insect populations have shown shifts in phenology, higher order consumers have responded to a lesser degree, increasing costs to individuals and ultimately impacting populations (Visser et al. 1998, Both and Visser 2001, Both et al. 2006).

In migratory birds, numerous accounts highlight phenological shifts in response to changes in seasonality (Crick et al. 1997, Hüppop and Hüppop 2003, Jenni and Kéry 2003, Lehikoinen et al. 2004, Stervander et al. 2005, Jonzén et al. 2006,

* Fontaine, J. J., R. J. Stutzman, and L. Z. Gannes. 2015. Leaps, chains, and climate change for western migratory songbirds. Pp. 3–15 in E. M. Wood and J. L. Kellermann (editors). Phenological synchrony and bird migration: changing climate and seasonal resources in North America. Studies in Avian Biology (no. 47), CRC Press, Boca Raton, FL.

Tøttrup et al. 2006), but considerable variation remains among species and even populations (Inouye et al. 2000, Both and Visser 2001, Strode 2003, Gordo et al. 2005, Rubolini et al. 2007, Weidinger and Král 2007, Wilson 2007, Møller et al. 2008, Askeyev et al. 2010, Both 2010). Recent research has begun to address the discrepancies, but has largely focused on aspects related to conditions at breeding grounds alone (Gordo 2007, Lehikoinen and Sparks 2010, but see Kanuscak et al. 2004, Both 2010). Due to their implicit dependence on spatially, temporally, and climatically separated habitats, understanding phenological responses of migratory species requires careful consideration of the effects of climate change across multiple temporal and geographic scales (Calvert et al. 2009). Moreover, because conditions during any period of the migratory cycle have implications for subsequent periods (Visser et al. 2004, Bearhop et al. 2005, Lehikoinen et al. 2006), particular attention must be paid to how changing climatic conditions at locations throughout the migratory cycle are interrelated to understanding how individuals and species will respond (Hedenström et al. 2007, Both 2010). For example, warming spring conditions at breeding grounds may favor earlier arrival and reproduction to optimize resource availability for the young (Crick et al. 1997, Both and te Marvelde 2007), but the benefits must be weighed against the cost of advancing migration and the corresponding timing of stopover events (Alerstam 1991, Both 2010). Shifts in the timing or duration of migration must ensure that birds are able to obtain adequate energy resources while avoiding predation and environmental hazards to arrive ultimately at breeding locations at the optimal time (Alerstam and Lindström 1990, Moore et al. 1995, Klaasen 1996, Gannes 2002, Moore et al. 2005). Given the degree of heterogeneity in climate change across landscapes and differences in the responses of local communities (IPCC 2001, Rosenzweig et al. 2008), changes in resource phenology due to climate change may differ greatly at stopover and breeding habitats, leading to differential selection, which may limit the response of individuals and populations to conditions at breeding grounds (Fontaine et al. 2009, Both 2010). However, despite the potential importance of migration in limiting populations (Moore et al. 1995, 2005) and clear evidence that climate

change is both spatially and temporally heterogeneous (IPCC 2001, Rosenzweig et al. 2008), we know surprisingly little about how climates are changing across the vast geographical range that migrants occupy (Fontaine et al. 2009, Both 2010), which species and populations are most at risk (Møller et al. 2008), and how migratory patterns and behaviors may influence responses to changing climatic conditions (Petersen 2009).

As an important step in addressing these issues, we examined how the expression of two common migratory patterns, chain and leapfrog migration (Figure 1.1), either constrain or facilitate responses to heterogeneity in climate change for songbirds migrating across western North America. We developed a series of theoretical models that consider variation in the timing and distance of migration to assess how discordance in climate change between migratory regions in the desert Southwest and breeding regions throughout western North America may impact species that express different migratory patterns. Relating spatial and temporal variation in climate change to the timing, distance, and patterns of migration may help to elucidate the overall costs of climate change to individuals and to identify species and populations of particular conservation concern.

METHODS

We gathered unadjusted data from the US Historical Climatology Network (Williams et al. 2007), Alaska Climate Research Center (2009), and Canadian National Climate Data and Information Archive (2009) for 143 weather stations representing 13 states, two provinces and two territories (Figure 1.2). To minimize missing data, we limited our analysis to monthly climate data from 1954 to 2006 for the months of March–June. We were interested in assessing the potential for mismatches between migratory phenology and resource phenology across the spring migratory period, and we limited our analysis to changes in temperature alone. Precipitation clearly plays an important role in the dynamics of the arid systems, particularly as it pertains to productivity and diversity (Sharifi et al. 1988, Bowers 2005, Miranda et al. 2011), but the role of precipitation in phenological events is more ambiguous (Cleland et al. 2007). While precipitation may influence phenology (Llorens and Penuelas 2005,

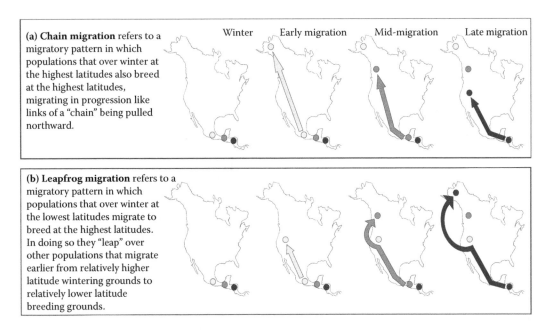

Figure 1.1. Songbird species exhibit a variety of migratory strategies, including (a) chain migration and (b) leapfrog migration. In western North America, species as diverse as Yellow Warblers (*Setophaga petechia*), White-crowned Sparrow (*Zonotrichia leucophrys*), Fox Sparrow (*Passerella iliaca*), and Wilson's Warbler (*Cardellina pusilla*) exhibit leapfrog migration, while American Coot (*Fulica americana*), Sharp-shinned Hawk (*Accipiter striatus*), Hermit Thrush (*Catharus guttatus*), and American Redstart (*Setophaga ruticilla*) are all known to exhibit chain migration. Unfortunately, the migratory strategy of many species remains unknown.

Crimmins et al. 2010), most examples outside of studies of monsoon cycles suggest interactions with temperature (Post and Stenseth 1999, Inouye 2008, Crimmins et al. 2010). Even in highly arid environments, temperature is a more reliable predictor of phenological events (Penuelas et al. 2002, Gordo and Sanz 2005). Moreover, the interannual and geographic variability of precipitation in arid systems is highly dynamic and unpredictable, making precipitation an unreliable climatic condition upon which to evolve a general migratory strategy. Thus, for each station we recorded the latitude, longitude, elevation, and average minimum monthly temperature for each month over the 52 years. We focused on minimum temperature because, across a wide array of ecosystems, changes in minimum temperature are known to influence plant phenology (Crimmins et al. 2008), and thereby the cues some songbirds use to make settlement decisions (McGrath et al. 2009). More importantly, arthropod phenology, and thus the majority of food resources for migratory songbirds, exhibits a threshold response to temperature whereby development and proliferation cease below a given temperature (reviewed by Honěk

1996). Changes in minimum temperature, rather than average temperature, or other climatic conditions such as precipitation are therefore a more relevant estimation of the potential influence of climate change on songbird food resource phenology across the wide range of biomes present in western North America.

Utilizing complete case regression analysis (Haitovsky 1968), we estimated the slope for the change in minimum temperature over the 52-year period for each month, at each climate station (hereafter referred to as the rate of warming). We assigned months as either spring migration (March–May) or breeding arrival (April–June) based on generalities about when western songbirds migrate and arrive at breeding locations. Arrival, which is strongly correlated with the onset of breeding (Moore et al. 2005, but see Ahola et al. 2004), appears sensitive to changing climatic conditions (Jonzén et al. 2006) and has important fitness implications (Moore et al. 2005, Decker and Conway 2009). We also divided the climate stations into latitudinal bands based on generalities about where western songbirds are during migration (desert Southwest: 26°–35°,

Figure 1.2. The distribution of 143 weather stations representing 13 states, two provinces, and two territories. All analyses were performed on unadjusted data from weather stations managed by the US Historical Climatology Network, Alaska Climate Research Center, and Canadian National Climate Data and Information Archive. Data were divided into four latitudinal bands representing high (▲, n = 23), mid (●, n = 35), and low (★, n = 50) latitude breeding populations as well as the migratory zone (◆, n = 35) used by all populations during the spring migration.

n = 35) versus breeding (western North America: 37°–72°), and further subdivided breeding areas into low (37°–48°, n = 50), mid (49°–59°, n = 35), and high (≥60°, n = 23) latitudinal bands to examine more closely how migration distance may influence the relationship between warming at migratory and breeding locations. Latitudinal bands were selected because they explained significant variation in the rate of warming across latitudes and because they represent important geopolitical zones with the low band representing the continental United States, the mid band the Canadian provinces and the panhandle of Alaska, and the high band the northern territories and the rest of Alaska (Figure 1.2). Choosing latitudinal bands based on climatological and geopolitical importance not only may help elucidate important biological phenomena, but also may facilitate management responses in geographically specific regions. To estimate the importance of spatial autocorrelation, we calculated Moran's I for the 16 possible latitudinal band months.

Utilizing the latitudinal bands to represent potential breeding populations, we developed a series of models to examine the relative rate of warming at migration and breeding locations based on all possible months of migration and arrival across western North America. First, we tested for spatial and temporal patterns using a global GLM (generalized linear model) that included month as a factor; latitude, longitude, and elevation as covariates; and the rate of warming as the dependent variable. We then assigned each station to a latitudinal band and added this categorical variable to the model to ensure that the general patterns continued to be representative.

After testing for the overall effect of month and latitudinal band, we compared the relative rate of warming of migratory habitats in the desert Southwest to breeding habitats throughout western North America by estimating the difference in the rate of warming between latitudinal bands (breeding − $\bar{x}$ migration). This comparison produces eight new values for each of the breeding location weather stations representing the difference in the rate of warming for each of three possible migration and arrival months, but excluding the difference between May migration and April arrival as an impossible event. For example, to compute the difference in the rate of warming for a bird that migrates through the desert Southwest in April and arrives at a low latitude breeding area in May, we subtract the average rate of warming for the migratory zone of the desert Southwest in April from each of the low latitude climate stations. By repeating this process we were able to estimate the difference in the rate of warming for each breeding habitat weather station for all possible months of migration and arrival. Using the differences as the dependent variable, we developed an additional global GLM model that included migration month, arrival month, and latitudinal band as factors and latitude, longitude, and elevation as covariates to test for the overall effect of each categorical variable and the potential interactions on the relative rate of warming between migration and breeding habitats.

Last, to test the potential for leapfrog or chain migration either to constrain or to facilitate responses to spatial and temporal variation in climate change, we developed a simple model representing each migration type. We made assumptions about how the timing of migration and migration distance interact, such that chain migration is represented by March migrants arriving at high latitude breeding locations in April, April migrants arriving at mid latitudes in May, and May migrants arriving at low latitudes in June (Figure 1.1). In contrast, leapfrog migration is represented by March migrants arriving at low latitude breeding locations in April, April migrants arriving at mid latitudes in May, and May migrants arriving at high latitudes in June. Although the timing of migration and arrival events is theoretical, it is based on the limited information available on migratory patterns of species that stop over in the Southwest (Finch and Yong 2000, Skagen et al. 2005, Paxton et al. 2007, Carlisle et al. 2009, McGrath et al. 2009, Delmore et al. 2012) and general information about the timing of the breeding season at different latitudes throughout North America (Cooper et al. 2005). Using the values from the differences between breeding and migratory locations for each weather station, we tested whether the average difference

in the rate of warming was greater for leapfrog or chain migration using a global GLM that included migration type as a factor and longitude, latitude, and elevation as covariates.

RESULTS

After breaking the 143 weather stations into latitudinal bands and considering the rate of warming for each of the months, only 4 of 16 latitudinal band months showed significant ($P < 0.05$) spatial autocorrelation as measured by Moran's index. Given the continuing debate over the benefits of correcting for spatial autocorrelations (Diniz-Filho et al. 2003, Hawkins et al. 2007, Beale et al. 2010) and the relatively weak correlations we found (Moran's I varied from 0.35 to −0.05), all further tests were run on uncorrected data. Over the 52-year period, we found that the rate of warming increased seasonally ($F_{3,428} = 18.72$, $P < 0.001$) and was sensitive to geographic location, but not elevation ($F_{1,428} = 2.18$, $P = 0.141$), with higher latitude ($F_{1,428} = 11.66$, $P = 0.001$) and more westerly locations ($F_{1,428} = 62.95$, $P < 0.001$) warming at a faster rate. When we divided the data into latitudinal bands, the latitudinal effect was lost ($F_{1,428} = 0.67$, $P = 0.415$), and replaced by the significant effect of the bands ($F_{3,428} = 7.95$, $P < 0.001$), indicating that much of the latitudinal variation was represented in the latitudinal bands. Longitude ($F_{1,428} = 66.68$, $P < 0.001$) and month ($F_{3,428} = 17.20$, $P < 0.001$) continued to influence the rate of warming, but there was also a significant interaction between month and latitudinal band ($F_{6,428} = 9.50$, $P < 0.001$).

Based on the estimated marginal means after controlling for elevation, longitude, and latitude, the average increase in minimum temperature for the 35 climate stations located in the desert Southwest was March: $2.35°C \pm 0.28°C$; April: $1.74°C \pm 0.28°C$; and May: $2.40°C \pm 0.28°C$. Using these estimates, we calculated the difference in the rate of warming for all climate stations throughout the three latitudinal bands. Differences in the rate of warming between migration and breeding locations were sensitive to temporal patterns of migration ($F_{2,863} = 33.16$, $P < 0.001$) and arrival ($F_{2,863} = 80.57$, $P < 0.001$). There continued to be a longitudinal pattern ($F_{1,863} = 137.79$, $P < 0.001$), but overall, the model showed limited spatial sensitivity (latitude: $F_{1,863} = 3.17$, $P = 0.075$;

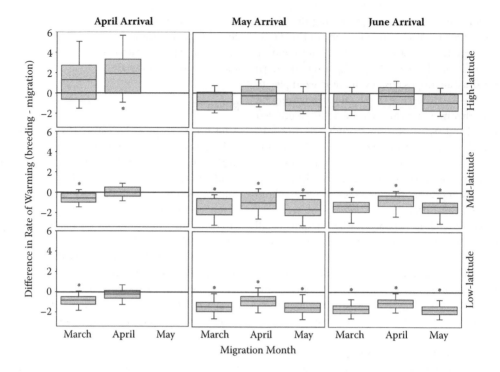

Figure 1.3. Significant differences in the rate of warming between migration and breeding habitats were shown by 15 of 24 hypothetical populations. When compared to the average rate of warming during migration, the rate of warming for breeding habitats was most consistently different for populations breeding at low latitudes and showed the smallest differences for populations breeding at high latitude and migrating in April. Box plots represent 5th, 25th, median, 75th, and 95th percentiles for the relationship between climate change at migration and breeding locations (breeding − $\bar{x}$ migration), where zero indicates breeding and migration locations are changing at the same rate, negative numbers indicate migration locations are warming faster, and positive numbers indicate breeding locations are warming faster. Boxes marked with asterisks are significantly different from zero at P ≤ 0.001 level according to independent one-sample t-tests.

elevation: $F_{1,863} = 2.02$, $P = 0.155$; latitudinal band: $F_{2,863} = 2.45$, $P = 0.087$). There was, however, a strong interaction between spatial and temporal components resulting in 15 of 24 theoretical populations experiencing discordance in the relative rate of warming between breeding and migratory locations (Figure 1.3; migration month by arrival month by latitudinal band: $F_{17,863} = 2.35$, $P = 0.002$). Moreover, this interaction resulted in a significant difference in the relative rates of warming experienced by the different migration patterns with the theoretical populations expressing leap-frog migration experiencing greater discordance in the rate of warming between migration and breeding locations (Figure 1.4; migration type: $F_{1,215} = 5.21$, $P = 0.024$; longitude: $F_{1,215} = 40.40$, $P < 0.001$; latitude: $F_{1,215} = 0.06$, $P = 0.814$; elevation: $F_{1,215} = 0.47$, $P = 0.492$).

DISCUSSION

Avian migration is among the most well studied of phenological characteristics, yet despite our wealth of knowledge, we still have only a cursory understanding of the mechanisms underlying the ability of migratory birds to respond to changing climatic conditions (Møller et al. 2008). While we might assume that warmer spring temperatures will impose strong selection for earlier arrival at breeding locations, which in turn would lead to changes in the timing or duration of spring migration, such assumptions fail to consider the importance of selection during migration and, ultimately, the trade-offs that occur between advancing breeding phenology and advancing migratory phenology (Ahola et al. 2004, Kanuscak et al. 2004, Both 2010). Given that migratory

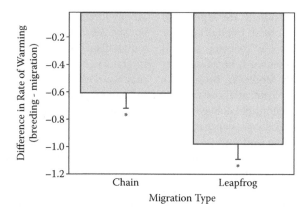

Figure 1.4. The relative discordance in climate change differs between chain and leapfrog migration. Migration type significantly influenced the relative difference in warming between breeding and spring migration habitats, with leapfrog migration seemingly experiencing greater discordance. Columns represent estimated marginal means (± SE) of the difference in warming at migration and breeding locations (breeding − $\bar{x}$ migration) after correcting for longitude, latitude, and elevation, where zero indicates breeding and migration locations are changing at the same rate, negative numbers indicate *migration* locations are warming faster, and positive numbers indicate *breeding* locations are warming faster. Columns marked with asterisks are significantly different from zero at $P \leq 0.001$ level according to independent one-sample t-tests.

birds are dependent upon spatially, temporally, and climatically separated habitats, successful management of migratory species requires careful consideration of the effects of climate change across multiple temporal and geographic scales (Calvert et al. 2009).

As an important step in understanding the potential implications of climate change on western migratory songbirds, we considered how spatial and temporal variation in climate change may interact with patterns and processes of migration to constrain or facilitate individual and population responses to changing climatic conditions. Our findings show that despite consistent increases in temperature throughout western North America, the relative rate of warming varied widely among locations and months, leading to significant differences in the rate of warming at migration versus breeding locations for 15 of 24 theoretical populations. While the long-term impact of climate change on migratory bird populations remains unclear, differences in the rate of climate change between migratory and breeding locations are likely to have significant fitness consequences that may constrain responses to climate change per se (Ahola et al. 2004). If we assume that current migratory patterns are adaptive, such that populations have evolved to

optimize the timing and duration of migration to coincide with peaks in local phenology (McGrath et al. 2009), then discordance in climate change makes adapting to new conditions more challenging as individuals are forced to balance differential selection pressures among locations. For example, if warming (and thus local phenology) is advancing more rapidly at migratory locations than breeding locations, then individuals must:

1. Migrate earlier to optimize food availability en route (McGrath et al. 2009), but arrive at breeding grounds when food is limited and risk of severe weather is high (Decker and Conway 2009),

2. Time migration to optimize food availability at breeding grounds (Martin 1987, Moore et al. 2005), but after food availability has peaked en route,

3. Extend the migratory period to optimize arrival at all locations, but be exposed to increase risk of predation and severe weather (Moore et al. 2005), or

4. Alter migratory routes to optimize local phenology en route, but be exposed to novel habitats, food resources, and predators.

For each scenario, individuals are exposed to trade-offs that have important implications for migratory bird populations by reducing survival en route, reproductive potential at breeding locations, or potentially both. Moreover, because the trade-offs are in addition to potential costs associated with advancing migratory phenology per se, the potential impacts are likely greater than if climate change were occurring uniformly.

There are important implications for how spatial and temporal variation in climate change may affect the different theoretical populations that we examined. For example, populations breeding at the highest latitudes consistently show the least discordance in the rate of climate change they experience, while low latitude populations show the greatest discordance. This finding may offer an explanation as to why there appears to be a latitudinal gradient in the phenological response of migratory bird populations to changing climates (Sparks et al. 2005, Hüppop and Winkel 2006, Rubolini et al. 2007). While this phenomenon may simply reflect the corresponding latitudinal gradient in warming (IPCC 2001, Rosenzweig et al. 2008), our findings suggest that high latitude populations not only are under the strongest selection from warming at breeding locations, but also are the least constrained by patterns of warming during migration as local phenology is presumably advancing at a similar rate between migratory and breeding habitats. In contrast, populations breeding at low latitudes face the weakest strength of selection due to limited warming at breeding locations, while also experiencing the greatest discordance in the rate of warming between migratory and breeding locations. The apparent latitudinal gradient in the degree of discordance in warming between high and low latitude breeding populations may facilitate responses to climate change at high latitudes while simultaneously constraining responses at low latitudes. The resulting paradox is that even though climate change is less extreme, birds breeding at low latitudes may face more extreme costs relating to the relative timing and duration of migration and breeding, which may ultimately constrain their ability to respond to even minor changes in climate. As for high latitude breeding populations, there appears to be a much tighter correlation between the rate of warming at migratory and breeding habitats, but there may also be greater variation in the degree of discordance

among locations within the high latitudinal band. For species that exhibit limited site fidelity or have a high dispersal potential, as may be expected for migratory birds, such variation in discordance may again ultimately constrain adaptive responses to changing climatic conditions.

Although the discordance between climate change at breeding and migratory locations was most extreme for populations breeding at low latitudes, within this region, populations arriving earlier faced less discordance. In contrast, high-latitude populations showed the opposite pattern with populations arriving later showing the least degree of discordance. In concert, the patterns are significant because they highlight the potential for strong directional selection on arrival timing beyond simply considering warming at breeding locations alone. Indeed, in the case of populations breeding at high latitudes, individuals arriving earlier face the greatest discordance in climate, thus, we might expect selection to favor delayed arrival—opposite what is expected from models that consider the effects of warming at breeding locations independently from warming at migratory locations. In contrast, the seasonal increase in the relative degree of discordance for populations breeding at low latitudes favors earlier arrival by migrants and therefore acts additively to what is expected from models that consider climate change at breeding locations alone.

Last, when we considered how patterns of migration might affect the discordance in the rate of warming between migration and breeding locations, we found that migration pattern may play an important role, with the relative rates of warming differing significantly between leapfrog and chain migration. Given the simplicity of our models, any relationship between migratory pattern and the potential costs of spatial and temporal variation in climate change is suggestive that the type of migration pattern a species exhibits has significant implications on the ability of that species to respond to changing climatic conditions. Indeed, our findings suggest that species exhibiting chain migration may be more resilient in the face of changing climates, a hypothesis that to our knowledge has not been previously tested and thus requires further exploration.

Here, we demonstrate that rates of warming vary substantially among locations occupied during the migratory cycle of western songbirds, and in doing so we highlight a nonintuitive source of selection that may shape the phenology of multiple avian life-history traits. While in some instances spatial and temporal variation in climate change appears to facilitate responses to warming climates, in other instances it clearly constrains both evolutionary and phenotypic adaptation. Such complexity likely underlies the significant variation in the ability of migratory birds to respond to climate change, as has been previously noted by many authors (Crick et al. 1997, Inouye et al. 2000, Both and Visser 2001, Hüppop and Hüppop 2003, Jenni and Kéry 2003, Strode 2003, Lehikoinen et al. 2004, Gordo et al. 2005, Stervander et al. 2005, Jonzén et al. 2006, Tøttrup et al. 2006, Weidinger and Král 2007, Wilson 2007, Møller et al. 2008).

Variation in the response of migratory birds to climatic trends will likely continue to challenge conservation efforts (Parmesan 2007), but by exploring how spatial and temporal variation in climate change impacts migratory birds and how behavioral and life-history strategies both constrain and facilitate responses to climate change, there exists the possibility that research efforts may elucidate the overall costs to individuals and help identify species and populations of particular conservation concern. Although our findings only illustrate theoretical populations, they represent a key first step into developing proactive management strategies to mitigate climate change impacts. For example, in many species there are significant differences in the phenology, duration, and distance of migration among age classes and between sexes (reviewed by Cristol et al. 1999). In the American West, taxa as diverse as songbirds (e.g., Hermit Thrush, *Catharus guttatus*), raptors (e.g., Red-tailed Hawk, *Buteo jamaicensis*), shorebirds (e.g., Western Sandpiper, *Calidris mauri*), and even waterfowl (e.g., Mallard, *Anas platyrhynchos*) exhibit age- and/or sex-specific migratory segregation (Pattenden and Boag 1989, Cristol et al. 1999, Mueller et al. 2000, Stouffer and Dwyer 2003, Bishop et al. 2004). Although such strategies are assumed to be adaptive (Cristol et al. 1999), differential costs of climate change among sexes or age classes due to differences in migration distance or phenology may lead to differential selection within the same population, potentially decoupling evolved adaptive strategies. Unfortunately, excluding Mallards and Northern Pintail (*Anas acuta*), few migratory birds are managed for age- or sex-specific demographic parameters. The resulting outcome could manifest in reduced effective population size or reproductive potential, or altered population or breeding age structure. Similarly, races or populations of the same species also experience different sources of selection due to variation in the rate of climate change, but potential impacts appear particularly high for species that exhibit leapfrog migration.

Differential impacts of climate change represent a unique challenge as populations are rarely managed independently, and while we know the migration patterns for some species, migratory patterns for many species continue to remain an enigma (Blanchard 1941, Phillips 1951, Ryder 1963, Bell 1997, Clegg et al. 2003, Smith et al. 2003, Norris et al. 2006). These problems are further exacerbated because although species are often the target, management actions are carried out locally. Our findings suggest that, without an understanding of the migratory patterns of the species, and more specifically, the local population, it may prove difficult to mitigate against climate change impacts through local management actions alone. Moreover, because most management actions, even species-specific actions, affect the entirety of the community, an understanding of the migratory patterns of various species representing a multitude of populations may be necessary truly to understand and manage climate change impacts on migratory bird communities.

ACKNOWLEDGMENTS

We thank J. Kellermann and E. Wood for their efforts in bringing together this volume of Studies in Avian Biology and K. Decker, S. Skagen, and two anonymous reviewers for comments on previous versions of this manuscript. This work was supported by funding from the US Geological Survey National Climate and Wildlife Science Center; T & E, Inc.; the Great Plains Landscape Conservation Cooperative; and the Rainwater Basin Joint Venture. Any use of trade, firm, or product names is for descriptive purposes only and does not imply endorsement by the US government. The Nebraska Cooperative Fish and Wildlife Research Unit is supported by a cooperative agreement among the US Geological Survey, the Nebraska Game and Parks Commission, the University of Nebraska, the US Fish and Wildlife Service, and the Wildlife Management Institute.

LITERATURE CITED

Ahola, M., T. Laaksonen, K. Sippola, T. Eeva, K. Rainio, and E. Lehikoinen. 2004. Variation in climate warming along the migration route uncouples arrival and breeding dates. Global Change Biology 10:1610–1617.

Alaska Climate Research Center. [online]. 2009. Climatological data—Monthly time series. <http://climate.gi.alaska.edu/Climate/Location/TimeSeries/index.html> (1 June 2009).

Alerstam, T. 1991. Bird migration. Cambridge University Press, Cambridge, UK.

Alerstam, T., and A. Lindström. 1990. Optimal bird migration: The relative importance of time, energy, and safety. Pp. 331–351 in E. Gwinner (editor), Bird migration: Physiology and ecophysiology. Springer–Verlag, Berlin, Germany.

Askeyev, O., T. Sparks, I. Askeyev, D. Tishin, and P. Tryjanowski. 2010. East versus West: Contrasts in phenological patterns? Global Ecology and Biogeography 19:783–793.

Beale, C. M., J. J. Lennon, J. M. Yearsley, M. J. Brewer, and D. A. Elston. 2010. Regression analysis of spatial data. Ecology Letters 13:246–264.

Bearhop, S., W. Fiedler, R. W. Furness, S. C. Votier, S. Waldron, J. Newton, G. J. Bowen, P. Berthold, and K. Farnsworth. 2005. Assortative mating as a mechanism for rapid evolution of a migratory divide. Science 310:502–504.

Bell, C. P. 1997. Leapfrog migration in the Fox Sparrow: Minimizing the cost of spring migration. Condor 99:470–477.

Bishop, M. A., N. Warnock, and J. Y. Takekawa. 2004. Differential spring migration by male and female Western Sandpipers at interior and costal stopover sites. Ardea 92:185–196.

Blanchard, B. D. 1941. The White-crowned Sparrows Zonotrichia leucophrys of the Pacific seaboard: Environment and annual cycle. University of California Publications in Zoology 46:1–178.

Both, C. 2010. Flexibility of timing of avian migration to climate change masked by environmental constraints en route. Current Biology 20:243–248.

Both, C., S. Bouwhuis, C. M. Lessells, and M. E. Visser. 2006. Climate change and population declines in a long-distance migratory bird. Nature 441:81–83.

Both, C., and L. te Marvelde. 2007. Climate change and timing of avian breeding and migration throughout Europe. Climate Research 35:93–105.

Both, C., and M. E. Visser. 2001. Adjustment to climate change is constrained by arrival date in a long-distance migrant bird. Nature 411:296–298.

Bowers, J. 2005. El Niño and displays of spring-flowering annuals in the Mojave and Sonoran Deserts. Journal of the Torrey Botanical Society 132:38–49.

Calvert, A. M., P. D. Taylor, and S. Walde. 2009. Cross-scale environmental influences on migratory stopover behavior. Global Change Biology 15:744–759.

Canadian National Climate Data and Information Archive. [online]. 2009. Canadian climate data online customized search. <http://climate.weatheroffice.ec.gc.ca> (1 June 2009).

Carlisle, J. D., S. K. Skagen, B. E. Kus, C. van Riper III, K. L. Paxton, and J. F. Kelly. 2009. Land bird migration in the American West: Recent progress and future research directions. Condor 111:211–225.

Clegg, S. M., J. F. Kelly, M. Kimura, and T. B. Smith. 2003. Combining genetic markers and stable isotopes to reveal population connectivity and migration patterns in a Neotropical migrant, Wilson's Warbler (Wilsonia pusilla). Molecular Ecology 12:819–830.

Cleland, E. E., I. Chuine, A. Menzel, H. A. Mooney, and M. D. Schwartz. 2007. Shifting plant phenology in response to global change. Trends in Ecology and Evolution 22:357–365.

Cooper, C. B., W. M. Hochachka, and A. A. Dhondt. 2005. Latitudinal trends in within-year reoccupation of nest boxes and their implications. Journal of Avian Biology 36:31–39.

Crick, H. Q. P., C. Dudley, D. E. Glue, and D. L. Thomson. 1997. UK birds are laying eggs earlier. Nature 388:526–527.

Crimmins, T. M., M. A. Crimmins, and C. D. Bertelsen. 2010. Complex responses to climate drivers in onset of spring flowering across a semi-arid elevation gradient. Journal of Ecology 98:1042–1051.

Crimmins, T. M., M. A. Crimmins, C. D. Bertelsen, and J. Balmat. 2008. Relationships between alpha diversity of plant species in bloom and climatic variables across an elevation gradient. International Journal of Biometeorology 52:353–366.

Cristol, D. A., M. B. Baker, and C. Carbone. 1999. Differential migration revisited: Latitudinal segregation by age and sex classes. Current Ornithology 15:33–88.

Decker, K. L., and C. J. Conway. 2009. Effects of an unseasonal snowstorm on Red-faced Warbler nesting success. Condor 111:392–395.

Delmore, K. E., J. W. Fox, and D. E. Irwin. 2012. Dramatic intraspecific differences in migratory routes, stopover sites and wintering areas, revealed using light-level geolocators. Proceedings of the Royal Society of London B 279:4582–4589.

Diniz-Filho, J. A. F., L. M. Bini, and B. A. Hawkins. 2003. Spatial autocorrelation and red herrings in geographical ecology. Global Ecology and Biogeography 12:53–64.

Finch, D. M., and W. Yong. 2000. Land bird migration in riparian habitats of the Middle Rio Grande: A case study. Studies in Avian Biology 20:88–98.

Fontaine, J. J., K. L. Decker, S. K. Skagen, and C. van Riper. 2009. Spatial and temporal variation in climate change: A bird's eye view. Climatic Change 97:305–311.

Gannes, L. Z. 2002. Mass change pattern of Blackcaps refueling during spring migration: Evidence for physiological limitations to food assimilation. Condor 104:231–239.

Gordo, O. 2007. Why are bird-migration dates shifting? A review of weather and climate effects on avian migratory phenology. Climate Research 35:37–58.

Gordo, O., L. Brotons, X. Ferrer, and P. Comas. 2005. Do changes in climate patterns in wintering areas affect the timing of spring arrival of trans-Saharan migrant birds? Global Change Biology 11:12–21.

Gordo, O., and J. J. Sanz. 2005. Phenology and climate change: A long-term study in a Mediterranean locality. Oecologia 146:484–495.

Haitovsky, Y. 1968. Missing data in regression analysis. Journal of the Royal Statistical Society Series B 30:67–82.

Hawkins, B. A., J. A. F. Diniz-Filho, L. M. Bini, P. De Marco, and T. M. Blackburn. 2007. Red herrings revisited: Spatial autocorrelation and parameter estimation in geographical ecology. Ecography 30:375–384.

Hedenström, A., Z. Barta, B. Helm, A. I. Houston, J. M. McNamara, and N. Jonzén. 2007. Migration speed and scheduling of annual events by migrating birds in relation to climate change. Climate Research 35:79–91.

Honěk, A. 1996. Geographical variation in thermal requirements for insect development. European Journal of Entomology 93:303–312.

Hüppop, O., and K. Hüppop. 2003. North Atlantic oscillation and timing of spring migration in birds. Proceedings of the Royal Society of London B 270:233–240.

Hüppop, O., and W. Winkel. 2006. Climate change and timing of spring migration in the long-distance migrant Ficedula hypoleuca in central Europe: The role of spatially different temperature changes along migration routes. Journal of Ornithology 147:344–353.

IPCC. 2001. Climate change 2001: The scientific basis. 881 pp. in J. T. Houghton, Y. Ding, D. J. Griggs, M. Noguer, P. J. van der Linden, X. Dai, K. Maskell, and C. A. Johnson (editors). Contribution of Working Group I to the Third Assessment Report of the Intergovernmental Panel on Climate Change. Cambridge University Press, Cambridge, U.K.

Inouye, D. W. 2008. Effects of climate change on phenology, frost damage, and floral abundance of montane wildflowers. Ecology 89:353–362.

Inouye, D. W., B. Barr, K. B. Armitage, and B. D. Inouye. 2000. Climate change is affecting altitudinal migrants and hibernating species. Proceedings of the National Academy of Sciences of the USA 97:1630–1633.

Jenni, L., and M. Kéry. 2003. Timing of autumn bird migration under climate change: Advances in long-distance migrants, delays in short-distance migrants. Proceedings of the Royal Society of London B 270:1467–1471.

Jonzén, N., A. Lindén, T. Ergon, E. Knudsen, J. O. Vik, D. Rubolini, D. Piacentini, C. Brinch, F. Spina, L. Karlsson, M. Stervander, A. Andersson, J. Waldenström, A. Lehikoinen, E. Edvardsen, R. Solvang, and N. C. Stenseth. 2006. Rapid advance of spring arrival dates in long-distance migratory birds. Science 312:1959–1961.

Kanuscak, P., M. Hromada, T. H. Sparks, and P. Tryjanowski. 2004. Does climate at different scales influence the phenology and phenotype of the River Warbler Locustella fluviatilis? Oecologia 141:158–163.

Klaasen, M. 1996. Metabolic constraints on long-distance migration in birds. Journal of Experimental Biology 199:57–64.

Lehikoinen, E., and T. Sparks. 2010. Changes in migration. Pp. 89–112 in A.P. Møller, W. Fiedler, and P. Berthold (editors), Effects of climate change on birds. Oxford University Press, Oxford, UK.

Lehikoinen, A., M. Kilpi, and M. Ost. 2006. Winter climate affects subsequent breeding success of Common Eiders. Global Change Biology 12:1355–1365.

Lehikoinen, E., T. H. Sparks, and M. Zalakevicius. 2004. Arrival and departure dates. Pp. 1–31 in A. P. Møller, W. Fiedler, and P. Berthold (editors), Birds and climate change (vol. 35), Elsevier, Amsterdam, Netherlands.

Llorens, L., and J. Penuelas. 2005. Experimental evidence of future drier and warmer conditions affecting flowering of two co-occurring Mediterranean shrubs. International Journal of Plant Sciences 166:235–245.

Martin, T. E. 1987. Food as a limit on breeding birds: A life-history perspective. Annual Review of Ecology and Systematics 18:453–487.

McGrath, L. J., C. van Riper, and J. J. Fontaine. 2009. Flower power: tree flowering phenology as a settlement cue for migrating birds. Journal of Animal Ecology 78:22–30.

Miranda, J. D., C. Armas, F. M. Padilla, and F. I. Pugnaire. 2011. Climatic change and rainfall patterns: Effects on semi-arid plant communities of the Iberian Southeast. Journal of Arid Environments 75:1302–1309.

Møller, A. P., D. Rubolini, and E. Lehikoinen. 2008. Populations of migratory bird species that did not show a phenological response to climate change are declining. Proceedings of the National Academy of Science of the USA 105:16195–16200.

Moore, F. R., S. A. Gauthreaux, P. Kerlinger, and T. R. Simons. 1995. Habitat requirements during migration: Important link in conservation. Pp. 121–144 in T. E. Martin and D. M. Finch (editors), Ecology and management of Neotropical migratory birds. Oxford University Press, New York, NY.

Moore, F. R., R. J. Smith, and R. Sandberg. 2005. Stopover ecology of intercontinental migrants. Pp. 251–261 in R. Greenberg and P. P. Marra (editors), Birds of two worlds: The ecology and evolution of migration. Johns Hopkins University Press, Baltimore, MD.

Mueller, H. C., N. S. Mueller, D. D. Berger, G. Allez, W. Robichaud, and J. L. Kaspar. 2000. Age and sex differences in the timing of fall migration of hawks and falcons. Wilson Bulletin 112:214–224.

Norris, D. R., P. P. Marra, P. K. Kyser, G. J. Bowen, L. M. Ratcliffe, J. A. Royle, and T. K. Kyser. 2006. Migratory connectivity in a widely distributed songbird, the American Redstart (*Setophaga ruticilla*). Ornithological Monographs 61:14–28.

Parmesan, C., and G. Yohe. 2003. A globally coherent fingerprint of climate change impacts across natural systems. Nature 421:37–42.

Parmesan, C. 2007. Influences of species, latitudes and methodologies on estimates of phenological response to global warming. Global Change Biology 13:1860–1872.

Pattenden, R. K., and D. A. Boag. 1989. Effects of body mass on courtship, pairing and reproduction in captive Mallards. Canadian Journal of Zoology 67:495–501.

Paxton, K. L., C. van Riper, T. C. Theimer, and E. H. Paxton. 2007. Spatial and temporal migration patterns of Wilson's Warbler (*Wilsonia pusilla*) in the Southwest as revealed by stable isotopes. Auk 124:162–175.

Penuelas, J., I. Fillella, and P. Comas. 2002. Changed plant and animal life cycles from 1952 to 2000 in the Mediterranean region. Global Change Biology 8:531–544.

Petersen, M. R. 2009. Multiple spring migration strategies in a population of Pacific Common Eiders. Condor 111:59–70.

Phillips, A. R. 1951. Complexities of migration: A review. Wilson Bulletin 63:129–136.

Post, E., and N. C. Stenseth. 1999. Climatic variability, plant phenology, and northern ungulates. Ecology 80:1322–1339.

Root, T. L., J. T. Price, K. R. Hall, S. H. Schneider, C. Rosenzweig, and J. A. Pounds. 2003. Fingerprints of global warming on wild animals and plants. Nature 421:57–60.

Rosenzweig, C., D. Karoly, M. Vicarelli, P. Neofotis, Q. Wu, G. Casassa, A. Menzel, T. L. Root, N. Estrella, B. Seguin, P. Tryjanowski, C. Liu, S. Rawlins, and A. Imeson. 2008. Attributing physical and biological impacts to anthropogenic climate change. Nature 453:353–358.

Rubolini, D., A. P. Møller, K. Rainio, and E. Lehikoinen. 2007. Intraspecific consistency and geographic variability in temporal trends of spring migration phenology among European bird species. Climate Research 35:135–146.

Ryder, R. A. 1963. Migration and population dynamics of American Coots in western North America. Proceedings of the International Ornithological Congress 13:441–453.

Sæther, B. E., W. J. Sutherland, and S. Engen. 2004. Climate influences on avian population dynamics. Pp. 185–209 in A. P. Møller, W. Fiedler, and P. Berthold (editors), Birds and climate change (vol. 35), Elsevier, Amsterdam, Netherlands.

Sharifi, M. R., F. C. Meinzer, E. T. Nilsen, P. W. Rundel, R. A. Virginia, W. M. Jarrell, D. J. Herman, and P. C. Clark. 1988. Effect of manipulation of water and nitrogen supplies on the quantitative phenology of *Larrea tridentata* (creosote bush) in the Sonoran Desert of California. American Journal of Botany 75:1163–1174.

Skagen, S. K., J. F. Kelly, C. van Riper III, R. L. Hutto, D. M. Finch, D. J. Krueper, and C. P. Melcher. 2005. Geography of spring land bird migration through riparian habitats in southwestern North America. Condor 107:212–227.

Smith, T. B., T. D. Meehan, and B. O. Wolf. 2003. Assessing migration patterns of Sharp-shinned Hawks *Accipiter striatus* using stable isotope and band encounter analysis. Journal of Avian Biology 34:387–392.

Sparks, T. H., F. Bairlein, J. G. Bojarinova, O. Huppop, E. A. Lehikoinen, K. Rainio, L. V. Sokolov, and D. Walker. 2005. Examining the total arrival distribution of migratory birds. Global Change Biology 11:22–30.

Stervander, M., K. Lindström, N. Jonzén, and A. Andersson. 2005. Timing of spring migration in birds: Long-term trends, North Atlantic oscillation and the significance of different migration routes. Journal of Avian Biology 36:210–221.

Stouffer, P. C., and G. M. Dwyer. 2003. Sex-biased winter distribution and timing of migration of Hermit Thrushes (*Catharus guttatus*) in eastern North America. Auk 120:836–847.

Strode, P. K. 2003. Implications of climate change for North American wood-warblers (Parulidae). Global Change Biology 9:1137–1144.

Tøttrup, A. P., K. Thorup, and C. Rahbek. 2006. Patterns of change in timing of spring migration in North European songbird populations. Journal of Avian Biology 37:84–92.

Visser, M. E., C. Both, and M. M. Lambrechts. 2004. Global climate change leads to mistimed avian reproduction. Pp. 89–110 in A. P. Møller, W. Fiedler, and P. Berthold (editors), Birds and climate change (vol. 35), Elsevier, Amsterdam, Netherlands.

Visser, M. E., A. J. van Noordwijk, J. M. Tinbergen, and C. M. Lessells. 1998. Warmer springs lead to mistimed reproduction in Great Tits (*Parus major*). Proceedings of the Royal Society of London Series B 265:1867–1870.

Walther, G. R., E. Post, A. Convey, A. Menzel, C. Parmesan, T. J. C. Beebee, J. Fromentin, O. Hoegh-Guldberg, and F. Bairlein. 2002. Ecological responses to recent climate change. Nature 416:389–395.

Weidinger, K., and M. Král. 2007. Climatic effects on arrival and laying dates in a long-distance migrant, the Collared Flycatcher *Ficedula albicollis*. Ibis 149:836–847.

Williams, C. N., M. J. Menne, R. S. Vose, and D. R. Easterling. [online]. 2007. United States Historical Climatology Network monthly temperature and precipitation data. ORNL/CDIAC-118, NDP-019. <http://cdiac.ornl.gov/epubs/ndp/ushcn/usa_monthly.html> (1 June 2009).

Wilson, W. H. 2007. Spring arrival dates of migratory breeding birds in Maine: Sensitivity to climate change. Wilson Journal of Ornithology 119:665–677.

CHAPTER TWO

Landbird Stopover in the Great Lakes Region[*]

INTEGRATING HABITAT USE AND
CLIMATE CHANGE IN CONSERVATION

David N. Ewert, Kimberly R. Hall, Robert J. Smith, and Paul G. Rodewald

Abstract. Millions of landbirds migrate annually through the Great Lakes region of North America, where protecting stopover habitat is a critical conservation concern, especially along highly developed coastlines. Observed and projected climate changes have the potential to alter the quality and phenology of resources and the migratory timing of landbirds. Based on our review of key migrant-stopover habitat relationships in the Great Lakes region, we propose guidelines that address projected shifts in stopover conditions. Given the high uncertainty associated with the rate and magnitude of future changes in phenology, our approach increases the likelihood of achieving the conservation outcome of providing stopover sites that will sustain migrants as conditions change. We suggest that current efforts focus on protecting areas near the Great Lakes and other bodies of water, as well as on sharing key information on stopover habitat needs broadly with other conservation planners and the general public. We propose that more attention be devoted to (1) identifying and protecting sites with ecological and topographic diversity, to retain variation in the phenology of resources during the migratory period; (2) enhancing and expanding the current network of stopover sites by increasing the size, habitat quality, and distribution of important stopover habitat; and (3) increasing protection of existing large habitat patches in the northern Great Lakes region, which are expected to be under increased development pressure as the climate warms.

Key Words: bird distribution, climate change, coastal, Great Lakes, migratory birds, phenology, stopover ecology.

One of the challenges of conserving migratory landbirds is deciding where and when to act: What conservation strategies will best sustain a highly mobile and ecologically diverse suite of species at scales ranging from local to intercontinental scales? To protect migrants effectively, we need to understand how current and projected stressors act on each phase of the life cycle (Faaborg et al. 2010), and implement strategies that support each phase. Traditionally, the migratory phase of the life cycle has not been considered in conservation planning for landbirds

[*] Ewert, D. N., K. R. Hall, R. J. Smith, and P. G. Rodewald. 2015. Landbird stopover in the Great Lakes region: integrating habitat use and climate change in conservation. Pp. 17–46 in E. M. Wood and J. L. Kellermann (editors), Phenological synchrony and bird migration: changing climate and seasonal resources in North America. Studies in Avian Biology (no. 47), CRC Press, Boca Raton, FL.

(Moore et al. 1995, Mehlman et al. 2005), yet recent studies suggest that landbirds may have higher mortality rates during migration compared to the stationary periods of the breeding and wintering season (Sillett and Holmes 2002). Although it is unclear if mortality during migration actually limits populations or varies along the migratory route, projected rapid changes in climate have strong potential to influence the fitness of migrating birds. The high level of physiological stress and risk associated with migration suggests that changes in temperature, storm intensity, or other climate-linked factors—such as the availability of arthropod prey—could strongly affect fitness (Klaassen et al. 2012). Consideration of climate change also emphasizes the need to think of the landbird life cycle holistically, as impacts that occur during migration may carry over from this stage to the next, similar to effects of droughts on migratory timing and condition of birds (Marra et al. 1998, Norris et al. 2004, Rockwell et al. 2012).

In addition to the physiological stress of migration, the spatial extent of the migratory process makes climate change-related impacts seem particularly likely in this phase of the life cycle. Birds are exposed to many environments in a short time along the migratory route, and migration represents a highly dynamic set of responses to both extrinsic factors like weather and food availability, and to intrinsic factors (Moore et al. 1995, Petit 2000). Further, climate change may influence shifts in breeding or wintering ranges, migration distances, and routes taken by migrants, any of which could influence timing of migration and overall exposure to climate change (Visser et al. 2009, Pulido and Berthold 2010, Knudsen et al. 2011, James and Abbott 2014). While the body of work linking breeding and wintering areas of some populations (Rubenstein et al. 2002, Ryder et al. 2011) and identifying migration routes of migrants is growing (Stutchbury et al. 2009, Stanley et al. 2012, McKinnon et al. 2013), anticipating how these interactions might be altered by climate change provides an additional challenge to the goal of developing conservation programs focused on the migration period.

Here, we provide guidance that specifically integrates habitat and climate change into conservation strategies for migratory birds in the Laurentian Great Lakes region of the United States and Canada (Figure 2.1), which supports millions of migrants during each spring and fall migration.

In this region and elsewhere, conservation actions to benefit birds typically focus on protecting habitat. However, choosing where to invest resources in stopover habitat is a challenge, as habitat use and quality vary across sites and may also vary considerably within a site due to factors such as season (fall or spring; Mehlman et al. 2005, Keller and Yahner 2007, Deutschlander and Muheim 2009, Ewert et al. 2011), weather, and variation in resource availability at the site (Bonter et al. 2009, France et al. 2012, Johnson 2013, Kirsch et al. 2013). Climate change is likely to increase variability in resource availability, as exposure of both migrants and their stopover habitats to climate change will depend on latitude, local factors like topographic position or proximity to smaller water bodies, and proximity to the Great Lakes—each of which can influence climatic variables (Scott and Huff 1996, Notaro et al. 2013a). We expect that both regional and local patterns of climate change will influence the phenology of plants and arthropod prey at stopover sites (Polgar and Primack 2011, Elmore et al. 2012), which may contribute to changes in the predictability or abundance of resources during this physiologically demanding time.

The urgency in addressing climate change in conservation planning for migratory landbirds is informed by global scale work suggesting that accelerating rates of change in temperature and related changes in other climatic factors will have significant consequences for the maintenance of biodiversity and ecosystems (Parmesan and Yohe 2003, Root et al. 2003, Thomas et al. 2004, Mora et al. 2013, Warren et al. 2013), and that climate drivers may supplant habitat loss as the primary threat to biodiversity (Bellard et al. 2012). Although evidence is strong that many birds around the globe are responding to climate change by starting spring migration earlier and leaving breeding grounds later (Root et al. 2003, Jonzén et al. 2006, Van Buskirk et al. 2009), much remains to be learned about drivers of variability across species and geographic regions, as well as the key mechanisms underlying these responses (Knudsen et al. 2011, Small-Lorenz et al. 2013, James and Abbott 2014).

Linking migratory processes to actions on the ground adds other sources of uncertainty. As in many other regions, it is difficult to adapt current conservation approaches for migrants that stop over in the Great Lakes basin, given the

Figure 2.1. Laurentian Great Lakes region showing states and provinces bordering the Great Lakes and the Great Lakes watershed boundary. (NOAA. [online]. 2004. Basin map of Great Lakes. Great Lakes Environmental Research Laboratory. Ann Arbor, MI. <http://www.flic kr.com/ph otos/noaa_glerl/4037600466/> [16 July 2013].)

uncertainties associated with projecting future climates, including the responses of migrants and their habitats, changes in land use, and adoption of management practices in response to climate change. In this chapter, we provide a foundation for advancing conservation strategies by examining stopover habitat use in the region in the context of climate change, with a focus on reducing risks related to changes in phenology. We frame our presentation of interactions between climate change and stopover ecology using the terminology of climate change vulnerability assessments (Glick et al. 2011). In this framing, vulnerability is a factor of the amount of exposure to changes in key climate drivers, as well as the amount of sensitivity shown by species or habitats to these drivers, modified by their capacity to adaptively respond or be resilient to change. Compared to other taxa, migratory birds are typically viewed as highly sensitive to changes in climate, especially changes in spring temperature and phenology, but they are also typically thought to have high potential to respond in ways that maintain

fitness, in part due to their high mobility (Foden et al. 2008, Young et al. 2011). However, typical assessment approaches are likely to underestimate risks because they do not evaluate climate impacts across the entire life cycle (Small-Lorenz et al. 2013), and little consideration is given to where migrants might go and how they might fare in new or modified sites. Habitats critical to all life stages continue to be lost, which emphasizes the need to accelerate conservation efforts.

For this risk assessment for the Great Lakes region, we first review observed and projected changes in climate, and associated phenological responses and trends. We then review habitat attributes associated with landbird migrants at both landscape and local (site) scales, with an emphasis on Great Lakes coastal regions, and we consider these attributes in the context of climate change. Last, we articulate conservation strategies and management options that should increase the probability that forest and shrub-dependent landbirds will successfully migrate through the Great Lakes region as climate conditions change.

FOCAL REGION: THE LAURENTIAN GREAT LAKES REGION

The Great Lakes are a dominant feature of the Upper Midwest landscape and play an important role in shaping bird migration patterns. The Great Lakes region, especially coastal areas, provides stopover sites for large numbers of landbird migrants (Peet 1908, Janssen 1976, Diehl et al. 2003, Ewert et al. 2006, Smith et al. 2007a, Bonter et al. 2009, Ewert et al. 2011, France et al. 2012, Johnson 2013). Migrating landbirds in this region occupy many habitats, including upland forests (Rodewald and Matthews 2005), shrublands (Petrucha et al. 2013), riparian corridors (Kirsch et al. 2013), and alder swamps (Nicholls et al. 2001), and they often show preferences for sites near water or wetlands (Ewert and Hamas 1996, Diehl et al. 2003, Ewert et al. 2011). Habitat protection and restoration are high priorities because many suitable stopover habitats have been lost as shorelines have been developed, and much of the southern half of the region has been urbanized or converted to agriculture. In addition, the Great Lakes states have experienced dramatic losses of wetlands, with estimates up to 85% in the southern part of the region (Illinois, Indiana, and Ohio—Mitsch and Gosselink 2007, Appendix A). The degree to which these natural systems have been altered by land conversion, invasive species, and pollution underscores the need to accelerate ongoing work to address interacting stressors as we develop conservation priorities for migratory stopover sites in light of climate change.

METHODS

Literature reviewed for this chapter was identified from comprehensive bibliographies on migratory stopover and related topics maintained by the authors. Publications included over 1,220 citations, of which approximately 110 are specific to the Great Lakes region and date back to 1893. To integrate climate change impacts, we drew from a literature database of approximately 200 citations developed by KRH over the past 5 years for assessing ecological risks from climate change in the Great Lakes region. To expand this synthesis, we also conducted targeted key word searches for papers on impacts or interactions that we had not reviewed previously. Key words included the following terms: stopover, stopover sites, landbirds,

landscape, resources, fragmentation, radar, Great Lakes, phenology, climate change, and invertebrates. Our conclusions are largely based on studies focused on songbirds in the Great Lakes region (although relatively few studies have been conducted around Lake Superior), supplemented by studies in eastern North America and western Europe that have comparable features. Similarly, we draw primarily from climate change impacts observed and projected for this region, with the caveat that few studies on phenological patterns have been conducted, so our work also references research conducted in other locations.

Observed and Projected Climate Change in the Great Lakes Region

Climate change is already leading to many environmental changes in the Great Lakes region. Here, we focus on increases in temperature because they are the main source of exposure to change in terrestrial and aquatic systems, associated with relatively high certainty, and a key driver of phenological change. We include water temperature changes in addition to air temperature because these changes could influence phenology of emergent arthropods, an important resource for spring migrating birds in the Great Lakes region (Smith et al. 2007a, Ewert et al. 2011). Changes in wind pattern, precipitation, and lake level are important aspects of regional climate change, but are associated with larger ranges of uncertainty and are not strongly tied to phenology. Nonetheless, we review them briefly as they are relevant to updating conservation approaches and priorities for migrating birds in the Great Lakes region.

Changes in Air Temperature

Since the 1880s, the linear trend in average global surface temperature suggests an increase of approximately 0.85°C in the Northern Hemisphere, and the last 30 years were likely the warmest period in the last 1,400 years (IPCC 2013). Consistent with other locations at similar latitudes, changes in the Great Lakes region exceed the global average. In the US portion of the region, average temperatures have increased 1°C–1.6°C from 1950 to 2009 (Hayhoe et al. 2010). An acceleration in the rate of temperature change for the midwestern United States is reflected in a mean increase of

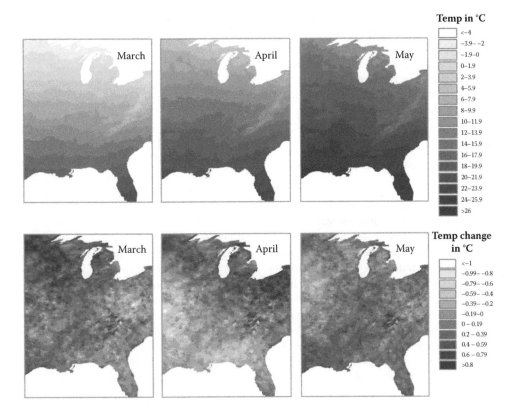

Figure 2.2. Mean observed temperature by month of spring migration from 1979 to 2010 and patterns of temperature change (bottom row) over the same time period for the eastern United States. In the bottom figure, darker colors indicate local warming trends, while white or light gray values indicate cooling. (Source: Prism data set [Gibson, W. P. et al. 2002. Proceedings of 13th AMS Conference on Applied Climatology, American Meteorological Society, Portland, OR, 13–16 May, 181–183] as available from ClimateWizard (Girvetz et al. 2009).

0.06°C per decade from 1900 to 2010 and a much faster increase of 0.24°C per decade from 1979 to 2010 (Pryor et al. 2013). This rate of change is expected to continue. Average projections across multiple global circulation models (GCMs) suggest further increases of 2°C–3°C (for a range of low- to high-emission scenarios) by the middle of this century and increases of 3°C–5°C by the 2080s (Hayhoe et al. 2010, Pryor et al. 2013).

Determining how changes in air temperature may influence the timing of bird migration and phenological conditions at stopover sites is complicated because patterns of change vary with the spatial and temporal scale of observation (Elmore et al. 2012). The most obvious patterns in temperature experienced by birds flying toward the Great Lakes region in spring correlate with latitudinal gradients, though the full range of this gradient is not likely to be experienced by birds because conditions tend to warm over time as birds are

moving north (Figure 2.2). Broadly, the entire gradient is warming, though temperatures in the northern Great Lakes region are warming more quickly than the southern ones (Hayhoe et al. 2010, Dobrowski et al. 2013, Pryor et al. 2013). For example, Michigan, Minnesota, and Wisconsin are among the 10 fastest warming states in the United States (Tebaldi et al. 2012); yet, even within these rapidly warming northern states, considerable heterogeneity can be found in the magnitude, temporal pattern, and even direction of temperature change (Figure 2.2; see Dobrowski et al. 2013 for annual patterns). Warmer winters and, to a lesser degree, extension of warmer temperatures into fall are extending growing seasons in most of the Midwest (Pryor et al. 2013). By 1999, the frost-free period had increased by more than 1 week in most parts of the region (CCSP 2009, Andresen et al. 2012) and it is projected to increase by an additional 3 weeks by the middle of this century (Pryor et al. 2013). At local scales, factors such as topographic variation, proximity to the Great Lakes and other water bodies, and the influence of urban heat islands contribute to local temperature regimes and thus habitat conditions along migration routes (Dobrowski 2011, Fontaine et al. 2009, Andresen et al. 2012, Elmore et al. 2012).

Changes in Great Lakes Water Temperature and Ice Cover

Changes in air temperature typically correlate with changes in surface water temperatures, but recent work in the Great Lakes region indicates that rapid loss of ice cover and earlier stratification with separation of water into layers by temperature of the upper lakes (Lakes Superior, Michigan, and Huron) promotes rates of increases in surface water temperatures that are much higher than the changes in air temperature (Austin and Colman 2008). For Lake Superior, which shows the strongest trend, open water temperatures in summer have increased 3.5°C over 100 years, with the summer surface water temperature increasing at a rate of 1.1°C ± 0.6°C per decade from 1979 to 2006 (Austin and Colman 2008). Changes in surface water temperatures in the upper Great Lakes are triggering systemwide impacts, including increases in wind and current speeds, and further increases in the duration of the stratified period (Austin and Colman 2007, 2008; Desai et al. 2009; Dobiesz and Lester 2009; Wang et al. 2011). The clearest indication that aquatic system phenology is advancing in response to spring warming comes from estimates of stratification start dates and duration. For the upper lakes, duration of stratification has been increasing at a rate of 4–14 days/decade since the 1970s (Austin and Colman 2007). Projections for the end of this century for Lakes Huron and Michigan suggest possible increases in the stratified period of 40–50 days by the 2050s and 60–90 days by the 2080s relative to the 1980s (Trumpickas et al. 2009). Given the recent sharp declines (73%) in total annual ice cover on the Great Lakes from 1973 to 2010 (Wang et al. 2011), it seems likely that little ice will be on the Great Lakes by the middle of this century in most years—possibly precipitating further changes in lake dynamics, changes in the degree of lake-influenced weather experienced in Great Lakes coastal zones, and changes in aquatic species' phenology.

Interacting Climate Factors: Changes in Precipitation, Great Lakes Water Levels, and Wind

While temperature changes are the dominant driver of phenological changes, precipitation-related impacts may also interact with changes in temperature to influence habitat quality at stop-over sites in the Great Lakes region. Although seasonal and annual averages for precipitation remain highly variable, showing no clear trends over the past century, heavy rain events have become more common (Hayhoe et al. 2010). Compared to past averages (1948–1978), over the last 30 years the central United States has experienced significant increases of up to 40% in the frequency of rainfall events that fall in the top two categories ("very heavy" and "extreme") of precipitation events (Groisman et al. 2012). GCM outputs for total annual precipitation in the Great Lakes region project a variety of scenarios ranging from strong increases to strong decreases, but most models indicate that precipitation in the winter–early spring period is likely to increase (Hayhoe et al. 2010, Pryor et al. 2013). Further rise in temperature is also projected to lead to continued increases in storm intensity, an impact that is of higher certainty than other precipitation-related projections because warmer air can hold more water (Groisman et al. 2004).

While storms during migration or flooding of habitats can influence birds as they migrate, related droughts may be even more important drivers of food availability along migration routes. Projections suggest increases in the duration of periods between rain events, leading to summer and fall droughts (Mishra et al. 2010, Trenberth 2011). Further, because warmer conditions should increase evapotranspiration rates, plants may experience increased water stress even if precipitation rates remain similar. Interactions between temperature and precipitation are also important considerations for understanding changes in ecosystem dynamics in spring, through changes in snow depth and timing. Changes in snow are challenging to predict because of the uncertainty associated with future precipitation amounts, combined with the likelihood that while the proportion of precipitation that falls as snow rather than rain is likely to decrease, reductions in ice cover on the Great Lakes suggest that lake-effect snowfall may increase (Kunkel et al. 2009; Notaro et al. 2013a, b; Pryor et al. 2013). While migrants are not typically directly impacted by snow, changes in snowmelt timing and the resultant changes in albedo (reflectance) can promote changes in soil temperature that may show somewhat different patterns from temporal trends in air temperatures. These interactions may contribute to diverse phenological responses of plant species and in rates of ecosystem processes in habitats used by migrants (Groffman et al. 2012).

Coastal stopover habitats are affected by the dynamics of water levels in the Great Lakes. Surface water of the Great Lakes comprises about one-third of the total area of the Great Lakes basin and water levels vary by season and decade and in association with global multidecadal cycles, but less so for Lakes Ontario and Superior, where outflows are managed (Gronewald et al. 2013). The main drivers of change in Great Lakes water levels are direct inputs from precipitation and evaporation and runoff from land; the relative importance of each varies as weather conditions change (Gronewald et al. 2013). Changes in some climatic factors are likely to drive water levels in opposite directions, contributing to high uncertainty in future projections (Angel and Kunkel 2010, Lofgren et al. 2012, Gronewald et al. 2013). Specifically, higher air and water temperatures contribute to increased evaporation rates, and reductions in ice cover increase the time period during which evaporation can occur—all potentially driving water levels lower. Conversely, if we see increases in winter and spring precipitation and increases in runoff from peak storms, these changes could lead to higher inputs to the lakes and thus lake level increases. Given that the Great Lakes water levels could increase or decrease (Angel and Kunkel 2010, Lofgren et al. 2012, Gronewald et al. 2013), conservation plans should include protection strategies for assuring that coastal habitats will be available under a range of future water levels, including higher and lower extreme values.

As with precipitation, changes in wind have been observed and are likely, but they are highly variable and hard to predict. High wind energy is a characteristic of coastal habitats, as the gradient in temperature between water and land drives movement of air, which in turn drives currents that shape aquatic habitats and temperature regimes. Wind patterns also greatly influence conditions for migration, including both migrant passage (Liechti 2006) and the spatial patterns of emergent aquatic insects (Corkum et al. 2006), which may be important food resources for spring migrants using Great Lakes coastal habitats during stopover (Smith et al. 2007a, Ewert et al. 2011).

Correlated with higher summer surface water temperatures, summer wind speed increases of up to 5% per decade on Lake Superior have been documented since the 1980s (Austin and Colman 2007, Desai et al. 2009). Austin and Coleman (2007) also detected increases in wind speed over Lakes Michigan, Superior, and Huron. More than a decade ago, work focusing on summer wind direction from 1980 to 1999 across the Great Lakes demonstrated a significant change in wind direction from southwest to south around 1990, which is consistent with a southward shift in the dominant summer storm track (Waples and Klump 2002). At a regional scale, research on wind patterns in the context of planning for risks to people, as well as potential changes in wind as a resource for energy generation, suggests that, at least in the next 50 years, average wind patterns will remain within the historic range of variability (Pryor and Barthelmie 2011, Pryor et al. 2012). From the perspective of planning for bird migration, we have additional challenges related to understanding how changes will influence fitness, and a mismatch with respect to how attributes of wind are mapped and studied. For example, birds typically migrate at heights above the wind resource utilized by wind turbines, though they move through these elevations when taking off and landing. Thus, while we know changes in wind during migration could be very important, given these uncertainties and data gaps, our strategy for addressing risks related to changes in wind focuses on targeted research and monitoring.

Changes in Phenology

Understanding how migrant landbirds are likely to be at risk during migration requires that we consider the changes in climate drivers within the context of the sensitivity and adaptability of both birds and stopover habitats. Our review focuses on changes in phenology such as shifts in the timing of seasonal events, which are a common response to changes in temperature, especially spring temperature (Bradley et al. 1999, Gibbs and Breisch 2001, Root et al. 2003, Parmesan 2006, Cleland et al. 2007). In the eastern United States, as in many other parts of the world, strong evidence suggests that many species of landbird migrants are arriving earlier in spring and departing later in fall, consistent with warming trends (Marra et al. 2005, Mills 2005, Miller-Rushing et al. 2008, Swanson and Palmer 2009, Van Buskirk et al. 2009, Hurlbert and Liang 2012). In the following, we review broad changes in migratory bird phenology and phenological changes in habitats and prey to set the stage for integrating these types of changes into conservation planning to protect stopover habitats in the Great Lakes region.

Whereas change in day length is the dominant cue that determines the approximate date to initiate migration (Berthold 1996), temperature and precipitation can influence when birds initiate migration or the rate at which birds travel along migratory routes (Marra et al. 2005, Jonzén et al. 2007, Tøttrup et al. 2008, Chapter 5, this volume, Smith and Carey, unpubl. data). Even for species where day length likely plays an important role in timing, higher food availability in winter has been linked to earlier spring migration (Studds and Marra 2007, Bridge et al. 2010), supporting the hypothesis that responses to changing conditions, including those en route (Smith and Moore 2005), may result from climate-related influences on food availability. Other landbird species, however, appear to be more consistent in spring departure dates despite varying environmental conditions, while other aspects of migration (route, fall timing) have been found to be more variable (Stanley et al. 2012). In addition to shifts in departure dates and passage rates, some species may show changes in arrival time at any one site that correlate with a reduction in migration distance. Work by Visser et al. (2009) on birds in Europe and by Heath et al. (2012) on American Kestrels (*Falco sparverius*) in the western United States suggest phenological advances can occur as a result of birds wintering farther north as conditions warm. Similarly, northern range shifts and range expansions have also been observed for many species that winter in North America, including many that migrate or breed in the Great Lakes region (La Sorte and Thompson

2007, McDonald et al. 2012). Associating change in arrival timing at the migratory destination with relative vulnerability of different migratory species is difficult as some species have shifted breeding ranges north (Hitch and Leberg 2007, Zuckerberg 2009), which may reduce the extent to which they are exposed to changes in resource timing on the breeding grounds relative to their migration timing. Given the variation in specific habitats used, distances traveled, body sizes, food preferences, and reproductive strategies of migrant landbirds, we expect species- and population-specific sensitivity to different aspects of climate change, as well as variation in capacities to respond facultatively or to adapt via a change in gene frequencies. Understanding sensitivities and vulnerabilities to climate change is also difficult because, e.g., simple phenological metrics such as arrival date can be influenced by factors that are independent of migratory timing, including changes in breeding and wintering distributions (Knudsen et al. 2011, James and Abbott 2014). As selection increases for shorter migration distances, populations may even show rapid reductions in the tendency to migrate (Pulido and Berthold 2010). In addition to interspecific variation, the highly dynamic nature of most species' habitat use and movement through stages of the annual cycle makes it challenging to understand whether an observed change, or lack of change, confers a fitness benefit or cost (Knudsen et al. 2011). Given that a great deal of uncertainty still exists regarding both sensitivity and the extent to which observed responses represent successful evolutionary adaptations, in our review we do not attempt to rank climate vulnerability across subsets of species with similar traits, such as long-distance versus short-distance migrants.

A key driver in the timing of migration is likely to be changes in the phenology of resources that birds require during migration or that allow them to deposit lipids needed to begin or continue migration. Air temperature, precipitation, and related factors like snowmelt timing can strongly influence the emergence of flowers and growth of leaves, which determine food available for terrestrial arthropods that are important prey for migratory birds in the spring (McGrath et al. 2009, Strode 2009, Polgar and Primack 2011, Groffman et al. 2012). While strong evidence suggests a general trend of advancing spring plant phenology, and delays in the onset of fall both globally and in the Great Lakes region, these changes tend to be highly variable across species, space, and time (Bradley et al. 1999, Parmesan and Yohe 2003, Root et al. 2003, Strode 2003, Schwartz et al. 2006, Cleland et al. 2007, Willis et al. 2008, Primack et al. 2009, Cleland et al. 2012, Cook et al. 2012a).

As for passage dates of migrating birds, changes in leaf-out or blooming dates in plants may be a result of multiple mechanisms. While temperature, often expressed as growing degree-days, is a strong predictor of spring plant phenology (e.g., budburst, flowering dates), climatic events such as spring frost or weather events in other seasons (i.e., drought or warm winters) can lead to different responses by species exposed to the same climate in spring (Rathcke and Lacey 1985, Fenner 1998, Parmesan 2006, Cleland et al. 2007, Polgar and Primack 2011). For example, warming in fall and winter can delay dormancy or fulfillment of chilling requirements for plants, leading to a delay in spring phenology even under warmer spring conditions (Cook et al. 2012b).

Relative to studies of the phenology of leaf-out and flowering, few studies of fruiting phenology have been conducted (Rathcke and Lacey 1985). In the temperate zone, timing of fruiting in animal-dispersed plant species appears less linked to climate factors than to spring phenological traits, and it has likely evolved in response to several factors, including the need for plants to have sufficient time (after flowering) to build up reserves that can be allocated to fruit production, and the timing of abundant seed dispersers, such as migratory birds (Rathcke and Lacey 1985, Fenner 1998). While timing of onset of fruiting may be less sensitive to climate, warmer temperatures can promote faster ripening, and extreme conditions such as frosts early in the growing season and high temperatures in fall can cause significant reductions in fruit production (Rathcke and Lacey 1985). Typically, the types of fruits produced in the fall, when birds are migrating, are viable for weeks or months, which is longer than summer fruits, presumably due at least in part to reduction in the impact of pathogens and decomposers (i.e., bacteria, fungi) when conditions are cooler (Cipollini and Stiles 1992, 1993). While not a phenological response of the plants, timing of available fruit could change if increasing fall temperatures contribute to increases in the effectiveness of bacteria and fungi to compete with birds for fruit resources.

As with temperature changes on land, climate change-related increases in water temperatures could influence the timing or availability of food resources for migratory birds. In the Great Lakes region, many species feed on emergent aquatic arthropods (e.g., midges—Diptera: Chironomidae) during spring migration (Smith et al. 2007, Ewert et al. 2011, MacDade et al. 2011). Increasing water temperature reduces dissolved oxygen levels and hence productivity of aquatic food webs, and it can impact the development rate and behavior of aquatic insects. As temperatures increase within a species' range of tolerance, shifts from one life stage to the next typically occur faster, suggesting earlier emergence phenology (Hassall and Thompson 2008, Eggermont and Heiri 2012). However, because timing of emergence is also influenced by factors other than temperature, such as photoperiod, food availability for preemergence developmental stages, and water temperature, it is difficult to predict abundance of aquatic emergent insects with temperature change. Thus, our planning should allow for unexpected dynamics of these important resources for migrants (Root and Schneider 2006, Groffman et al. 2012).

Linking Phenological Change and Risk to Migrants

The fact that many bird species are responding to changes in temperature by shifting their phenologies and ranges has been interpreted as evidence for resilience to climate change. However, three aspects of changing migration phenology could represent increasing risk to migrants.

First, many have suggested and some have documented the potential for critical mismatches in the timing of birds and the availability of key resources (Knudsen et al. 2011). For example, documented mismatches include early arrival but not earlier breeding of Pied Flycatchers (*Ficedula hypoleuca*) in Finland due to a lack of phenological change at the breeding site (Ahola et al. 2004), while arrival of the same species in the Netherlands has not changed and has become late relative to peak food abundance (Both et al. 2005). While we are not aware of any evidence of increased stress to migrants as a result of phenology shifts within the region, work by Marra et al. (2005) suggested that birds migrating toward the Great Lakes region have shown reduced rates of timing shift relative to plants along their route. Similar patterns have been observed in other northern regions (Saino et al. 2011, Ovaskainen et al. 2013). Although a variety of mechanisms can result in changes of migration timing (departure date, rate of travel, location of breeding or wintering grounds), it is difficult to say at what point change in migration timing relative to various phenological metrics of resources might be associated with reduced fitness, given that risks are associated with arriving too early (i.e., cold snaps, lack of food) or too late (Knudsen et al. 2011). Theoretical work on optimal arrival timing suggests reduced rate of change relative to plant resources should be favored by selection in most cases, and that we should expect variation in responses across species (Jonzén et al. 2007, Johansson and Jonzén 2012). However, increases in the potential for mismatches between migrant timing and food resources seems particularly likely in northern Great Lakes coastal zones where aquatic and terrestrial systems are experiencing high, but different, rates of temperature change. Further, in coastal ecosystems, many other factors that can influence phenology are simultaneously changing (i.e., wind direction and speed, start and duration of the stratified period).

Second, changes in climate may lead to species vulnerabilities through increases in the frequency of extreme events or increased impact from typical conditions due to a shift in phenology. An example of this type of risk is somewhat counterintuitive; spring warming can lead to increased frost impacts. In years with warmer springs, many plants respond with earlier budburst and leaf development. When early spring warm periods are followed by a frost, sensitive plant tissues are damaged (Gu et al. 2008, Hufkens et al. 2012, Winkler et al. 2012, Augspurger 2013), potentially reducing food resources in spring (arthropods associated with new plant tissues) or fall (fruit; Rathcke and Lacey 1985, Augspurger 2009, Groffman et al. 2012). Clearly, frost damage is not a new risk factor. Over evolutionary time, the risk of frost damage has likely played a major role in shaping the phenology of species in the temperate zone (Rathcke and Lacey 1985, Inouye 2000), including migratory birds (Brown and Brown 2000). However, this risk appears to be increasing in the Great Lakes region. In an Illinois forest, Augspurger (2013) has found that

the risks of significant frost damage to plant tissues have increased from an annual probability of 0.03 (average for 1889–1979) to 0.21 (1980–2012). While it is unclear how general or persistent this high increase in frost risk might be, projections suggest that periods of cold conditions in spring will persist in the Great Lakes region through the end of the next century, in spite of overall trends toward warmer average temperatures (Kodra et al. 2011). Proximity to water, which can buffer air temperature fluctuations, as well as topographical factors that influence movement of cold air, can influence frost risk (Gu et al. 2008, Elmore et al. 2012, Hufkens et al. 2012). Identification of higher and lower frost-risk areas could inform land protection priorities.

Third, while many species may be adapting now, the rate of climate change is expected to continue to increase, perhaps exceeding migrants' phenotypic plasticity (Gienapp et al. 2008), the primary mechanism believed responsible for recently observed changes in migration timing (Knudsen et al. 2011, though see Van Buskirk et al. 2012 for evidence that the role of microevolution may be underestimated). Over evolutionary time, variability in resources and stressors encountered during migration and at breeding sites has selected for traits promoting flexible phenotypic responses in many species. However, the rate of change in environmental conditions is rapidly increasing, especially at higher latitudes. These rates of change may already exceed some species' capacities to respond. In northern Europe, Saino et al. (2011) studied 117 bird species, demonstrating that even as most species arrive earlier, they arrive at a higher degree-day, a key driver of plant phenology, than before. Species with the greatest mismatch with respect to the onset of spring are also among the set showing the largest population declines. Similarly, abundance of resources may decline. Of over 400 plant species studied in Massachusetts, most showed advanced flowering phenology by an average of 7 days over 100 years, associated with a 2.4°C increase in local temperatures. Failure by some plant families to shift flowering time was associated with strong declines in abundance (Willis et al. 2008).

These three types of risks related to changing phenology represent the vulnerabilities that need to be considered for conservation approaches of stopover habitat.

Integrating Climate Change with Habitat Use

A comprehensive approach to protecting habitat requires understanding how birds are using habitats in the Great Lakes at both landscape and site scales (Buler et al. 2007, Deppe and Rotenberry 2008, Pennington et al. 2008), even as we recognize that these patterns may be shifting in response to climatic drivers. Habitat selection theory predicts that migrating birds initiate the process of selecting a stopover site at a broad spatial scale, likely while descending from nocturnal migration (Hutto 1985a, Bowlin et al. 2005, Chernetsov 2005, Mukhin et al. 2008), and then refine their selection through a period of exploration after making landfall (Hutto 1985a, Aborn and Moore 1997, Moore and Aborn 2000, Jenni and Schaub 2003, Chernetsov 2005). Here, we provide an overview of the state of knowledge regarding what and how broad- and local-scale features shape habitat use by migrant landbirds within the Great Lakes basin and speculate how climate change might influence these systems, and thus how migrants will use these habitats in the future.

Landscape-Scale Features

The distribution and abundance of migrants in a region can be influenced by broad-scale features of the landscape, such as proximity to geographical impediments to migration (e.g., ocean, large lake, desert) and land cover (McCann et al. 1993, Moore et al. 1995, Buler et al. 2007, Buler and Moore 2011, Ewert et al. 2011, France et al. 2012). In addition to geographic barriers, landscape-level factors include habitat availability and indices of fragmentation such as patch size, patch isolation, and composition of the areas that surround patches (Buler et al. 2007, Keller and Yahner 2007, Deppe and Rotenberry 2008, Bonter et al. 2009), as well as proximity to water, as species often concentrate near or in lacustrine, wetland, and riparian habitats (Ewert and Hamas 1996, Diehl et al. 2003, Ewert et al. 2011).

Climate change may contribute to further losses of habitats (or habitat components) as conditions in the northern parts of the Great Lakes region become more suitable for farming, and as additional pressure is put on aquatic systems to provide water to support human communities. Even in intact ecosystems, increasing temperatures can make habitats for migrants or their emergent aquatic insect prey scarcer or fragmented as some patches dry out, an impact that may accelerate if lake levels decline.

Habitat Availability

Numerous studies have documented negative relationships between migrant abundance and the amount of developed area, such as urban or agricultural land, and positive relationships between migrant abundance and forested or other natural areas within a region. For example, Mizrahi et al. (2006) found negative relationships between migrant abundance and urban areas in Pennsylvania with extreme habitat loss/fragmentation, whereas Bonter et al. (2009) detected strong negative relationships between migrant abundance and agricultural lands within the Great Lakes basin. Buler et al. (2007) provided evidence that hardwood forest was important in explaining distributions of insectivorous migrants in coastal Mississippi and Louisiana during both spring and fall migration, suggesting that amount and type of forest cover at the landscape level was an indicator of habitat quality for insectivorous landbird migrants. Within the Great Lakes region, as noted before, the extent of intact natural habitats varies widely. For example, within coastal areas (here defined as a 20-km band around each of the Great Lakes), an assessment of landbird stopover habitat availability at a landscape scale (5 km) shows that highest availability is in the northern part of the region and along the eastern shores of Lakes Erie and Ontario (Great Lakes migratory bird stopover portal, Ewert et al. 2012), suggesting that the greatest immediate need for conservation and protection is in the southern half of the region (Figure 2.3).

Patch Size, Isolation, and Composition of the Surrounding Matrix

Field studies in combination with assessments of radar-derived imagery indicate that the size of habitat fragments (patches) influences migrant densities within that patch, which in turn may affect how migrants use and benefit from use of that patch. Although more forested areas generally hold more migrants within highly fragmented

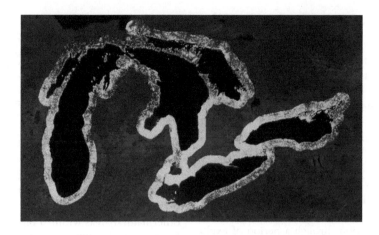

Figure 2.3. Landscape-scale assessment of availability of stopover habitat for landbird migrants within a 20-km zone around Lakes Michigan, Huron, Erie, and Ontario and connecting waters. Within the assessment zone, each 1-km pixel is shaded from white (no or little habitat) to dark gray (more habitat) to indicate the proportion of area within a 5-km radius window that represents habitat. (Source: Great Lakes Migratory Bird Stopover Portal: www.glmigratorybirds.org)

landscapes of the southern half of the Great Lakes region (Bonter et al. 2009), landbird migrants can occur at relatively high densities in remaining small patches (Diehl 2003, Chartier 2006, Bonter et al. 2009, France et al. 2012, Johnson 2013, Smith et al., unpubl. data).

Birds concentrating in small patches may depress food resources, in turn reducing the rate at which they meet nutritional requirements and ultimately delaying passage. For example, Moore and Wang (1991) demonstrated that spring migrants depressed resources while passing through a small chenier in coastal Louisiana, while Kelly et al. (2002) found that mass gain in fall migrating Wilson's Warblers (*Cardellina pusilla*) in southwestern New Mexico declined as a function of their abundance. Further, Smith et al. (unpubl. data), working in southwest Michigan, found that small patches within a predominantly agricultural matrix held both the fewest substrate arthropods and the highest number of long-distance migrants relative to larger patches surrounded by a more forested matrix. However, even as birds using these small habitat remnants may face increasing competition for available food resources, it is clear these patches are used by migrants and can provide important habitat where birds can rest and gain mass (Seewagen and Slayton 2008). Due to their limited area, small patches are likely to have the least topographic variation and plant and insect species diversity, suggesting that even without climate change, phenological variation at

a given site will likely be lower than at larger sites. In addition, rapid shifts in phenology in birds or prey are more likely to result in increased risk of phenological mismatch or high resource losses during frost events.

In addition to reductions in patch size, consequences of habitat loss include increasingly isolated patches. Although little information exists describing the influence of patch isolation on migrant landbirds, movement of landbirds during the breeding season is inhibited by fragmentation and isolation (Desrochers and Hannon 1997). Similarly, Creegan and Osborne (2005) demonstrated that woodland birds are increasingly reluctant to venture into gaps between patch fragments as distance between woodlots increases, while Bélisle and St. Clair (2001) suggest that even small gaps in forest cover may negatively influence movement of breeding birds. More recently, Cochran and Wikelski (2005) demonstrated that migrating *Catharus* thrushes tended to remain within an area of approximately 100 m in diameter. Buler (2006), working in habitats adjacent to the Gulf Coast, reported that spring migrating Ovenbirds (*Seiurus aurocapilla*) exhibited limited movement within a patch during the day, regardless of site size or presumed quality. Matthews and Rodewald (2010) also found that Swainson's Thrushes (*Catharus ustulatus*) showed high site tenacity within urban forest stopover locations in the highly fragmented, Columbus, Ohio, metropolitan area during spring migration. Although a few birds departed from the smallest forest patches, 93% of radio-tracked thrushes remained within a patch until presumed migratory departure. Collectively, these studies suggest a reluctance in some species to move between patches during the day, although at least some individuals in the Lake Erie region move up to 30 km at night within a broader stopover region (Taylor et al. 2011).

If migrant landbirds are influenced by patch isolation and are increasingly reluctant to move between patches, then the amount of isolation within a landscape will likely influence habitat use and the rate at which energy stores are replenished because birds must allocate more time and energy searching for suitable habitat, which may depress mass gain rates and increase time spent en route (Moore et al. 1995). Further, reluctance to move between patches may concentrate migrants, increasing competition for limited resources while movement between blocks may increase exposure to predators. Consequently, a migrant's ability to gain access to blocks of habitat where fat stores can be safely replenished is restricted if fragments of suitable habitat are widely dispersed. Climate change has the potential to exacerbate impacts of patch isolation. Isolated patches are most likely to show net reductions in plant and insect diversity as conditions change, because developed areas around patches are likely to represent barriers to species from farther south that might shift into a system. The effect could be compounded if climate change leads to increases in land converted to agriculture or to land used for supporting alternative energy sources such as wind farms.

The matrix surrounding available habitat may also influence migrant use of remaining patches, either through concentrating landbirds within the habitat remnant and/or through possibly influencing resource abundance within the remaining patch. For example, France et al. (2012), working along the south shore of Lake Ontario, found that migrants concentrated within suitable habitat that was surrounded and isolated by an unsuitable habitat matrix, principally developed land and agricultural cover. Further, arthropod diversity in forested habitats is often orders of magnitude greater than in agricultural fields (Pimentel et al. 1992); consequently, resource abundance in habitat remnants within an agricultural or otherwise developed matrix may be depressed because fewer

arthropods are available to disperse among habitat remnants within a developed matrix. Patch size, degree of patch isolation, and the matrix surrounding a habitat remnant are highly interrelated and all may interact to produce negative consequences to migrants as climate changes if patches become smaller, more isolated, and increasingly surrounded by unsuitable habitat.

Proximity to Water

From a regional and landscape perspective, migrant abundance is often greatest near large bodies of water, including near-shore areas of the Gulf of Mexico (Buler et al. 2007, Buler and Moore 2011), Atlantic coast (McCann et al. 1993), and Great Lakes region (Bonter et al. 2009, Ewert et al. 2011, France et al. 2012, Johnson 2013), where birds concentrate following or preceding flights over inhospitable habitat or where locally abundant food supplies occur. For example, Bonter et al. (2009) used WSR-88D imagery from six radar stations within the Great Lakes basin to demonstrate that large numbers of spring migrants departed from near-shore habitats along the south shore of Lake Erie, and that the amount of core water area within a landscape (large water bodies such as inland lakes) was ~15 times greater in areas with high levels of migrant activity than in areas with relatively few migrants. Ewert et al. (2011), working along the north shore of Lake Huron within Michigan's eastern Upper Peninsula, demonstrated that large numbers of both spring and fall migrants concentrated within near-shore habitats, coincident with shoreline/inland differences in emergent aquatic midges (Diptera: Chironomidae), at least in spring. France et al. (2012) found that both species richness and abundance of spring and fall migrant landbirds were higher in habitats adjacent to Lake Ontario compared to inland sites. Johnson (2013), working within the western Lake Erie basin, identified proximity to the coast as one of the strongest predictors of the distribution of Blackpoll Warblers (Setophaga striata) and Black-throated Green Warblers (S. virens) during spring migration and that proximity to the coast, where emergent aquatic insects were relatively abundant, was the lone habitat variable in the top model for a guild of 24 transient warbler species.

Local-Scale Features

Migrants select habitats at local scales (Bairlein 1983, Moore et al. 1995, Moore and Aborn 2000, Moore et al. 2005, Rodewald and Brittingham 2007, Newton 2008) and their distributions tend to be associated positively with food abundance both among (Hutto 1985a, b; Hatch and Smith 2009; Cohen et al. 2012; Smith et al., unpubl. data) and within (Martin and Karr 1986, Buler et al. 2007, Ewert et al. 2011, Johnson 2013) habitats. For example, within black oak (*Quercus velutina*)-dominated habitats in southwestern Michigan, the presence of water (e.g., riparian, lacustrine, wooded palustrine habitats) was positively associated with both resource and migrant abundance at a local scale (Smith et al., unpubl. data). Indeed, food availability is thought to be a primary determinant of fine-scale habitat use patterns during migration (Hutto 1985a, Moore et al. 1995, Buler et al. 2007). Consequently, factors that influence food availability within a stopover site can play a significant role in determining migrant abundance.

Many researchers and observers have identified a strong connection between migratory bird distributions and proximity to water. The idea that migrants concentrate near water to take advantage of abundant food resources is supported by studies of aquatic-borne insects and arthropods near the Great Lakes. Work in the Detroit River and Lake St. Clair region suggests that dispersing adult caddis and mayfly species are highly concentrated near the shore, showing an exponential decline in abundance with distance and mean dispersal distances ranging from 650 to 1850 m (Kovats et al. 1996). Also, substrate type may be another factor that could be used to predict landbird migrant habitat use and prioritize conservation work, as emergent midges were found to be more than 20 times more abundant over rocky rather than sandy substrates in the Lake Michigan littoral zone (Kornis and Janssen 2011).

Field studies suggest that migrant landbirds are attracted to riparian areas within (Wilson 2001, Pennington et al. 2008) and outside the Great Lakes area (Ohmart 1994, Skagen et al. 1998, Yong et al. 1998, Skagen et al. 2005), and wooded wetlands within (Nicholls et al. 2001) and outside the Great Lakes basin (Simons et al. 2000, Gauthreaux and Belser 2002), perhaps due to the enhanced productivity of these wetland areas relative to upland areas (but see Rodewald and Matthews 2005).

Riparian areas, wetlands, and especially mesic-forested wetlands that are periodically flooded have high primary productivity (Mitsch et al. 1991). High primary productivity may create a trophic cascade, with increased arthropod abundance (both terrestrial and aquatic) and, consequently, greater bird density (Forkner and Hunter 2000, Buler et al. 2007).

With expected changes in climate and phenology, habitat quality near either the Great Lakes or riparian systems could be reduced or become less predictable. Increases in water temperature are already occurring, and thus we should expect shifts and/or mismatches in aquatic system phenology, as well as high regional variability. As noted earlier, Great Lakes surface waters, especially in the upper lakes, are warming more rapidly than air temperatures. Stream and inland water bodies are likely also warming, though some streams will be more buffered than others to changing temperatures, depending on the amount of cold groundwater input. Similarly, at local scales, some systems are more protected from warming due to shading by vegetation or protection from wind, while others are exposed to relatively warm runoff from urban areas. Further, productivity of aquatic arthropods may increasingly fluctuate with more variable precipitation and water levels, while locations of key aquatic habitats, and swarming locations and behavior, can be influenced by changes in currents, as well as wind strength and direction.

Vegetation Structure and Species Composition

Vegetation structure and composition within a patch are also associated with migrant use. While fewer studies have focused on structural characteristics of habitats as they relate to en route habitat use compared to breeding habitats, migration studies typically identify positive associations between foliage density or structurally complex habitats and migrant abundance (Petit 2000). For example, foliage density is often positively associated with migrant density in the Midwest (Blake and Hoppes 1986, Martin and Karr 1986, Packett and Dunning 2009, Cashion 2011) and the southern United States (Somershoe and Chandler 2004, Buler et al. 2007), and increasing evidence suggests that migrants, including forest breeding species, use structurally complex shrubland (Moore et al. 1990, Smith and Hatch 2008) and edge habitats during spring (Rodewald and Brittingham 2007) and fall (Moore et al. 1990, Suthers et al. 2000, Rodewald and Brittingham 2004, Smith and Hatch, unpubl. data) in eastern North America.

Use of structurally complex, early successional, and edge habitats during spring has largely been attributed to higher foliage density and associated invertebrate abundance (Blake and Hoppes 1986, Martin and Karr 1986, Hatch and Smith 2009). For fall migrants, the importance of fruit and fruiting shrubs, characteristic of early successional and edge habitats, is well documented (White and Stiles 1992, Parrish 1997, Suthers et al. 2000, McGranahan et al. 2005), including the positive dietary effects of utilizing a mixed fruit and invertebrate diet (Bairlein 2003, Podlesak and McWilliams 2006, Smith et al. 2007b). Migrant use of structurally complex habitats may also be related to predation pressure, such as fall migrating Blue-gray Gnatcatchers (Polioptila caerulea) and American Redstarts (Setophaga ruticilla) on the Gulf of Mexico coast (Cimprich et al. 2005). While more research is needed to better understand why birds are using these habitats, it is clear that early successional and edge habitats are important for both spring and fall migrants and that including them as part of an overall conservation strategy is important, irrespective of climate change.

Migrant landbirds also tend to be associated positively or negatively with particular plant species in the Great Lakes states (Strode 2009, Wood et al. 2012) and elsewhere (McGrath et al. 2009), likely due to differential distribution of food resources available from particular plant species or structure of foraging substrates. Previous research in the Great Lakes states demonstrated that landbird migrants forage on oaks (Quercus spp.) and elms (Ulmus spp.) and avoid maples (Acer spp., Graber and Graber 1983, Strode 2009, Wood et al. 2012), possibly because oak and elm species support greater invertebrate abundance (Graber and Graber 1983, Wood et al. 2012). Within coastal Lake Michigan habitats, migrants are negatively associated with density of American beech (Fagus grandifolia; Smith et al., unpubl. data), presumably because beech ranks low in terms of invertebrate diversity relative to other common plant species within the study area. Plant physiognomy may also influence suitability by affecting how a migrant moves through vegetation and perceives and captures prey, and what prey are available (Rodewald and Brittingham 2004, 2007).

For example, bird species that characteristically glean from nearby substrates should forage more readily in vegetation with short petioles, whereas species that employ more flight-based maneuvers should forage more efficiently in plants with longer petioles (Holmes and Robinson 1981, Wood et al. 2012).

Given the significance of plant species and plant community composition on the ability for a landbird to meet the exigencies of migration, climate-induced range shifts in individual plant species and the resulting changes in plant community composition will undoubtedly affect the quality of remaining habitat for migrants, perhaps especially in small, isolated patches. In some cases, benefits may exist; e.g., many species of oaks are expected to increase in abundance in this region as temperatures warm, although maples are expected to become even more dominant (Iverson et al. 2008, Scheller and Mladenoff 2008, Ravenscroft et al. 2010). Understanding these preferences can help to inform how we prioritize sites for protection and help to identify which of the many possible climate-related changes are likely to pose risks or benefits to migrants.

Non-native trees and shrubs are an increasingly common component of early successional and edge habitat and new non-native species that degrade stopover habitat, such as kudzu (Pueraria lobata), are likely to become more frequent in the Great Lakes region as the barrier of cold winters is reduced with climate change (Hall and Root 2012). Even so, surprisingly few studies have examined the use of exotics by landbirds during migration. Parrish (1997) and Johnson et al. (1985) demonstrated that fall migrating landbirds incorporate fruits of exotic shrubs into their diet and Suthers et al. (2000) found higher numbers of fall landbird migrants in habitats with high fruit abundance (containing both non-native and native species), but these studies did not explicitly evaluate the significance or the fitness consequences of exotic shrubs on migrating landbirds.

Berries of many exotic shrubs may be nutritionally poor (Ingold and Craycraft 1983, Herrera 1987) and non-native shrub species tend to hold fewer arthropods than native plants (Smith and Hatch 2008, Tallamy and Shropshire 2008). However, fruit is also an important source of dietary carotenoid, a known antioxidant important in health and condition of birds (Baeta et al. 2008, Schaefer et al. 2008). Consequently, even as the nutritional quality of non-native fruit may be low for migrating landbirds, exotic fruits may be a primary source of carotenoids. For example, fruits of non-native honeysuckle (*Lonicera* spp.) and autumn olive (*Elaeagnus umbellata*) appear to be disproportionately used by fall migrating land-birds relative to native species in northeastern Pennsylvania (R. Smith and M. Hatch, unpubl. data). Clearly, non-native species have a poten-tially diverse set of consequences for avian com-munities, and determining how exotics affect songbird populations has important management implications as plant communities change in response to climate change, especially as differ-ent suites of non-native and native plant species form new communities. Interestingly, one of the mechanisms through which exotics appear to be able to gain competitive advantage is through dif-ferences in phenology: Many invaders in northern forests leaf out earlier and keep leaves longer in the season (Smith 2013). A competitive advantage may be enhanced as growing seasons lengthen (Wolkovich and Cleland 2011).

Behavioral Flexibility at a Local Scale

Foraging ecology and habitat use of migrating landbirds are highly variable (Graber and Graber 1983; Hutto 1985b; Loria and Moore 1990; Martin and Karr 1990; Smith et al. 1998, 2004) due to the variety of vegetation structures encountered and variation in resource quality and quantity as well as changes in competitive pressures and community composition along a migratory route (Moore et al. 1995). Morse (1971) suggested such behavioral plasticity may be adaptive in that it permits a migrant to utilize a diverse array of habitat types successfully and respond to novel circumstances (Greenberg 1990, Petit et al. 1990). On the other hand, migrants have been shown to adjust aspects of their foraging behavior, even as other components remain essentially unchanged (Loria and Moore 1990, Martin and Karr 1990, Moore et al. 1995). For example, a close associa-tion exists between morphology and habitat use (Winkler and Leisler 1985, Leisler 1990); hence, morphology circumscribes a migrant's ability to exploit habitat opportunistically (Bairlein 1992). Even as migratory bird populations may have the evolutionary potential to respond to environmen-tal change (Berthold 1988, 1990), flexibility in

the behavior of migratory birds may be insuffi-cient to accommodate large-scale, rapid shifts in habitat structure (Wiedenfeld 1992, Moore et al. 1995) or vegetation community composition due to climate change. Thus, even though migratory birds may be highly mobile, their responses are only advantageous if individuals ultimately find the resources they need. If habitats become a poor match for migrant needs, survivorship during migration could decrease.

Climate Change and Phenology in the Great Lakes Region: Conservation Implications for Stopover Sites

In this section, we integrate general principles for stopover habitat conservation applicable to the Great Lakes region with proposed conserva-tion strategies that we expect will enhance the capacity of landbirds to migrate through the area, given anticipated phenological and other climate change factors.

Landbirds typically use several stopover sites between wintering and breeding grounds (Stutchbury et al. 2009, 2011), but ensuring a suc-cessful migration requires maintaining a network of suitable stopover sites throughout the migration route (Petit 2000, Mehlman et al. 2005, Moore and Woodrey 2005), perhaps especially near bar-riers such as the Great Lakes and in highly altered landscapes (Petit 2000, Mehlman et al. 2005, Faaborg et al. 2010). Based on our understanding of habitat selection at both the landscape and site scales, the following key principles should inform our efforts to establish and protect a network of stopover sites:

1. Protect areas of natural habitat, especially those of high quality, near barriers such as the Great Lakes and in highly altered agricultural landscapes.

2. Maintain, restore, or create tall, structur-ally diverse forests and other habitats such as wetlands, grasslands, shrublands, and forests at different successional stages (Petit 2000).

3. Ensure protection of small, isolated habitat patches, especially in highly urban or agricultural landscapes (Moore et al. 1995), that may serve as "stepping stones" (Petit 2000) or "fire escapes" (small, scattered

spaces where birds can find shelter and at least some, but perhaps limited food supplies *sensu* Mehlman et al. 2005).

At site scales, efforts to build a network should focus on areas with diverse plant assemblages, especially of these species known to contribute positively to migrant fitness, and in arrangements that promote connectivity across the landscape for both migrants and the plant and invertebrate species they need to survive.

We update these conservation principles here by incorporating climate change and potential shifts in phenology (Table 2.1). We provide additional criteria for evaluating habitat and use a hedge betting approach to increase the chances that habitat is available for migrants when they arrive within the Great Lakes region. Our approach should increase stopover opportunities in locations where habitat is most limited, such as highly fragmented agricultural regions, and anticipate where we might see the most ecological change, such as northern-conifer dominated sites or coastal zones, and the largest climate-related changes in land use such as more agriculture at northern latitudes.

Our criteria for selection of stopover sites for protection have an important focus on sites that are topographically diverse (Table 2.1). These locations should promote diversity of plant and invertebrate phenology and should hedge bets with respect to maintaining habitat value under changes in precipitation and water availability, an important factor since both spring high-water periods and summer and autumn droughts are expected to increase in frequency with climate change. Further, regions or sites with higher topographic diversity are likely to sustain a higher diversity of plant and prey species, even as conditions change, due to the availability of a higher variety of microclimates (Anderson and Ferree 2010).

Stopover Sites and Climate Change: Research Priorities

Here, we suggest several areas of research that will help inform more specific conservation strategies. First, in order to move from broad strategies to site-specific management tactics, we need information on likely changes in distribution of both native and exotic species that could affect

the viability of a site for migrants (Moore and Woodrey 1993, Ewert and Hamas 1996, Petit 2000, Mehlman et al. 2005, Faaborg et al. 2010, Wood et al. 2012). Similarly, information focused on where phenological changes may have the greatest impact—especially at sites that consistently support many migrants and/or enhance fitness and sites in highly altered agricultural and urban landscapes that provide refugia (Faaborg et al. 2010, Knudsen et al. 2011)—should be priority research foci.

For coastal regions, research is needed to improve our understanding of how multiple climate and land-use–related impacts in Great Lakes coastal zones may combine to influence the abundance of emergent aquatic insects during spring migration. While we have focused primarily on phenological changes, periods of low water levels of the Great Lakes may also affect migration of landbirds. For example, in coastal areas with low gradient offshore bathymetry adjacent to forests, dispersal of insects from near-shore hatching areas may be decreased, thereby reducing the quality of near-shore forests to migrants. Changes in insect abundance may exacerbate the already decreasing productivity of aquatic insects due to water temperature changes and alteration of the current area, as well as the distribution and abundance of some Great Lakes coastal wetlands (Herdendorf 2004, Albert et al. 2005, Mortsch et al. 2006). Other areas of concern include places with steep topography or bathymetry where wetlands may be lost, or where development limits conservation options regardless of changes in water level. Conservation work might then focus on high-productivity coastal areas where future suitable foraging habitat is expected to be within dispersal distance of aquatic-borne insects.

Further, as research techniques continue to improve, our approaches can be fine-tuned with analyses of comparative mortality at stopover sites within the Great Lakes region and throughout migration routes to better understand where migrants are most susceptible and where conservation programs are most needed.

Stopover Sites and Climate Change: Building Support for Conservation

In addition to revised guidelines for stopover site protection and new research, other activities need

TABLE 2.1
Conservation strategies to cope with phenological change.

Habitat factor	Climate update	Rationale
Landscape scale		
Protect contiguous blocks of habitat, favoring forest-dominated areas with a high diversity of habitat types.	Incorporate topographic and hydrologic diversity into criteria for selecting priority areas. Consider potential for change in habitat values as plant species' ranges shift, especially in northern areas. Seek opportunities to protect intact blocks of habitat in the northern half of the region, both near and farther away from coasts.	Phenological mismatches between migrants and key resources may be less likely if resource phenology is more varied. Topographic variation should increase the variety of microclimates within the area, promoting phenological diversity. Hydrologic diversity (e.g., groundwater vs. surface water-fed streams) can promote variation in water temperatures and increase the range of aquatic insect emergence timing. Boreal forest tree species that are common in the northern part of the region are expected to be less common; northern areas are also experiencing larger increases in temperature. Efforts to monitor and manage forests may need to be increased to assess and maintain diverse forested systems. Currently, much of the northern part of the region is forested, with low development pressure. However, climate change may drive shifts in land use (i.e., such as expansion of agriculture) and we should anticipate these changes as we frame our long-term strategies for land protection.
Focus conservation strategies on protecting areas near the Great Lakes shoreline, as these are heavily used by migrants.	Increase attention to areas with moderate bathymetry, given uncertainty in future water levels and fluctuations, and with forest and shrub lands close to shore.	Coastal areas will remain as concentration areas for migrants. Coastal areas with rocky or gravelly substrates and moderate bathymetry changes may be productive for aquatic insects, and near-shore woody plant foraging sites will minimize distances to aquatic insect concentrations and associated arthropods near the shoreline.
Identify and protect key "islands" of habitat in regions dominated by nonhabitat.	Increase protection of isolated and ecologically diverse patches to maximize phenological diversity, given increasing climate variability. Favor strategies that promote topographic/hydrologic diversity across protected habitat patches.	Habitat patches in a matrix of agriculture or developed land are likely to continue to be critical, especially as storm intensities increase, and some migrants experience increased stress from multiple pathways. Diversity and ecological function at isolated patches may be particularly vulnerable as plant and prey species that are lost are less likely to be replaced by dispersers due to low habitat connectivity; thus, more and larger sites may be needed to maintain the same level of benefit to migrants. Phenological mismatches may be less likely if nearby patches vary in terms of proximity to water and topographic position both of which contribute to variation in plant phenology and prey availability.
Prioritize regions with more surface water.	In addition to the resource value of habitats with water, regional surface water may buffer these regions from frost damage.	Surface waters moderate air temperature fluctuations and can reduce spring frost risk to adjacent vegetation. Areas with a higher proportion of surface water may be of increased value in spring migration periods if regions show an early warming trend followed by cold periods that damage new leaves.

Habitat factor	Climate update	Rationale
	Site scale	
Protect sites with structural diversity and diversity of plant species, with emphasis on plants that provide key food resources.	Favor sites with variation in resource phenology. Consider frost tolerance, and early versus late phenology of dominant plants when assessing habitat value. Consider potential for change in habitat values as plant species' ranges shift, especially in northern areas.	Phenological "mismatches" between migrants and key resources may be less likely if resource phenology is more varied. Plant species vary in their typical timing and, within a site, topographic or hydrological variation may promote wider variation in resource timing. More combinations of species and site types should lead to more phenological variation. In sites where frost risk is high, frost-tolerant species may be more likely to provide resources consistently, even when plant phenology advances. Plants that leaf out later may be less likely to experience damage. This factor may be less important in areas where temperatures are buffered by proximity to open water. Northern forests are expected to show declines in several conifer species' richness and aspen–birch associations. Sites dominated by these species may show declines in forest cover or regeneration; sites that already support some species that are expected to do well (many hardwoods and especially oaks may increase) may be more resource rich over the long term.
Protect sites where some portion is adjacent to or includes surface waters.	In addition to the resource value of habitats near water, surface waters may buffer adjacent vegetation from frost damage.	Surface waters moderate air temperature fluctuations and can reduce spring frost risk to adjacent vegetation. Larger water bodies (e.g., rivers or lakes, as long as they are not frozen) are likely to promote a larger area of reduced frost risk than smaller surface waters (i.e., a small stream or wetland).

to be continued. For example, an ever-increasing need is to foster partnerships to conserve habitats in ways that explicitly address a range of spatial scales (site to landscape, regional, rangewide) and in the context of interactions throughout the life cycle. In many cases, this may be most effectively done by considering guilds or possibly species of conservation concern with stopover-specific habitat requirements (Box 2.1).

Conservation projects that enhance passage of landbirds through the Great Lakes region should be integrated with other projects focused on ecological restoration, native species protection, and those with societal benefits such as improvement of water and air quality. Because direct and indirect costs associated with climate change are likely to increase, thus placing more financial strain on

the public and private sectors, stopover site conservation projects are most likely to be supported by the public if they serve multiple purposes (Hutto 2000, Mabey and Watts 2000, Karieva and Marvier 2012). Conservation may be facilitated if projects also provide economic benefits, such as shoreline management, water quality protection, sustainable timber management, and ecotourism (Xie 2012).

To facilitate an integrated approach, tools and guidance are needed to help practitioners incorporate the needs of migrants in land use and land management decisions, especially in the highly developed coastal zones. The Great Lakes migratory bird stopover portal (www.glmigratorybirds. org) is one example of a conservation planning tool designed to provide information to meet this

BOX 2.1   Kirtland's Warbler
(Setophaga kirtlandii) and
Climate Change

An example of stopover conservation considering life-cycle interactions

The Kirtland's Warbler is an endangered species that breeds only in the Great Lakes region, and it has been relatively well studied during all phases of the life cycle, except migration (Mayfield 1992, Wunderle et al. 2010, Wunderle et al. 2014). Although rarely detected in migration, the greatest number of records occurs within the Great Lakes region, where Kirtland's Warblers have been most frequently reported in shrub/scrub habitat, often near Great Lakes shorelines (Petrucha et al. 2013). If water levels recede, this could affect their ability to access midges and other aquatic emergent insects, an effect that would increase as distance between water and suitable shrub/scrub habitat increases. Carryover effects from both the breeding and wintering seasons may be another important factor that could affect the ability of Kirtland's Warblers to migrate successfully, as increasingly dry conditions are predicted to precede fall (Mishra et al. 2010, Trenberth 2011) migration, e.g., these projected changes could also affect birds during migration with carryover effects to both the breeding and wintering grounds resulting in birds migrating relatively late and in relatively poor condition. Birds arriving late on the breeding grounds often suffer lower productivity (Rockwell et al. 2012). Consequently, predicted changes in climate may increase stress on migrating Kirtland's Warblers (and other species), lower productivity, and perhaps lower survivorship. Further loss of stopover sites or stopover site quality could exacerbate carryover effects experienced immediately prior to or during migration, especially if the Great Lakes shorelines support a relatively high proportion of migrating Kirtland's Warblers.

need. We also must engage in dialog on how to protect birds in regions, such as the Great Lakes, that have many locations suitable for wind power (Pocewicz et al. 2013). Last, successful conservation activities must be effectively communicated to the public, along with results of research that highlight the extraordinary abilities of birds to migrate (Hutto 2000), to gain the public support needed to generate the funding and other resources required to achieve conservation goals for migratory birds in the Great Lakes region and elsewhere.

ACKNOWLEDGMENTS

We thank the Upper Midwest and Great Lakes Landscape Conservation Cooperative of the US Fish and Wildlife Service for partial funding of time for David Ewert and Kimberly Hall to develop this chapter. Additional funding was provided by the University of Scranton, Pennsylvania Wild Resource Conservation Program, US Fish and Wildlife Service Upper Mississippi River and Great Lakes Region Joint Venture, and the US Fish and Wildlife Service Region 3 for Robert Smith. Research support to Paul Rodewald was provided by the Ohio Division of Wildlife, US Fish and Wildlife Service, Upper Mississippi River and Great Lakes Region Joint Venture, Ohio Lake Erie Commission, and the Ohio State University. Kristin E. France contributed to development of the framework for this chapter. Carol Bocetti commented on the Kirtland's Warbler text and Michael Ward provided valuable insight on potential interactions of climate change and agriculture relative to stopover sites. Two anonymous reviewers and the editors of this volume, Eric Wood and Jherime Kellermann, provided constructive comments that improved the manuscript. We thank Sarah Burgess for assisting with preparation of the manuscript.

LITERATURE CITED

Aborn, D. A., and F. R. Moore. 1997. Pattern of movement by Summer Tanagers (*Piranga rubra*) during migratory stopover: A telemetry study. Behaviour 134:1–24.

Ahola, M., T. Laaksonen, K. Sippola, T. Eeva, K. Rainio, and E. Lehikoinen. 2004. Variation in climate warming along the migration route uncouples arrival and breeding dates. Global Change Biology 10:1610–1617.

Albert, D. A., D. A. Wilcox, J. W. Ingram and T. A. Thompson. 2005. Hydrogeographic classification for Great Lakes wetlands. Journal of Great Lakes Research 31 (Supplement 1):129–146.

Anderson, M. G., and C. E. Ferree. 2010. Conserving the stage: climate change and the geophysical underpinnings of species diversity. PLoS One 5:e11554.

Andresen, J., S. Hilberg, and K. Kunkel. 2012. Historical climate and climate trends in the Midwestern USA, 18 pp. in J. Winkler, J. Andresen, J. Hatfield, D. Bidwell, and D. Brown (coordinators), US National Climate Assessment Midwest Technical Input Reports. Available from the Great Lakes Integrated Sciences and Assessments (GLISA) Center. <http://glisa.msu.edu/docs/NCA/MTIT_Historical.pdf>

Angel, J. R., and K. E. Kunkel. 2010. The response of Great Lakes water levels to future climate scenarios with an emphasis on Lake Michigan–Huron. Journal of Great Lakes Research 36:51–58.

Augspurger, C. K. 2009. Spring 2007 warmth and frost: phenology, damage and refoliation in a temperate deciduous forest. Functional Ecology 23:1031–1039.

Augspurger, C. K. 2013. Reconstructing patterns of temperature, phenology, and frost damage over 124 years: spring damage risk is increasing. Ecology 94:41–50.

Austin, J. A., and S. M. Colman. 2007. Lake Superior summer water temperatures are increasing more rapidly than regional air temperatures: a positive ice–albedo feedback. Geophysical Research Letters 34:L06604.

Austin, J. A., and S. M. Colman. 2008. A century of warming in Lake Superior. Limnology and Oceanography 53:2724–2730.

Baeta, R., B. Faivre, S. Motreuil, M. Gaillard, and J. Moreau. 2008. Carotenoid trade-off between parasitic resistance and sexual display: An experimental study in the Blackbird (Turdus merula). Proceedings of the Royal Society B 275:427.

Bairlein, F. 1983. Habitat selection and associations of species in European passerine birds during southward, post-breeding migrations. Ornis Scandinavica 14:239–245.

Bairlein, F. 1992. Morphology-habitat relationships in migrating songbirds. Pp. 356–369 in J. M. Hagan and D. W. Johnston (editors), Ecology and conservation of Neotropical migrant landbirds. Smithsonian Institution Press, Washington, DC.

Bairlein, F. 2003. Nutritional strategies in migratory birds. Springer–Verlag, Berlin, Germany.

Bélisle, M., and C. C. St. Clair. 2001. Cumulative effects of barriers on movements of forest birds. Conservation Ecology 5:art9.

Bellard, C., C. Bertelsmeier, P. Leadley, W. Thuiller, and F. Courchamp. 2012. Impacts of climate change on the future of biodiversity. Ecology Letters 15:365–377.

Berthold, P. 1988. The control of migration in European warblers. Pp. 215–249 in Acta XIX Ornithological Congress.

Berthold, P. 1990. Genetics of migration. Pp. 269–280 in E. Gwinner (editor), Bird migration. Springer–Verlag, Berlin.

Berthold, P. 1996. Chapter 2: control mechanisms and ecophysiology. Pp. 34–232 in Control of bird migration. Chapman and Hall, London.

Blake, J. G., and G. Hoppes. 1986. Influence of resource abundance on use of tree-fall gaps by birds in an isolated woodlot. Auk 103:328–340.

Bonter, D. N., S. A. Gauthreaux, Jr., and T. M. Donovan. 2009. Characteristics of important stopover locations for migrating birds: remote sensing with radar in the Great Lakes basin. Conservation Biology 23:440–448.

Both, C., R. G. Bijlsma, and M. E. Visser. 2005. Climatic effects on timing of spring migration and breeding in a long-distance migrant, the Pied Flycatcher Ficedula hypoleuca. Journal of Avian Biology 36:368–373.

Bowlin, M. S., W. W. Cochran, and M. C. Wikelski. 2005. Biotelemetry of New World thrushes during migration: physiology, energetics, and orientation in the wild. Integrative and Comparative Biology 45:295–304.

Bradley, N. L., A. C. Leopold, J. Ross, and W. Huffaker. 1999. Phenological changes reflect climate change in Wisconsin. Proceedings of the National Academy of Sciences of the USA 96:9701–9704.

Bridge, E. S., J. F. Kelly, P. E. Bjornen, C. M. Curry, P. H. C. Crawford, and J. M. Paritte. 2010. Effects of nutritional condition on spring migration: do migrants use resource availability to keep pace with a changing world? Journal of Experimental Biology 213:2424–2429.

Brown, C. R., and M. B. Brown. 2000. Weather-mediated natural selection on arrival time in Cliff Swallows (Petrochelidon pyrrhonata). Behavioral Ecology and Sociobiology 47:339–345.

Buler, J. J. 2006. Understanding habitat use by landbirds during migration along the Mississippi Gulf Coast using a scale-dependent approach. Ph.D. dissertation, The University of Southern Mississippi, Hattiesburg, MS.

Buler, J. J., and F. R. Moore. 2011. Migrant–habitat relations during stopover along an ecological gradient: extrinsic constraints and conservation implications. Journal of Ornithology 152 (Supplement 1):101–112.

Buler, J. J., F. R. Moore, and S. Woltmann. 2007. A multi-scale examination of stopover habitat use by birds. Ecology 88:1789–1802.

CCSP. [online]. 2009. Global climate change impacts in the United States. US Climate Change Science Program, Unified Synthesis Product. US Climate Science Program. Washington, DC. <http://www.globalchange.gov/publications/reports/scientific-assessments/us-impacts> (15 May 2013).

Cashion, E. B. 2011. Avian use of riparian habitats and the Conservation Reserve Program: Migratory stopover in agroecosystems. M.S. thesis, Ohio State University, Columbus, OH.

Chartier, N. A. 2006. The influence of woodlot size and location in suburban and rural matrices on bird species richness and individual abundance. M.S. thesis, Eastern Michigan University, Ypsilanti, MI.

Chernetsov, N. 2005. Spatial behavior of medium and long-distance migrants at stopovers studied by radio tracking. Annals of the New York Academy of Sciences 1046:242–252.

Cimprich, D. A., M. Woodrey, and F. R. Moore. 2005. Passerine migrants respond to variation in predation risk during stopover. Animal Behaviour 69:1173–1179.

Cipollini, M. L., and E. W. Stiles. 1992. Relative risks of fungal rot for temperate ericaceous fruits—effects of seasonal variation on selection for chemical defense. Canadian Journal of Botany 70:1868–1877.

Cipollini, M. L., and E. W. Stiles. 1993. Fruit rot, antifungal defense, and palatability of fleshy fruits for frugivorous birds. Ecology 74:751–762.

Cleland, E. E., I. Chuine, A. Menzel, H. A. Mooney, and M. D. Schwartz. 2007. Shifting plant phenology in response to global change. Trends in Ecology and Evolution 22:357–365.

Cleland, E. E., J. M. Allen T. M. Crimmins, J. A. Dunne, S. Pau, S. E. Travers, E. S. Zavaleta, and E. M. Wolkovich. 2012. Phenological tracking enables positive species responses to climate change. Ecology 93:1765–1771.

Cochran, W. W., and M. Wikelski. 2005. Individual migratory tactics of New World *Catharus* thrushes. Pp. 274–289 in R. Greenberg and P. P. Marra (editors), Birds of two worlds: the ecology and evolution of migration. Johns Hopkins University Press, Baltimore, MD.

Cohen, E. B., F. R. Moore, and R. A. Fischer. 2012. Experimental evidence for the interplay of exogenous and endogenous factors on the movement ecology of a migrating songbirds. PloS One 7:e41818.

Cook, B. I., E. M. Wolkovich, T. J. Davies, T. R. Ault, J. L. Betancourt, J. M. Allen, K. Bolmgren, E. E. Cleland, T. M. Crimmins, N. J. B. Kraft, L. T. Lancaster, S. J. Mazer, G. J. McCabe, B. J. McGill, C. Parmesan, S. Pau, J. Regetz, N. Salamin, M. D. Schwartz, and S. E. Travers. 2012a. Sensitivity of spring phenology to warming across temporal and spatial climate gradients in two independent databases. Ecosystems 15:1283–1294.

Cook, B. I., E. M. Wolkovich, and C. Parmesan. 2012b. Divergent responses to spring and winter warming drive community level flowering trends. Proceedings of the National Academy of Sciences of the USA 109:9000–9005.

Corkum, L. D., J. J. H. Ciborowski, and D. M. Dolan. 2006. Timing of Hexagenia (*Ephemeridae ephemeroptera*) mayfly swarms. Canadian Journal of Zoology 84:1616–1622.

Creegan, H. P., and P. E. Osborne. 2005. Gap-crossing decisions of woodland songbirds in Scotland: An experimental approach. Journal of Applied Ecology 42:678–687.

Deppe, J. L., and J. T. Rotenberry. 2008. Scale-dependent habitat use by fall migratory birds: vegetation structure, floristics and geography. Ecological Monographs 78:461–487.

Desai, A. R., J. A. Austin, V. Bennington, and G. A. McKinley. 2009. Stronger winds over a large lake in response to weakening air-to-lake temperature gradient. Nature Geoscience 2:855–858.

Desrochers, A., and S. J. Hannon. 1997. Gap crossing decisions by forest songbirds during the post-fledging period. Conservation Biology 11:1204–1210.

Deutschlander, M. E., and R. Muheim. 2009. Fuel reserves affect migratory orientation of thrushes and sparrows both before and after crossing an ecological barrier near their breeding grounds. Journal of Avian Biology 40:85–89.

Diehl, R. H. 2003. Landscape associations of birds during migratory stopover. Ph.D. dissertation. The University of Illinois, Urbana-Champaign, IL.

Diehl, R. H., R. P. Larkin, and J. E. Black. 2003. Radar observations of bird migration over the Great Lakes. Auk 120:278–290.

Dobiesz, N. E., and N. P. Lester. 2009. Changes in midsummer water temperature and clarity across the Great Lakes between 1968 and 2002. Journal of Great Lakes Research 35:371–384.

Dobrowski, S. Z. 2011. A climatic basis for microrefugia: the influence of terrain on climate. Global Change Biology 17:1022–1035.

Dobrowski, S. Z., J. Abatzoglou, A. K. Swanson, J. A. Greenberg, A. R. Mynsberge, Z. A. Holden, and M. K. Schwartz. 2013. The climate velocity of the contiguous United States during the 20th century. Global Change Biology 19:241–251.

Eggermont, H., and O. Heiri. 2012. The chironomid–temperature relationship: expression in nature and palaeoenvironmental implications. Biological Reviews 37:430–456.

Elmore, A. J., S. M. Guinn, B. J. Minsley, and A. D. Richardson. 2012. Landscape controls on the timing of spring, autumn, and growing season length in mid-Atlantic forests. Global Change Biology 18:656–674.

Ewert, D. N., P. J. Doran, K. R. Hall, A. Froehlich, J. Cannon, J. B. Cole, and K. E. France. 2012. On a wing and a (GIS) layer: prioritizing migratory bird stopover habitat along Great Lakes shorelines. Final report to the Upper Midwest/Great Lakes Landscape Conservation Cooperative.

Ewert, D. N., and M. J. Hamas. 1996. Ecology of migratory landbirds during migration in the Midwest. Pp. 200–208 in F. R. Thompson III (editor), Management of midwestern landscapes for the conservation of migratory birds. USDA Forest Service General Technical Report NC-187. USDA Forest Service, North Central Forest Experiment Station, St. Paul, MN.

Ewert, D. N., M. J. Hamas, R. J. Smith, M. E. Dallman, and S. W. Jorgensen. 2011. Distribution of migratory landbirds along the northern Lake Huron shoreline. Wilson Journal of Ornithology 123:9–14.

Ewert, D. N., G. J. Soulliere, R. D. Macleod, M. C. Shieldcastle, P. G. Rodewald, J. Fujimura, E. Shieldcastle, and R. J. Gates. 2006. Migratory bird stopover site attributes in the western Lake Erie basin. Final report to the George Gund Foundation.

Faaborg, J., R. T. Holmes, A. D. Anders, K. L. Bildstein, K. M. Dugger, S. A. Gauthreaux, Jr., P. Heglund, K. A. Hobson, A. E. Jahn, D. H. Johnson, S. C. Latta, D. J. Levey, P. P. Marra, C. L. Merkford, E. Nol, S. I. Rothstein, T. W. Sherry, T. S. Sillett, F. R. Thompson III, and N. Warnock. 2010. Conserving migratory land birds in New World. Do we know enough? Ecological Applications 20:398–418.

Fenner, M. The phenology of growth and reproduction in plants. 1998. Perspectives in Plant Ecology, Evolution and Systematics 1:78–91.

Foden, W., G. Mace, J.-C. Vié, A. A. Angulo, S. Butchart, L. DeVantier, H. Dublin, A. Gutsche, S. Stuart, and E. Turak. 2008. Species susceptibility to climate change impacts. Pp. 1–11 in J.-C. Vié, C. Hilton-Taylor, and S. N. Stuart (editors), The 2008 review of the IUCN red list of threatened species. IUCN, Gland, Switzerland.

Fontaine, J. J., K. L. Decker, S. K. Skagen, and C. van Riper. 2009. Spatial and temporal variation in climate change: A bird's-eye view. Climatic Change 97:305–311.

Forkner, R. E., and M. D. Hunter. 2000. What goes up must come down? Nutrient addition and predation pressure on oak herbivores. Ecology 81:1588–1600.

France, K. E., M. Burger, T. G. Howard, M. D. Schlesinger, K. A. Perkins, M. MacNeil, D. Klein, and D. N. Ewert. 2012. Final report for Lake Ontario migratory bird stopover project. Prepared by the Nature Conservancy for the New York State Department of Environmental Conservation.

Gauthreaux, S. A. Jr., and C. G. Belser. 2002. Radar ornithology and the conservation of migratory birds. Pp. 871–875 in Bird Conservation Implementation and Integration in the Americas (C. J. Ralph and T. D. Rich, editors). USDA Department of Agriculture, Forest Service, General Technical Report PSW-GTR-19, Albany, CA.

Gibbs, J. P., and A. R. Breisch. 2001. Climate warming and calling phenology of frogs near Ithaca, New York, 1900–1999. Conservation Biology 15:1175–1178.

Gienapp, P., C. Teplitsky, J. S. Alho, J. A. Mills, and J. Merilä. 2008. Climate change and evolution: disentangling environmental and genetic responses. Molecular Ecology 17:167–178.

Girvetz, E. H., C. Zganjar, G. T. Raber, E. P. Maurer, P. Kareiva, and J. J. Lawler. 2009. Applied climate-change analysis: the Climate Wizard tool. PLoS One 4:e8320.

Glick, P., B. A. Stein, and N. A. Edelson (editors). 2011. Scanning the conservation horizon: a guide to climate change vulnerability assessment. National Wildlife Federation, Washington, DC.

Graber, J. W., and R. R. Graber. 1983. Feeding rates of warblers in spring. Condor 85:139–150.

Greenberg, R. 1990. Ecological plasticity, neophobia, and resource use in birds. Studies in Avian Biology 13:431–437.

Groffman, P. M., L. E Rustad, P. H. Templer, J. L. Campbell, L. M. Christenson, N. K. Lany, A. M. Socci, M. A. Vadeboncoeur, P. G. Schaberg, G. F. Wilson, C. T. Driscoll, T. J. Fahey, M. C. Fisk, C. L. Goodale,

M. B. Green, S. P. Hamburg, C. E. Johnson, M. J. Mitchell, J. L. Morse, L. H. Pardo, and N. L. Rodenhouse. 2012. Long-term integrated studies show complex and surprising effects of climate change in the northern hardwood forest. BioScience 62:1056–1066.

Gronewald, A. D., V. Fortin, B. Lofgren, A. Clites, C. A. Stow, and F. Quinn. 2013. Coasts, water levels, and climate change: a Great Lakes perspective. Climatic Change 120:697–711.

Gu, L., P. J. Hanson, W. Mac Post, D. P. Kaiser, B. Yang, R. Nemani, S. G. Pallardy, and T. Meyers. 2008. The 2007 eastern US spring freezes: increased cold damage in a warming world? BioScience 58:253–262.

Hall, K. R. [online]. 2012. Climate change in the Midwest: impacts on biodiversity and ecosystems. In J. Winkler, J. Andresen, J. Hatfield, D. Bidwell, and D. Brown (Coordinators). US National Climate Assessment Midwest Technical Input Report. Available from the Great Lakes Integrated Sciences and Assessment (GLISA) Center, <http://glisa.umich.edu/docs/NCA/MTIT_Biodiversity.pdf>

Hall, K. R., and T. L. Root. 2012. Climate change and biodiversity in the Great Lakes region: from fingerprints of change to helping safeguard species. Pp. 63–95 in T. Dietz and D. Bidwell (editors), Climate change in the Great Lakes region: navigating an uncertain future. Michigan State University Press, East Lansing, MI.

Hassall, C., and D. J. Thompson. 2008. The effects of environmental warming on Odonata: a review. International Journal of Odonatology 11:131–153.

Hatch, M. I., and R. J. Smith. 2009. Absence of protandry in a population of Gray Catbirds Dumetella carolinensis. Ibis 151:771–774.

Hayhoe, K., J. VanDorn, T. Croley, II, N. Schlegal, and D. Wuebbles. 2010. Regional climate change projections for Chicago and the US Great Lakes. Journal of Great Lakes Research 36:7–21.

Heath, J. A., K. Steenhof, and M. A. Foster. 2012. Shorter migration distances associated with higher winter temperatures suggest a mechanism for advancing nesting phenology of American Kestrels Falco sparverius. Journal of Avian Biology 43:376–384.

Herdendorf, C. E. 2004. Morphometric factors in the formation of Great Lakes coastal wetlands. Aquatic Ecosystem Health & Management 7:179–197.

Herrera, C. M. 1987. Vertebrate-dispersed plants of the Iberian Peninsula: A study of fruit characteristics. Ecological Monographs 57:305–331.

Hitch, A. T., and P. L. Leberg. 2007. Breeding distributions of North American bird species moving north as a result of climate change. Conservation Biology 21:534–539.

Holmes, R. T., and S. K. Robinson. 1981. Tree species preferences of foraging insectivorous birds in a northern hardwoods forest. Oecologia 48:31–35.

Hufkens, K., M. A. Friedl, T. F. Keenan, O. Sonnentag, A. Bailey, J. O'Keefe, and A. D. Richardson. 2012. Ecological impacts of a widespread frost event following early spring leaf-out. Global Change Biology 18:2365–2377.

Hurlbert, A. H., and Z. F. Liang. 2012. Spatiotemporal variation in avian migration phenology: citizen science reveals effects of climate change. PLoS One 7:e31662.

Hutto, R. L. 1985a. Habitat selection by nonbreeding, migratory land birds. Pp. 455–476 in M. L. Cody (editor), Habitat selection in birds. Academic Press, San Diego, CA.

Hutto, R. L. 1985b. Seasonal changes in the habitat distribution of transient insectivorous birds in southeastern Arizona: Competition mediated? Auk 102:120–132.

Hutto, R. L. 2000. On the importance of en route periods to the conservation of migratory landbirds. Studies in Avian Biology 20:109–114.

Ingold, J. L., and M. J. Craycraft. 1983. Avian frugivory on honeysuckle (Lonicera) in southwestern Ohio in fall. Ohio Journal of Science 83:256–258.

Inouye, D. W. 2000. The ecological and evolutionary significance of frost in the context of climate change. Ecology Letters 3:457–463.

IPCC. 2013. Summary for policymakers. In T. F. Stocker, D. Qin, G.-K. Plattner, M. M. B. Tignor, S. K. Allen, J. Boschung, A. Nauels, Y. Xia, V. Bex, and P. M. Midgley (editors), Climate change 2013: the physical science basis. Contribution of Working Group I to the Fifth Assessment Report of the Intergovernmental Panel on Climate Change, Cambridge University Press, Cambridge, UK.

Iverson, L. R., A. M. Prasad, S. N. Matthews, and M. Peters. 2008. Estimating potential habitat for 134 eastern trees species under six climate scenarios. Forest Ecology and Management 254:390–406.

James, A. R. M., and K. C. Abbott. 2014. Phenological and geographical shifts have interactive effects on migratory bird populations. American Naturalist 183:40–53.

Janssen, R. B. 1976. The spring migration, April 1–May 31, 1976, Western Great Lakes Region. American Birds 30:844–846.

Jenni, L., and M. Schaub. 2003. Behavioural and physiological reactions to environmental variation in bird migration: a review. Pp. 155–171 in P. Berthold, E. Gwinner, and E. Sonnenschein (editors), Avian migration. Springer, Heidelberg, Germany.

Johansson, J., and N. Jonzén. 2012. Effects of territory competition and climate change on timing of arrival to breeding grounds: a game-theory approach. American Naturalist 179:463–474.

Johnson, P. L. 2013. Migratory stopover of songbirds in the western Lake Erie basin. M.S. thesis, The Ohio State University, Columbus, OH.

Johnson, R. A., M. F. Willson, and J. N. Thompson. 1985. Nutritional values of wild fruits and consumption by migrant frugivorous birds. Ecology 66:819–827.

Jonzén, N., A. Hedenstrom, and P. Lundberg. 2007. Climate change and the optimal arrival of migratory birds. Proceedings of the Royal Society of London B 274:269–274.

Jonzén, N., A. Lindén, T. Ergon, E. Knudsen, J. O. Vik, D. Rubolini, D. Piacentini, C. Brinch, F. Spina, L. Karlsson, M. Stervander, A. Andersson, J. Waldenstrom, A. Lehikoinen, E. Edvardsen, R. Solvang, and N. C. Stenseth. 2006. Rapid advance of spring arrival dates in long-distance migratory birds. Science 312:1959–1961.

Karieva, P., and M. Marvier. 2012. What is conservation science? BioScience 62:962–969.

Keller, G. S., and R. H. Yahner. 2007. Seasonal forest-patch use by birds in fragmented landscapes of south-central Pennsylvania. Wilson Journal of Ornithology 119:410–418.

Kelly, J. F., L. S. DeLay, and D. M. Finch. 2002. Density-dependent mass gain by Wilson's Warblers during stopover. Auk 119:210–213.

Kirsch, E. M., P. J. Heglund, B. R. Gray, and P. McKann. 2013. Songbird use of floodplain and upland forests along the Upper Mississippi River corridor during spring migration. Condor 115:115–130.

Klaassen, M., B. J. Hoye, B. A. Nolet, and W. A. Buttemer. 2012. Ecophysiology of avian migration in the face of current global hazards. Philosophical Transactions of the Royal Society B 367:1719–1732.

Knudsen, E., A. Lindén, C. Both, N. Jonzén, F. Pulido, N. Saino, W. J. Sutherland, L. A. Bach, T. Coppack, T. Ergon, P. Gienapp, J. A. Gill, O. Gordo, A. Hedenström, E. Lehikoinen, P. P. Marra, A. P. Møller, A. L. K. Nilsson, G. Péron, E. Ranta, D. Rubolini, T. H. Sparks, F. Spina, C. E. Studds, S. A. Sæther, P. Tryjanowski, and N. Chr. Stenseth. 2011. Challenging claims in the study of migratory birds and climate change. Biological Reviews 86:928–946.

Kodra, E., K. Steinhaeuser, and A. R. Ganguly. 2011. Persisting cold extremes under 21st century warming scenarios. Geophysical Research Letters 38:L08705.

Kornis, M. S., and J. Janssen. 2011. Linking emergent midges to alewife (Alosa pseudoharengus) preference for rocky habitat in Lake Michigan littoral zones. Journal of Great Lakes Research 37:561–566.

Kovats, Z., J. A. N. Ciborowski, and L. Corkum. 1996. Inland dispersal of adult aquatic insects. Freshwater Biology 36:265–276.

Kunkel, K. E., L. Ensor, M. Palecki, D. Easterling, D. Robinson, K. G. Hubbard, and K. Redmond. 2009. A new look at the lake-effect snowfall trends in the Laurentian Great Lakes using a temporally homogenous data set. Journal of Great Lakes Research 35:23–29.

La Sorte, F. A. and F.R. Thompson. 2007. Poleward shifts in winter ranges of North American birds. Ecology 88:1803–1812.

Leisler, B. 1990. Selection and use of habitat by wintering migrants. Pp. 156–174 in E. Gwinner (editor), Bird migration. Springer-Verlag, Berlin, Germany.

Liechti, F. 2006. Birds: blowin' by the wind? Journal of Ornithology 147:202–211.

Lofgren, B. A., T. S. Hunter, and J. Wilbarger. 2012. Effects of using air temperature as a proxy for potential evapotranspiration in climate change scenarios of Great Lakes basin hydrology. Journal of Great Lakes Research 37:744–752.

Loria, D. E., and F. R. Moore. 1990. Energy demands of migration on Red-eyed Vireos, Vireo olivaceus. Behavioral Ecology 1:24–35.

Mabey, S. E., and B. D. Watts. 2000. Conservation of landbird migrants: addressing local policy. Studies in Avian Biology 20:99–108.

MacDade, L. S., P. G. Rodewald, and K. A. Hatch. 2011. Contribution of emergent insects to refueling in spring migrant songbirds. Auk 128:127–137.

Marra, P. P., C. M. Francis, R. S. Mulvihill, and F. R. Moore. 2005. The influence of climate on the timing and rate of spring bird migration. Oecologia 142:307–315.

Marra, P. P., K. A. Hobson, and R. T. Holmes. 1998. Linking winter and summer events in a migratory bird by using stable-carbon isotopes. Science 282:1884–1886.

Martin, T. E., and J. R. Karr. 1986. Patch utilization by migrating birds: Resource oriented? Ornis Scandinavica 17:165–174.

Martin, T. E., and J. R. Karr. 1990. Behavioral plasticity of foraging maneuvers of migratory warblers: Multiple selection periods for niches? Studies in Avian Biology 13:353–359.

Matthews, S. N., and P. G. Rodewald. 2010. Urban forest patches and stopover duration of migratory Swainson's Thrushes. Condor 112:96–104.

Mayfield, H. F. 1992. Kirtland's Warbler (*Dendroica kirtlandii*). Birds of North America, No. 19.

McCann, J. M., S. E. Mabey, L. J. Niles, C. Bartlett, and P. Kerlinger. 1993. A regional study of coastal migratory stopover habitat for Neotropical migrant landbirds: Land management implications. Transactions of the North American Wildlife and Natural Resources Conference 58:348–407.

McDonald, K. W., C. J. W. McClure, B. W. Rolek, and G. E. Hill. 2012. Diversity of birds in eastern North America shifts north with global warming. Ecology and Evolution 2:3052–3060.

McGranahan, D. A., S. Kuiper, and J. M. Brown. 2005. Temporal patterns in use of an Iowa woodlot during the autumn bird migration. American Midland Naturalist 153:61–70.

McGrath, L. J., C. van Riper III, and J. J. Fontaine. 2009. Flower power: tree flowering phenology as a settlement cue for migrating birds. Journal of Animal Ecology 78:22–30.

McKinnon, E. A., K. C. Fraser, and B. J. Stutchbury. 2013. New discoveries in land bird migration using geolocators, and a flight plan for the future. Auk 130:211–222.

Mehlman, D. W., S. E. Mabey, D. N. Ewert, C. Duncan, B. Abel, D. Cimprich, R. D. Sutter, and M. Woodrey. 2005. Conserving stopover sites for forest-dwelling migratory landbirds. Auk 122:1281–1290.

Miller-Rushing, A. J., T. L. Lloyd-Evans, R. B. Primack, and P. Satzinger. 2008. Bird migration times, climate change, and changing population sizes. Global Change Biology 14:1959–1972.

Mills, A. M. 2005. Changes in the timing of spring and autumn migration in North American migrant passerines during a period of global warming. Ibis 147:259–269.

Mishra, V., K. A. Cherkauer, and S. Shukla. 2010. Assessment of drought due to historic climate variability and projected future climate change in the midwestern United States. Journal of Hydrometeorology 11:46–68.

Mitsch, W. J. and J.G. Gosselink. 2007. Wetlands, 4th edition. John Wiley and Sons, Inc., Hoboken, NJ.

Mitsch, W. J., J. R. Taylor, and K. B. Benson. 1991. Estimating primary productivity of forested wetland communities in different hydrologic landscapes. Landscape Ecology 5:75–92.

Mizrahi, D. S., P. Hodgetts, V. Elia, and K. Peters. 2006. Oases along the flyway: preserving critical stopover habitat for migrating songbirds in Pennsylvania. Final report submitted to the Pennsylvania Game Commission, New Jersey Audubon Society, Harrisburg, PA.

Moore, F. R., and D. A. Aborn. 2000. Mechanisms of en route habitat selection: how do migrants make habitat decisions during stopover? Studies in Avian Biology 20:34–42.

Moore, F. R., S. A. J. Gauthreaux, P. Kerlinger, and T. R. Simons. 1995. Habitat requirements during migration: important link in conservation. Pp. 121–144 in T. E. Martin and D. M. Finch (editors), Ecology and management of Neotropical migratory birds. Oxford University Press, New York, NY.

Moore, F. R., P. Kerlinger, and T. R. Simons. 1990. Stopover on a Gulf Coast barrier island by spring trans-gulf migrants. Wilson Bulletin 102:487–500.

Moore, F. R., and Y. Wang. 1991. Evidence of food-based competition among passerine migrants during stopover. Behavioral Ecology and Sociobiology 28:85–90.

Moore, F. R., and M. S. Woodrey. 1993. Stopover habitat and its importance in the conservation of landbird migrants. Proceedings of the Forty-seventh Annual Conference, Southeastern Association of Fish and Wildlife Agencies 47:447–459.

Moore, F. R., M. S. Woodrey, J. J. Buler, S. Woltmann, and T. R. Simons. 2005. Understanding the stopover of migratory birds: a scale-dependent approach. Pp. 684–689 in C. J. Ralph and T. D. Rich (editors), USDA Forest Service Technical Report PSW-GTR-191. Pacific Southwest Research Station, Albany, CA.

Mora, C., A. G. Frazier, R. J. Longman, R. S. Dacks, M. M. Walton, E. J. Tong, J. J. Sanchez, L. R. Kaiser, Y. O. Stender, J. M. Anderson, C. M. Ambrosino, I. Fernandez-Silva, L. M. Giuseffi, and T. W.

Giambelluca. 2013. The projected timing of climate departure from recent variability. Nature 502:183–187.

Morse, D. H. 1971. The insectivorous bird as an adaptive strategy. Annual Review of Ecology and Systematics 2:177–200.

Mortsch, L., J. Ingram, A. Hebb, and S. Doka (editors). 2006. Great Lakes coastal wetland communities: Vulnerability to climate change and adaptation strategies. Final report submitted to the Climate Change Impacts and Adaptation Program, Natural Resources Canada. Environment Canada and the Department of Fisheries and Oceans, Toronto, ON.

Mukhin, A., N. Chernetsov, and D. Kishinev. 2008. Acoustic information as a distance cue for habitat recognition by nocturnally migrating passerines during landfall. Behavioral Ecology 19:716–723.

Newton, I. 2008. The migration ecology of birds. Elsevier, Amsterdam, Netherlands.

Nicholls, T. H., L. Egeland, J. Elias, and M. J. Robertsen. 2001. Habitat relationships of migratory songbirds in north-central Wisconsin. Passenger Pigeon 63:157–178.

Nolan, V., Jr. 1978. The ecology and behavior of the Prairie Warbler (Dendroica discolor). Ornithological Monographs 26:1–596.

Norris, D. R., P. P. Marra, T. K. Kyser, T. W. Sherry, and L. M. Ratcliffe. 2004. Tropical winter habitat limits reproductive success on the temperate breeding grounds in a migratory bird. Proceedings of the Royal Society of London B 271:59–64.

Notaro, M., K. Holman, A. Zarrin, E. Fluck, S. Vavrus, and V. Bennington. 2013a. Influence of the Laurentian Great Lakes on regional climate. Journal of Climate 26:789–804.

Notaro, M., K. Holman, A. Zarrin, E. Fluck, S. Vavrus, and V. Bennington. 2013b. Simulation of heavy lake-effect snowstorms across the Great Lakes Basin by RegCM4: synoptic climatology and variability. Monthly Weather Review 141:1990–2014.

Ohmart, R. D. 1994. The effects of human-induced changes on the avifauna of western riparian habitats. Studies in Avian Biology 15:273–285.

Ovaskainen, O., S. Skorokhodova, M. Yakovleva, A. Sukhov, A. Kutenkov, N. Kutenkova, A. Shcherbakov, E. Meyke, and M. del Mar Delgado. 2013. Community-level phenological response to climate change. Proceedings of the National Academy of Sciences of the USA 110:13434–13439.

Packett, D. L., and J. B. J. Dunning. 2009. Stopover habitat selection by migrant landbirds in a fragmented forest–agricultural landscape. Auk 126:579–589.

Parmesan, C. 2006. Ecological and evolutionary responses to recent climate change. Annual Review of Ecology Evolution and Systematics 37:637–669.

Parmesan, C., and G. Yohe. 2003. A globally coherent fingerprint of climate change impacts across natural systems. Nature 421:37–42.

Parrish, J. D. 1997. Patterns of frugivory and energetic condition in nearctic landbirds during autumn migration. Condor 99:681–697.

Peet, M. M. 1908. The fall migration of birds at Washington Harbor, Isle Royale, in 1905. Michigan Geological Survey Annual Report for 1908:97–119.

Pennington, D. N., J. Hansel, and R. B. Blair. 2008. The conservation value of urban riparian areas for landbirds during spring migration: land cover, scale, and vegetation effects. Biological Conservation 141:1235–1248.

Petit, D. R. 2000. Habitat use by landbirds along Neararctic–Neotropical migration routes: implications for conservation of stopover habitats. Studies in Avian Biology 20:15–33.

Petit, D. R., K. E. Petit, and L. J. Petit. 1990. Geographic variation in foraging ecology of North American insectivorous birds. Studies in Avian Biology 13:254–263.

Petrucha, M. E., P. W. Sykes, Jr., P. W. Huber, and W. W. Duncan. 2013. Spring and fall migrations of Kirtland's Warbler (Setophaga kirtlandii). North American Birds 66:382–427.

Pimentel, D., U. Stachow, D. A. Takacs, H. W. Brubaker, A. R. Dumas, J. J. Meaney, J. A. S. O'Neil, D. E. Onsi, and D. B. Corzilius. 1992. Conserving biological diversity in agricultural/forestry systems. BioScience 42:354–362.

Pocewicz, A., W. A. Estes-Zumpf, M. D. Andersen, H. E. Copeland, D. A. Keinath, and H. R. Griscom. 2013. Modeling the distribution of migratory bird stopovers to inform landscape-scale siting of wind development. PLoS One 8:e75363.

Podlesak, D. W., and S. R. McWilliams. 2006. Metabolic routing of dietary nutrients in birds: effects of diet quality and macronutrient composition revealed using stable isotopes. Physiological and Biochemical Zoology 79:534–549.

Polgar, C. A., and R. B. Primack. 2011. Leaf-out phenology of temperate woody plants: from trees to ecosystems. New Phytologist 191:926–941.

Primack, R. B., I. Ibanez, H. Higuchi, S. Don Lee, A. J. Miller-Rushing, A. M. Wilson, and J. A. Silander. 2009. Spatial and interspecific variability in phenological responses to warming temperatures. Biological Conservation 142:2569–2577.

Pryor, S. C., and R. J. Barthelmie. 2011. Assessing climate change impacts on the near-term stability of the wind energy resource over the United States. Proceedings of the National Academy of Sciences of the USA 108:8167–8171.

Pryor, S. C., R. J. Barthelmie, and J. T. Schoof. 2012. Past and future wind climates over the contiguous USA based on the North American Regional Climate Change Assessment Program Model Suite. Journal of Geophysical Research-Atmospheres 117:art16.

Pryor, S. C., R. J. Barthelmie, and J. T. Schoof. 2013. High-resolution projections of climate-related risks for the Midwestern USA. Climate Research 56:61–79.

Pulido, F., and P. Berthold. 2010. Current selection for lower migratory activity will drive the evolution of residency in a migratory bird population. Proceedings of the National Academy of Sciences of the USA 107:7341–7346.

Rathcke, B., and E. P. Lacey. 1985. Phenological patterns of terrestrial plants. Annual Review of Ecology and Systematics 16:179–214.

Ravenscroft, C., R. M. Scheller, D. J. Mladenoff, and M. A. White. 2010. Forest restoration in a mixed-ownership landscape under climate change. Ecological Applications 20:327–346.

Rockwell, S. M., C. I. Bocetti, and P. P. Marra. 2012. Winter climate on non-breeding grounds explains spring arrival date and reproductive success in an endangered migratory bird, the Kirtland's Warbler (Setophaga kirtlandii). Auk 129:744–752.

Rodewald, P. G., and M. C. Brittingham. 2004. Stopover habitats of landbirds during fall: use of edge-dominated and early-succession forests. Auk 121:1040–1055.

Rodewald, P. G., and M. C. Brittingham. 2007. Stopover habitat use by spring migrant landbirds: the roles of habitat structure, leaf development, and food availability. Auk 124:1063–1074.

Rodewald, P.G., and S. N. Matthews. 2005. Landbird use of riparian and upland forest stopover habitats in an urban landscape. Condor 107:259–268.

Root, T. L., J. T. Price, K. R. Hall, S. H. Schneider, C. Rosenzweig, and J. A. Pounds. 2003. Fingerprints of global warming on wild animals and plants. Nature 421:57–60.

Root, T. L., and S. H. Schneider. 2006. Conservation and climate change: the challenges ahead. Conservation Biology 20:706–708.

Rubenstein, D. R., C. P. Chamberlain, R. T. Holmes, M. P. Ayres, J. R. Waldbauer, G. R. Graves, and N. C. Tuross. 2002. Linking breeding and wintering ranges of a migratory songbird using stable isotopes. Science 295:1062–1065.

Ryder, T. B., J. W. Fox, and P. P. Marra. 2011. Estimating migratory connectivity of Gray Catbirds (Dumetella carolinensis) using geolocator and mark-recapture data. Auk 128:448–453.

Saino, N., R. Ambrosini, D. Rubolini, J. von Hardenberg, A. Provenzale, K. Huppop, O. Huppop, A. Lehikoinen, E. Lehikoinen, K. Rainio, M. Romano, and L. Sokolov. 2011. Climate warming, ecological mismatch at arrival and population decline in migratory birds. Proceedings of the Royal Society of London B 278:835–842.

Schaefer, H. M., K. J. McGraw, and C. Catoni. 2008. Birds use fruit colour as honest signal of dietary antioxidant rewards. Functional Ecology 22:303–310.

Scheller, R. M., and D. J. Mladenoff. 2008. Simulated effects of climate change, fragmentation, and inter-specific competition on tree species migration in Northern Wisconsin, USA. Climate Research 36:191–202.

Schwartz, M. D., R. Ahas, and A. Aasa. 2006. Onset of spring starting earlier across the Northern Hemisphere. Global Change Biology 12:343–351.

Scott, R. W., and F. A. Huff. 1996. Impacts of the Great Lakes on regional climate conditions. Journal of Great Lakes Research 22:845–863.

Seewagen, C. L., and E. J. Slayton. 2008. Mass changes of migratory landbirds during stopovers in a New York City park. Wilson Journal of Ornithology 120:296–303.

Sillett, T. S., and R. T. Holmes. 2002. Variation in survivorship of a migratory songbird throughout its annual cycle. Journal of Animal Ecology 71:296–308.

Simons, T. R., S. M. Pearson, and F. R. Moore. 2000. Application of spatial models to the stopover ecology of trans-gulf migrants. Studies in Avian Biology 20:4–14.

Skagen, S. K., J. F. Kelly, C. I. van Riper, R. L. Hutto, D. M. Finch, D. J. Krueper, and C. P. Melcher. 2005. Geography of spring landbird migration through riparian habitats in southwestern North America. Condor 107:212–227.

Skagen, S. K., C. P. Melcher, W. H. Howe, and F. L. Knopf. 1998. Comparative use of riparian corridors and oases by migrating birds in southeastern Arizona. Conservation Biology 12:896–909.

Small-Lorenz, S. L., L. A. Culp, T. B. Ryder, T. C. Will, and P. P. Marra. 2013. A blind spot in climate change vulnerability assessments. Nature Climate Change 3:91–93.

Smith, L. M. 2013. Extended leaf phenology in deciduous forest invaders: mechanisms of impact on native communities. Journal of Vegetation Science 24:979–987.

Smith, R., M. Hamas, M. Dallman, and D. Ewert. 1998. Spatial variation in foraging of the Black-throated Green Warbler along the shoreline of northern Lake Huron. Condor 100:474–484.

Smith, R. J., M. J. Hamas, D. N. Ewert, and M. E. Dallman. 2004. Spatial foraging differences in American Redstarts along the shoreline of northern Lake Huron during spring migration. Wilson Bulletin 116:48–55.

Smith, R. J., and M. I. Hatch. 2008. A comparison of shrub-dominated and forested habitat use by spring migrating landbirds in Northeastern Pennsylvania. Condor 110:682–693.

Smith, R. J., and F. R. Moore. 2005. Arrival timing and seasonal reproductive performance in a long-distance migratory landbird. Behavioral Ecology and Sociobiology 57:231–239.

Smith, R. J., F. R. Moore, and C. A. May. 2007. Stopover habitat along the shoreline of northern Lake Huron, Michigan: Emergent aquatic insects as a food resource for spring migrating landbirds. Auk 124:107–121.

Somershoe, S. G., and C. R. Chandler. 2004. Use of oak hammocks by Neotropical migrant songbirds: The role of area and habitat. Wilson Bulletin 116:56–63.

Stanley, C. Q., M. MacPherson, K. C. Fraser, E. A. McKinnon, and B. J. M. Stutchbury. 2012. Repeat tracking of individual songbirds reveals consistent migration timing but flexibility in route. PLoS One 7:e40688.

Stiles, E. W. 1980. Patterns of fruit presentation and seed dispersal in bird-disseminated woody plants in the eastern deciduous forest. American Naturalist 11:670–686.

Strode, P. K. 2003. Implications of climate change for North American Wood Warblers (Parulidae). Global Climate Biology 9:1137–1144.

Strode, P. K. 2009. Spring tree species use by migrating Yellow-rumped Warblers in relation to phenology and food availability. Wilson Journal of Ornithology 121:457–468.

Studds, C. E., and P. P. Marra. 2007. Linking fluctuations in rainfall to nonbreeding season performance in a long-distance migratory bird, *Setophaga ruticilla*. Climate Research 35:115–122.

Stutchbury, B. J. M., E. A. Gow, T. Done, M. MacPherson, J. W. Fox, and V. Afanasyev. 2011. Effects of post-breeding moult and energetic condition on timing of songbird migration into the tropics. Proceedings of the Royal Society B 278:131–137.

Stutchbury, B. J. M., J. R. Hill III, P. M. Kramer, J. Tautin, J. W. Fox, and V. Afanasyev. 2009. Tracking long-distance songbird migration by using geolocators. Science 323:896.

Suthers, H. B., J. M. Bickal, and P. G. Rodewald. 2000. Use of successional habitat and fruit resources by songbirds during autumn migration in central New Jersey. Wilson Bulletin 112:249–260.

Swanson, D. L., and J. S. Palmer. 2009. Spring migration phenology of birds in the northern prairie region is correlated with local climate change. Journal of Field Ornithology 80:351–363.

Tallamy, D. W., and K. J. Shropshire. 2008. Ranking Lepidopteran use of native versus introduced plants. Conservation Biology 23:941–947.

Taylor, P. D., S. A. Mackenzie, B. G. Thurber, A. M. Calvert, A. M. Mills, L. P. McGuire, and C. G. Guglielmo. 2011. Landscape movements of migratory birds and bats reveal an expanded scale of stopover. PLoS One 6:e27054.

Tebaldi, C., D. Adams-Smith, and N. Heller. [online]. 2012. The heat is on: US temperature trends. Climate central report. <http://www.climatecentral.org/wgts/heat-is-on/HeatIsOnReport.pdf> (accessed 27 June 2013).

Thomas, C. D., A. Cameron, R. E. Green, M. Bakkenes, L. J. Beaumont, Y. C. Collingham, B. F. N. Erasmus, M. F. de Siqueira, A. Grainger, L. Hannah, L. Hughes, B. Huntley, A. S. van Jaarsveld, G. F. Midgley, L. Miles, M. A. Ortega-Huerta, A. T. Peterson, O. L. Phillips, and S. E. Williams. 2004. Extinction risk from climate change. Nature 427:145–148.

Tøttrup, A. P., K. Thorup, K. Rainio, R. Yosef, E. Lehikoinen, and C. Rahbek. 2008. Avian migrants adjust migration in response to environmental conditions en route. Biology Letters 4:685–688.

Trenberth, K. E. 2011. Changes in precipitation with climate change. Climate Research 47:123–138.

Trumpickas, J., B. J. Shuter, and C. K. Minns. 2009. Forecasting impacts of climate change on Great Lakes surface water temperatures. Journal of Great Lakes Research 35:454–463.

Van Buskirk, J., R. S. Mulvihill, and R. C. Leberman. 2009. Variable shifts in spring and autumn migration phenology in North American songbirds associated with climate change. Global Change Biology 15:760–771.

Van Buskirk, J., R. S. Mulvihill, and R. C. Leberman. 2012. Phenotypic plasticity alone cannot explain climate-induced change in avian migration timing. Ecology and Evolution 2:2430–2437.

Visser, M. E., A. C. Perdeck, J. H. van Balen, and C. Both. 2009. Climate change leads to decreasing bird migration distances. Global Change Biology 15:1859–1865.

Wang, J., X. Bai, H. Hu, A. Clites, M. Colton, and B. Lofgren. 2011. Temporal and spatial variability of Great Lakes ice cover, 1973–2010. Journal of Climate 25:1318–1329.

Waples, J. T., and J. V. Klump. 2002. Biophysical effects of a decadal shift in summer wind direction over the Laurentian Great Lakes. Geophysical Research Letters 29:43/1–43/4.

Warren, R., J. VanDerWal, J. Price, J. A. Welbergen, I. Atkinson, J. Ramirez-Villegas, T. J. Osborn, A. Jarvis, L. P. Shoo, S. E. Williams, and J. Lowe. 2013. Quantifying the benefit of early climate change mitigation in avoiding biodiversity loss. Nature Climate Change 3:678–682.

White, D. W., and E. W. Stiles. 1992. Bird dispersal of fruits of species introduced into eastern North America. Canadian Journal of Botany 70:1689–1696.

Wiedenfeld, D. A. 1992. Foraging in temperate- and tropical-breeding and wintering male Yellow Warblers. Pp. 321–328 in J. M. Hagan and D. W. Johnson (editors), Ecology and conservation of Neotropical migrant landbirds. Smithsonian Institution Press, Washington, DC.

Willis, C. G., B. Ruhfel, R. B. Primack, A. J. Miller-Rushing, and C. C. Davis. 2008. Phylogenetic patterns of species loss in Thoreau's woods are driven by climate change. Proceedings of the National Academy of Sciences of the USA 105:17029–17033.

Wilson, T. E. 2001. Spring use of riparian habitat by migratory birds in Michigan's eastern Upper Peninsula. M.S. thesis, Central Michigan University, Mount Pleasant, MI.

Winkler, H., and B. Leisler. 1985. Morphological aspects of habitat selection in birds. Pp. 415–434 in M. L. Cody (editor), Habitat selection in birds. Academic Press, San Diego, CA.

Winkler, J., J. Andresen, J. Bisanz, R. Black, G. Guentchev, J. Nugent, K. Piromsopa, N. Rothwell, C. Zavalloni, J. Clark, H. Min, A. Pollyea, H. Prawiranata, and R. Torre. 2012. Michigan's tart cherry industry: vulnerability to climate variability and change. Pp. 104–116 in S. C. Pryor (editor), Climate change impacts, risk, vulnerability, and adaptation in the midwestern United States. Indiana University Press, Bloomington, IN.

Wolkovich, E. M., and E. E. Cleland. 2011. The phenology of plant invasions: a community ecology perspective. Frontiers in Ecology and the Environment 9:287–294.

Wood, E. M., A. M. Pidgeon, F. Liu, and D. J. Mladenoff. 2012. Birds see the trees inside the forest: the potential impacts of changes in forest composition on songbirds during spring migration. Forest Ecology and Management 280:176–186.

Wunderle, J. M., Jr., D. Currie, E. H. Helmer, D. N. Ewert, J. D. White, T. S. Ruzycki, B. Parresol, and C. Kwit. 2010. Kirtland's Warbler in anthropogenically disturbed early-successional habitats on Eleuthera, the Bahamas. Condor 112:123–137.

Wunderle, J. M. Jr., P.K. Ledrow, J.D. White, D. Currie, and D. N. Ewert. 2014. Sex and age differences in site fidelity, food resource tacking, and body condition of wintering Kirtland's Warbler (Setophaga Kirtlandii) in the Bahamas. Ornithological Monographs 80:1–62.

Xie, P. F. 2012. Socio-economic impacts of birdwatching along Lake Erie: A coastal Ohio analysis. Ohio Sea Grant OHSU-TS-061.

Yong, W., D. M. Finch, F. R. Moore, and J. F. Kelly. 1998. Stopover ecology and habitat use of migratory Wilson's Warblers. Auk 115:829–842.

Young, B., E. Byers, K. Gravuer, K. Hall, G. Hammerson, and A. Redder. [online]. 2011. Guidelines for using the NatureServe climate change vulnerability index. <https://connect.natureserve.org/sites/default/files/documents/Guidelines_NatureServeClimateChangeVulnerabilityIndex_r2.1_Apr2011.pdf> (Release 2.1, 7 April 2011).

Zuckerberg, B., A. M. Woods, and W. F. Porter. 2009. Poleward shifts in breeding bird distributions in New York State. Global Change Biology 15:1866–1883.

CHAPTER THREE

A Bird's-Eye View of the USA National Phenology Network[*]

AN OFF-THE-SHELF
MONITORING PROGRAM

Jherime L. Kellermann, Carolyn A. F. Enquist, Diana L. Humple, Nathaniel E. Seavy, Alyssa Rosemartin, Renée L. Cormier, and LoriAnne Barnett

Abstract. Phenology is central to the biology and ecology of organisms and highly sensitive to climate. Differential responses to climate change are impacting phenological synchrony of interacting species, which has been implicated in the decline of migratory birds that rely on seasonal resources. However, few studies explicitly measure phenology of seasonal habitat resources on the breeding and wintering grounds and at stopover sites. While avian monitoring methods are widely standardized, methods of monitoring resource phenology can be highly variable and difficult to integrate. The USA National Phenology Network (USA-NPN) has developed standardized plant and animal phenology protocols and a robust information management system to support a range of stakeholders in collecting, storing, and sharing phenology data, at the appropriate scale, to shed light on phenological synchrony. The USA-NPN's *Nature's Notebook* can be integrated into established research programs, ensuring that data will be comparable over time and across projects, taxa, regions, and research objectives. We use two case studies to illustrate the application of USA-NPN methods and protocols to established long-term landbird research programs. By integrating phenology into these programs, avian ecologists are increasing their ability to understand the magnitude and consequences of phenological responses to climate change.

Key Words: citizen science, climate change, phenological mismatch, phenology monitoring, phenology protocol, Ruby-throated Hummingbird, stopover habitat.

Phenology, the timing of reoccurring life cycle events, is integral to plant and animal physiology, ecology, population and community dynamics, and adaptive evolution as well as nutrient, carbon, and water cycles (Chuine 2010, Forrest and Miller-Rushing 2010, Pau et al. 2011). Phenology can be highly responsive to climate change and thus provides a valuable indicator of effects across spatial and temporal scales (Walther 2010, Yang and Rudolf 2010). Although not all

[*] Kellermann, J. L., C. A. F. Enquist, D. L. Humple, N. E. Seavy, A. Rosemartin, R. L. Cormier, and L. Barnett. 2015. A bird's-eye view of the USA National Phenology Network: an off-the-shelf monitoring program. Pp. 47–60 in E. M. Wood and J. L. Kellermann (editors), Phenological synchrony and bird migration: changing climate and seasonal resources in North America. Studies in Avian Biology (no. 47), CRC Press, Boca Raton, FL.

species are exhibiting shifts in their phenology, those that are responding often vary in their rate, direction, and magnitude of change (Parmesan 2007, Møller et al. 2008, Both et al. 2009). Visser and Both (2005) emphasized that understanding the significance of phenological changes in any particular species must be in context of the phenology of resources necessary to sustain demanding life-cycle periods. For example, bird migration is a seasonal period of extremely high physiological demand (McWilliams and Karasov 2001) that can have significant impacts on annual survival (Sillett and Holmes 2002).

Differential impacts of climate change on migratory birds and their seasonal resources are creating or exaggerating phenological mismatches (Jones and Cresswell 2010, Saino et al. 2011), especially in long-distance migrants (Coppack and Both 2002, Both et al. 2006). Migration includes three disinct stages: departure, the migratory journey, and arrival. Advances or delays in departure dates or duration of the migratory period in response to climate change can alter arrival dates (Marra et al. 2005, Balbontin et al. 2009, Both 2010), ultimately affecting reproductive success, fitness, and population dynamics (Møller 2001, Baker et al. 2004, Both et al. 2006). Migratory responses vary geographically and ecologically within and among bird populations and species, depending on sex, diet, migration distance, migration timing, brood size, population size, and phenotypic plasticity (Jenni and Kery 2003, MacMynowski and Root 2007, Vegvari et al. 2010, Gordo and Doi 2012). Selective pressures on birds to maintain phenological synchrony within seasonal communities under climate change may come from lower (e.g., food resources) or upper trophic levels (e.g., predators; Both et al. 2009) . Thus, documenting species phenology across trophic levels is critical to understanding avian responses to climate change during all stages of migration.

Changing phenology is altering seasonal vegetation condition and food availability, causing mismatch across trophic levels on breeding grounds and resulting in avian population declines (Jones and Cresswell 2010, Saino et al. 2011). Migratory birds time their arrival at breeding grounds to obtain high-quality territories and take advantage of seasonal food resources necessary to reproduce and fledge young successfully (Møller 2001). Important components of landbird breeding habitat quality include vegetation condition and food availability (Sherry and Holmes 1996, Smith and Moore 2005). Across the Northern Hemisphere, spring phenology of plants (e.g., flowering, leafing) has been advancing with warming trends during the past century (Menzel et al. 2006, Ellwood et al. 2013). Strong evolutionary selection for herbivorous insects to respond to cues that allow synchronization with host plant phenology (Bale et al. 2002, van Asch and Visser 2007) is also resulting in advancement of spring insect phenology, although mismatch may be occurring and even increasing (Parmesan 2007, Singer and Parmesan 2010).

Phenological changes in resources are also occurring along migration routes, which can affect stopover duration and frequency, arrival timing, body condition upon arrival, and reproductive success during the breeding season (Sandberg and Moore 1996, Smith and Moore 2005, Norris and Marra 2007, Seewagen et al. 2013). Over the course of successive stopover events, birds repeatedly encounter novel habitats and unpredictable environmental conditions (Nemeth and Moore 2007, Gillies and St. Clair 2010), where they attempt to maximize food acquisition and fat accumulation while balancing search-time costs (Aborn and Moore 1997, Paxton et al. 2008). The ability to identify patches of local food availability quickly based on habitat characteristics detectable across scales should shorten stopover duration and migration timing and improve fitness. Birds likely select stopover sites through a top-down hierarchical process (Chernetsov 2006), advancing from coarse- to fine-scale features (Buler et al. 2007). Plant phenology may be a direct indicator of food availability for birds that consume plant resources such as nectar or fruits (Smith et al. 2007, McKinney et al. 2012), or an index of availability of herbivorous insect prey synchronized with plant phenology (van Asch and Visser 2007, McGrath et al. 2009). Therefore, changes in plant phenology and phenological mismatch at stopover sites could affect avian fitness directly through decreased food abundance and indirectly though loss of habitat selection cues at stopover sites.

In contrast to arrival and stopover ecology, the dynamics of migratory departure are poorly understood; however, departure dates of migrants from both ends of their migratory range are undoubtedly being affected by climate-driven habitat condition and resource phenology. For example, American Redstarts (*Setophaga ruticilla*) wintering in tropical regions have later departure in dry years or when relegated to drier suboptimal habitats (Marra and Holberton 1998, McKellar et al. 2013) and show increased abundance on the breeding grounds in years following high plant productivity in wintering areas (Wilson et al. 2011). In many parts of Europe, timing of departure from breeding to wintering grounds is advancing (Cotton 2003). While long-distance migrants are advancing their fall departure in order to cross the Sahel before seasonal dry periods, multibrood species are taking advantage of longer growing seasons on breeding grounds and delaying their departure (Jenni and Kery 2003). Therefore, changes in departure may be related to changes in habitat and resources on the departure or arrival grounds or at stopover sites and depend on a species' migratory ecology and life history traits.

The complexity of avian phenological responses highlights the need for data on not only their temporal and spatial patterns, but also the critical seasonal resources upon which they depend. Most migration studies have been limited by logistics or funding to collect new phenological data, or to the availability of existing data at relevant ecological, spatial, or temporal scales. To fill this void, metrics of vegetation phenology for large-scale applications are often derived from remote sensing data, such as the normalized difference vegetation index (NDVI; Greenberg et al. 2008, Balbontin et al. 2009, Tottrup et al. 2010), which are available at large spatiotemporal scales but are limited in their spatial and ecological resolution relative to ground observations (White et al. 2009). However, the advent of the Internet combined with the growing popularity of citizen science campaigns is beginning to change the manner and scale at which we can collect and share data to address these problems at relevant scales and produce alternative metrics of vegetation phenology (Silvertown 2009, Dickinson et al. 2010).

OBSERVING PHENOLOGY FROM THE GROUND UP

Observational phenology monitoring can assist in downscaling and validation of remote sensing data (Kang et al. 2003, Kaheil et al. 2008) and supply detailed site-specific information on species and species interactions at fine resolutions relevant to individual organisms, populations, and communities (Schwartz 1999). The benefits of using observational phenology monitoring to address regional to global-scale changes include its generalizability across scales, its ability to represent ecological diversity, and its ability to reveal physical mechanisms (Schwartz 1994). However, understanding the complex dynamics of climate change impacts on bird migration phenology under changing climate conditions will require data on birds and seasonal resources across broad spatiotemporal scales, from individual breeding, wintering, and stopover sites to species' entire annual migratory range while transcending biogeographic and political boundaries (Kelly and Hutto 2005, MacMynowski and Root 2007, Taylor et al. 2011). Collection of sufficient data necessitates a monumental collaborative effort.

Historic plant phenology data sets exist for North America, including data from legendary naturalists such as Henry David Thoreau and Aldo Leopold (Miller-Rushing and Primack 2008) and the cloned lilac and honeysuckle phenology programs initiated in the mid-1950s (Schwartz et al. 2012). Europe has a somewhat richer trove of historic phenology data on a relatively wide range of species (Ahas et al. 2002), including data sets on grape harvest that span more than 500 years (Chuine et al. 2004, Menzel 2005). Despite the immense value of these data for examining long-term variation, trends, and extremes in biotic responses to climate (Bradley et al. 1999, Miller-Rushing and Primack 2008, Ellwood et al. 2013), they focus on either small geographic regions, a relatively narrow range of species, or species that may not be ecologically relevant to bird migration habitats (Marra et al. 2005).

Over the past decade, there has been a boon in the collection of ecological data by citizen scientists, especially with the advent of online science initiatives (Dickinson et al. 2010). New citizen science programs such as eBird have provided vast amounts of data on migratory birds in North America that are being used to explore spatiotemporal patterns of bird migration and responses to climate change at unprecedented scale and resolution (Sullivan et al. 2009, Fink et al. 2010, Hurlbert and Liang 2012). Arguably, additional observational data at similar spatiotemporal scales and ecological resolutions on the phenology of seasonal resources relevant to migratory birds at breeding, wintering, and stopover sites would be of significant value to understanding migratory habitat ecology, especially once integrated with bird monitoring data.

USA National Phenology Network and *Nature's Notebook*

In 2007, the USA-NPN (www.usanpn.org) was formed to track plant and animal phenology and ecological responses to climate change at a continental scale. The USA-NPN engages a diverse range of citizen scientist volunteers; federal, state, and nongovernmental organizations; professional research scientists; and educators to conduct monitoring as well as outreach and education. To guide this effort, the USA-NPN implemented *Nature's Notebook* in 2009, an online monitoring program comprising scientifically vetted protocols, observation guidelines, and interfaces for data entry and retrieval (the national phenology database, NPDb), in addition to a growing range of data products, educational materials, and support tools (Rosemartin et al. 2013, Denny et al. 2014).

Nature's Notebook provides all of the materials and tools necessary to implement phenology monitoring as an "off-the-shelf" package, ready for use and application to meet a wide range of monitoring and research goals and objectives (www. nn.usanpn.org). Through *Nature's Notebook*, people can create monitoring groups, register sites, plants, and animals to be observed; create and print standardized data sheets; and submit their observations. The methods and protocols explain and utilize species-specific phenophases for commonly occurring functional groups. Data can be collected throughout the year, and the methods and protocols help to minimize differences in terminology and phenological categories that regularly occur across projects using different protocols that later create challenges for data integration and analysis (Freeman et al. 2007).

All data submitted through *Nature's Notebook* are housed in the NPDb and, along with supporting metadata, are freely available online for download (www.usanpn.org/results/data). Registered sites are georeferenced and observers can record a range of supplemental details about the site such as slope, aspect, irrigation, landcover, development, and distance to nearest road or body of water. As of May 2014, there were 673 plant and 272 animal species available for observation with detailed description pages of the organisms and their phenophases (www.usanpn.org/nn/species_search).

The *Nature's Notebook* program employs "status" monitoring, whereby observers record the phenological status for a suite of species-specific phenophases on every observation date. Such repeated sampling can reveal trends throughout the annual life cycle of a species and has greater predictive power, in contrast to "event" monitoring, which typically only captures the date on which a phenological event first occurs (Denny et al. 2014). Observers can also record abundance or intensity measures for many plant phenophases such as the number of flowers per plant or the percent of flowers that are open (Table 3.1), which can provide data on the relative abundance of plant resources potentially available at a given location and time.

Nature's Notebook: Relevance to Bird Migration

While *Nature's Notebook* involves monitoring of a wide range of plants and animals and their species-specific phenophases, events most relevant to the study of bird migration and habitat ecology are the presence and feeding of bird species, plant phenophases related to bird food resources (e.g., flowers, fruits) or resources used directly by herbivorous invertebrate prey (e.g., leaf buds, young leaves), and the presence or emergence of invertebrate prey such as lepidopteran larvae. Table 3.1 lists some of the phenophases most frequently recorded through *Nature's Notebook* for birds, plants, and insects that could be applied to research and monitoring of bird migration and seasonal resources at breeding, wintering, and stopover habitats.

While other broad-scale monitoring programs that focus on a single taxonomic group provide extremely valuable phenological data that can be overlaid with data from other programs (e.g., eBird, Frogwatch), a significant benefit of *Nature's Notebook* is that it can provide colocated data for multiple taxa and phenophases using nationally standardized protocols. Because the protocols used in Nature's Notebook have been specifically developed to address phenological questions, their use will reduce the challenges that can arise when comparing phenological data collected with different methods (Freeman et al. 2007).

TABLE 3.1

*Bird, plant, and insect phenophases recorded through the USA-NPN's phenology monitoring program, Nature's Notebook,
relevant to study of bird migration, seasonal resources, and condition of stopover habitats.*

Taxa	Phenophase	Phenophase description	Abundance/intensity measures
Birds	Active individuals	One or more individuals seen moving or at rest	Number of birds in this phenophase
	Feeding	One or more individuals feeding; if possible, record species or substance being eaten	Number of birds in this phenophase
	Fruit/seed consumption	One or more individuals eating fruits, seeds, or cones of a plant; if possible, record plant name	Number of birds in this phenophase
	Insect consumption	One or more individuals seen eating insects; if possible, record insect or describe it	Number of birds in this phenophase
Plants	Flowers or flower buds	One or more fresh open or unopened flowers or flower buds visible	Number of flowers or flower buds
	Open flowers	One or more open, fresh flowers visible	Percentage open
	Pollen release	One or more flowers release visible pollen grains when gently shaken or blown onto a surface	Amount of pollen released
	Fruits	One or more fruits visible on the plant	Number of fruits
	Ripe fruits	One or more ripe fruits visible on the plant	Percentage ripe
	Recent fruit or seed drop	One or more mature fruits or seeds dropped or removed from the plant since last visit	Number mature fruits dropped seed
	Breaking leaf buds	One or more breaking leaf buds visible	Number of buds breaking
	Increasing leaf size	A majority of leaves have not yet reached full size and are still growing larger	Percentage of full size
Insects	Active caterpillars	One or more caterpillars (larvae) moving or at rest; when seen on a plant, record the name of the plant or describe it in the comments field	Number of individuals in this phenophase
	Caterpillars feeding	One or more caterpillars feeding; if possible, record species or substance being eaten	Number of individuals in this phenophase
	Flower visitation	One or more individuals visiting flowers or flying from flower to flower; if possible, record plant	Number of individuals in this phenophase
	Active subadults	One or more subadults moving or at rest	Number of individuals in this phenophase

Status of *Nature's Notebook* and the NPDb

As of 13 June 2014, over 3,629 registered observers had actively submitted over 3.5 million status records from 6,258 sites located across all 50 US states, the US Virgin Islands, and Puerto Rico. The species with the most observations in the NPDb recorded through *Nature's Notebook* include red maple (*Acer rubrum*), coyotebush (*Baccharis pilularis*), and quaking aspen (*Populus tremuloides*) for plants; bumblebees (*Bombus* spp.), monarch (*Danaus plexippus*), and red admiral (*Vanessa atalanta*) for insects; and American Robin (*Turdus migratorius*), Ruby-throated Hummingbird (*Archilochus colubris*), and Black-capped Chickadee (*Poecile atricapillus*) for birds. Up-to-date summaries and visualization of all species-specific data housed in the NPDb can be accessed and viewed graphically on the USA-NPN website using the data dashboard or the phenology visualization tool (www.usanpn.org).

Application of *Nature's Notebook* and NPDb Data

Phenological Mismatch in Ruby-Throated Hummingbirds

As an example of potential data applications for the NPDb, we examined phenological synchrony and overlap between flowering and movements of a nectarivorous bird (Miller-Rushing et al.

2010). We compared spring temperatures in 2011 and 2012, flowering times of 10 plant species, and migration and arrival times of Ruby-throated Hummingbird (RTHU, *Archilochus colubris*) at registered *Nature's Notebook* sites in the northeastern extent of their US breeding range in Maine. Hummingbird species can be highly responsive to variation in climate and habitat resources (Russell et al. 1994, McKinney et al. 2012, Courter et al. 2013), and central and eastern portions of the United States experienced record-breaking early spring temperatures and flowering phenology in 2012 (Ellwood et al. 2013).

We assessed flowering of nine plant species; three species of milkweed (*Asclepias* spp.), red columbine (*Aquilegia canadensis*), jewelweed (*Impatiens capensis*), Japanese knotweed (*Polygonum cuspidatum*), tulip tree (*Liriodendron tulipifera*), common lilac (*Aguilegia canadensis*), and common dandelion (*Taraxacum officinale*; a species that is not likely to be used by hummingbirds, but that may provide an indication of small flowering forb phenology). We first calculated the proportion of sites that recorded hummingbird presence out of all sites that were actively monitoring this bird species in

Maine during eleven 10-day periods from 1 April through 18 June in 2011 and 2012. We also calculated the proportion of individual plants of these species that had flowers during the same 10-day periods. We used these 10-day increments because all sites were surveyed at least once during that interval. We assessed phenological synchrony as the difference in date of mean and peak phenology between birds and flowering within and between years. We calculated annual overlap of hummingbird and flowering as the definite integral of the area shared by these plotted phenological response curves (Miller-Rushing et al. 2010) using R 3.0.1 (R Core Team 2014).

Ruby-throated Hummingbirds were monitored at 80 total sites (38 in 2011, 61 in 2012), and flowering phenology was monitored at 116 individual plants (65 in 2011, 88 in 2012). We found that mean flowering date was over 7 days earlier in 2012 than in 2011, while peak flowering advanced by nearly a month (Figure 3.1). In contrast, mean date of RTHU detection was about 14 days later in 2012 and peak date was about 10 days later. Therefore, the time between both mean and peak RTHU migration and plant flowering was

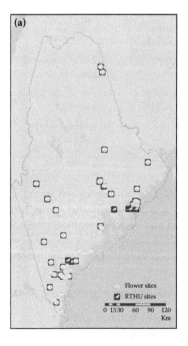
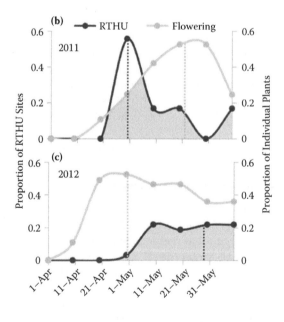

Figure 3.1. Map of registered USA National Phenology Network *Nature's Notebook* sites in (a) Maine, where the phenology of Ruby-throated Hummingbirds and relevant flowering plants was monitored in (b) 2011 and (c) 2012, and phenological synchrony and overlap of the proportion of sites with RTHU and proportion of individual nectar plants flowering per 10-day period from 1 April to 18 June, 2011–2012. Vertical dotted lines are the peak dates when the greatest proportion of sites reported hummingbirds (black) and the greatest proportion of individual plants were flowering (gray).

about 10 days later in 2012. Despite changes in synchrony, phenological overlap was almost two times greater in 2012 (1.74) than in 2011 (0.94), due to the earlier peak among hummingbirds in 2011. Analyses of long-term trends using historic records of the North American Bird Phenology Program for Ruby-throated Hummingbirds also showed the counterintuitive pattern of delayed phenology under early spring conditions of 2012. Courter et al. (2013) found that although Ruby-throated Hummingbirds have advanced their spring migration over the past 130 years, following warm winters their migration is delayed above 40° north following warm winters. Delays may be in response to the recent inability of plants to meet their winter chilling requirement in the southern United States, which in turn is reducing food resources, diminishing stopover habitat quality, and necessitating longer stopovers en route (Russell et al. 1994).

In contrast, our results suggest that phenological overlap actually increased despite drastic changes in synchrony. Overlap may be due to an even and prolonged, less concentrated or "peaky" distribution of bird migration and plant flowering. Nonetheless, it remains unclear whether there are fitness consequences to this interannual variation in phenology. Data on nesting and presence of fledglings and phenophases that also can be monitored through Nature's Notebook would shed light on the potential impacts on population and evolutionary dynamics. Furthermore, to improve the accuracy of phenological estimates, sites should be monitored at least twice a week during periods of rapid phenological change. Infrequent monitoring rates are a limitation of the current size of the NPDb. We anticipate, however, that as the number and frequency of observations grow, analysis at finer temporal resolutions will be more robust.

Integrating Nature's Notebook *into Established Research and Monitoring Efforts*

Ornithology enjoys a rich history of established, vetted, and standardized avian research and monitoring using methods such as point counts, area searches, nest monitoring, and mist-netting. These methods have been incorporated into national and international programs that have been collecting demographic and phenological data on bird populations for decades, such as the monitoring avian productivity and survivorship (MAPS) program, and the North American breeding bird survey (BBS; Sauer et al. 1994, DeSante et al. 1995). Vegetation monitoring is a common component of many landbird research and monitoring programs; however, these methods and protocols are often far more variable throughout the discipline, and often do not include a plant phenology component. Furthermore, some avian researchers may be uncertain of exactly how to monitor plant phenology at new or already established study sites.

For nearly 50 years, Point Blue Conservation Science (Point Blue, founded as Point Reyes Bird Observatory) has studied bird populations at the Palomarin Field Station in Point Reyes National Seashore, Marin County, California (Porzig et al. 2011). Program efforts include year-round monitoring of resident and migrant landbirds through constant-effort mist-netting and nest-monitoring programs, studying habitat associations, weather and vegetation monitoring, and a rigorous intern-training program. Long before researchers ever considered that these long-term data would be used to understand the consequences of climate change (MacMynowski et al. 2007, Goodman et al. 2012), researchers at Palomarin were already studying the variation in timing of avian annual life cycles—such as when birds breed, molt, and migrate (DeSante and Baptista 1989, Howell and Gardali 2003, Elrod et al. 2011). However, despite interest in the resources that plants provide for birds, relatively little standardized information on plant phenology was previously collected at the station.

Recent evidence of climate-change disruption to phenological relationships between birds and vegetation has highlighted the importance of incorporating phenological monitoring of plants into research and monitoring at Palomarin (Saino et al. 2011, Visser et al. 2012). With their primary expertise in avian ecology, station personnel looked to the broader scientific community for phenological monitoring methods. The broad scope, applications, and vetted and standardized plant phenology monitoring protocols of the USA-NPN and Nature's Notebook provided the tools necessary for integration into the station's long-term monitoring efforts. Furthermore, the USA-NPN's ample offerings of webinars and in-person training workshops assured station researchers that data quality is a high programmatic priority.

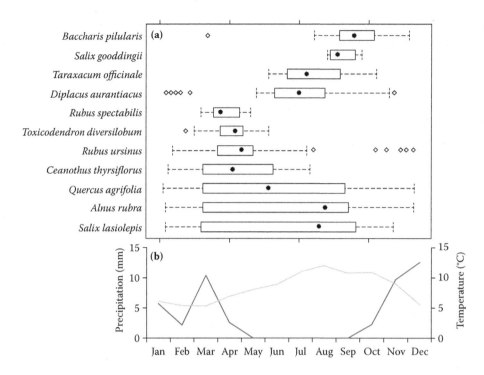

Figure 3.2. (a) The flowering phenology of plant species monitored at the Palomarin Field Station, California; associated study sites in the Point Reyes area using the USA-NPN *Nature's Notebook* program in 2012; and (b) mean daily precipitation (black line) and minimum temperature per month (gray line) for the Bolinas, California, region, 2012 (NOAA 2013). Box plots indicate median flowering date (black dot).

In 2012, Point Blue began collecting plant phenology data on 15 tree, shrub, and forb species at the Palomarin Field Station and four additional associated long-term bird monitoring sites in the Point Reyes area (in Point Reyes National Seashore, Golden Gate National Recreation Area, and Marin County Open Space District; hereafter, Palomarin study sites). The plant species selected are among the dominant species at the study sites that are ecologically important to the bird species studied. Plant phenology is monitored approximately every 7–10 days at some sites in conjunction with mist-netting visits, year-round except at two sites where no winter avian monitoring occurs. Although it requires a time commitment to conduct the phenology monitoring and ensure consistency among observers and proper data management, the efficiency of the protocols allows the plots to be monitored during normal mist-netting operational hours, given moderate bird capture rates.

We compared timing and duration of flowering among 11 plant species at Palomarin study sites using box plots generated in R 3.1.0 (R Core Team 2014). We found a wide range of plant phenological strategies from long to short durations (Figure 3.2). Additionally, while some species appear to respond to winter precipitation, flowering in the spring, others may respond to increasing temperatures, flowering in the late summer (Figure 3.2). These phenological patterns, along with other plant phenophases including leaf-out and fruiting, can now be compared with annual patterns of migratory bird phenophases including arrival, departure, passage, breeding, and molt. Furthermore, long-term monitoring could reveal the normal range of variation in plant and avian phenology, trends, and extreme events, as well as how these phenological patterns respond to climate variation and climate change (Gordo and Sanz 2010, Ellwood et al. 2013).

In addition to increased understanding of the ecological communities at these long-term study sites resulting from implementation of *Nature's Notebook*, incorporating phenology monitoring efforts at Palomarin has enhanced the intern training experience by broadening awareness and expertise to other ecological components of the systems they are studying and providing a better understanding of how ecological informatics, including citizen science projects like *Nature's Notebook*, are changing ecological research (Jeong et al. 2011, Hurlbert and Liang 2012, Miller-Rushing et al. 2012). Additionally, it has the potential to engage the visiting public and contributes to local, regional, and national efforts to track the influences of climate change.

DISCUSSION

Interannual and seasonal changes in habitat resources are central to understanding avian spatial and temporal distribution patterns, habitat ecology, reproductive success, and adaptive evolution in light of phenological synchrony across trophic levels (Both et al. 2009). Including vegetation phenology monitoring within bird monitoring programs can provide important insights into these dynamics. By moving beyond assessment of vegetation structure and composition alone, which may experience minimal change within and among years, phenology data provide information relevant to habitat quality that can change over short time periods of days to weeks. Furthermore, because phenology is highly sensitive to climate, phenology data can help develop a more mechanistic understanding of migratory bird habitat selection (Smith et al. 2007, McGrath et al. 2009, Carlisle et al. 2012).

Although a number of successful national scale bird monitoring programs involving both professional and citizen scientists exist, no contemporary programs for monitoring plant phenology have been implemented with comparable spatial scale and ecological complexity with equivalent success. The USA-NPN *Nature's Notebook* program is providing a unique range of data products and tools that can be directly applied, implemented, and integrated into bird monitoring and research programs for a wide group of users, including state and federal natural resource management agencies (Enquist et al. 2014) and nongovernmental conservation organizations, academic researchers, and citizen science and public education programs. Furthermore, despite the diversity of objectives, goals, and objectives represented by this wide range of stakeholders, the use of standardized data collection protocols and a shared database (NPDb) allows data to be integrated, comparable, and applicable to novel questions at both fine and coarse geographic scales.

Currently, *Nature's Notebook* is being applied by the Inventory and Monitoring Program of the US National Park Service through the California Phenology Project (www.usanpn.org/cpp/) and the Northeast Temperate Network (science.nature.nps.gov/im/units/netn/) to understand ecosystem responses to climate variability and climate change, inform natural resource management and decision making, and engage and educate the public. Existing efforts have produced extensive protocols and recommendations for implementing phenology monitoring using *Nature's Notebook* and these are freely available online through their websites. Building upon these early successes, the National Wildlife Refuge System of the US Fish and Wildlife Service is now implementing phenology via *Nature's Notebook* in its inventory and monitoring efforts. Detailed information about these programs and how to initiate an observation program is available through the USA-NPN website (www.usanpn.org).

Recommendations for Implementing Phenology Monitoring

Organizations should consider several factors when implementing or integrating *Nature's Notebook* into new or existing avian research and monitoring programs for migratory as well as resident birds. These include (1) selection of focal vegetation species for monitoring, (2) determining the number of sites to monitor and the number of individual plants of each species at each site, (3) the frequency of monitoring visits to each site, and (4) the phenological metrics of interest (e.g., onset, peak, duration) and methods for calculation.

Selecting focal plant species will depend on the objectives of the study, the bird species of interest, and the composition and diversity of vegetation communities at sites. Focal plant species should be relevant to bird species of interest, such as use as a foraging substrate or as a food resource; be sufficiently abundant to monitor multiple individuals at multiple sites; and capture flowering or other key phenophases throughout the study season (Figure 3.2).

Nature's Notebook is relatively easy to implement, but the amount of time required for observer training, data collection, data entry and management, and analysis should be carefully considered. As such, the number of sites and individual plants monitored will greatly depend on monetary and personnel resources for the study and the size and complexity of the study area. The greater the size or ecological complexity of a study area, the more sites or individual plants may be necessary to capture the phenological variation of the area. Similarly, the frequency of observations should also take into account the variation in plant phenology over time and space. During peak periods of change, such as spring leaf-out or flowering, one or two visits per week are preferable for capture of phase transitions, while during periods of less change, such as winter in northeastern temperate regions, one visit per month or less may be acceptable. For integration into established programs, phenology monitoring can be concurrent with other efforts, such as avian point counts, nest searching, or mist-netting, as at the Palomarin field station and associated study sites.

A wide variety of phenological metrics can be calculated from data collected using *Nature's Notebook* and available in the NPDb (Gerst et al., unpubl. ms). The onset, peak, and duration of phenophases within individuals or at sites are all common metrics. Each metric has important considerations related to sampling, analysis, and confounding factors. For example, estimates of onset may be sensitive to changes in population size and sampling frequency (Miller-Rushing et al. 2008, 2010). Therefore, desired metrics and expected analytical techniques should be considered before implementation of monitoring and data collection. Accordingly, the USA-NPN can provide an array of resources and support to facilitate project design, research, development, and implementation.

CONCLUSIONS

To understand long-term ecological responses to climate change that are critical for adaptive management and conservation of natural resources, we need multitaxa phenology monitoring across broad biogeographic regions (Parmesan 2007, Lawler 2009, Richardson et al. 2013, Enquist et al. 2014). As we have shown, the professional and amateur science communities have implemented and participated in a number of large bird-monitoring campaigns. These programs have primarily focused on landbirds and terrestrial systems; increased monitoring of freshwater and marine birds and habitats will provide additional insight into broader avian phenology dynamics. Unequivocally, colocated data on phenology of bird habitat and food resources will help us unravel the complexities of trophic cascades, phenological mismatch, and phenotypic plasticity and aid in assessing species vulnerability to environmental disturbance and change at spatiotemporal scales and ecological resolutions relevant to the interannual and seasonal ranges of migratory bird species across the globe.

ACKNOWLEDGMENTS

We thank all contributors to *Nature's Notebook* and the US Geological Survey, National Park Service, University of Arizona, and National Science Foundation for their support of the USA-NPN. Thanks to Point Reyes National Seashore, Golden Gate National Recreation Area, Marin County Open Space District, Karen A. and Keven W. Kennedy Foundation, March Foundation, Kimball Foundation, Makray Family Foundation, Lagunitas Brewing Company, and members and board of directors of Point Blue Conservation Science for their support of Palomarin; and interns and staff at the Palomarin and TomKat field stations, especially Carlene Henneman, Mark Dettling, Tom Gardali, and Geoff Deupel.

LITERATURE CITED

Aborn, D. A., and F. R. Moore. 1997. Pattern of movement by Summer Tanagers (*Piranga rubra*) during migratory stopover: A telemetry study. Behaviour 134:1077–1100.

Ahas, R., A. Aasa, A. Menzel, V. G. Fedotova, and H. Scheifinger. 2002. Changes in European spring phenology. International Journal of Climatology 22:1727–1738.

Baker, A. J., P. M. Gonzalez, T. Piersma, L. J. Niles, I. de Lima Serrano do Nascimento, P. W. Atkinson, N. A. Clark, C. D. T. Minton, M. K. Peck, and G. Aarts. 2004. Rapid population decline in Red Knots: fitness consequences of decreased refuelling rates and late arrival in Delaware Bay. Proceedings of the Royal Society of London B 271:875–882.

Balbontin, J., A. P. Møller, I. G. Hermosell, A. Marzal, M. Reviriego, and F. de Lope. 2009. Individual responses in spring arrival date to ecological conditions during winter and migration in a migratory bird. Journal of Animal Ecology 78:981–989.

Bale, J. S., G. J. Masters, I. D. Hodkinson, C. Awmack, T. M. Bezemer, V. K. Brown, J. Butterfield, A. Buse, J. C. Coulson, J. Farrar, J. E. G. Good, R. Harrington, S. Hartley, T. H. Jones, R. L. Lindroth, M. C. Press, I. Symrnioudis, A. D. Watt, and J. B. Whittaker. 2002. Herbivory in global climate change research: direct effects of rising temperature on insect herbivores. Global Change Biology 8:1–16.

Both, C. 2010. Flexibility of timing of avian migration to climate change masked by environmental constraints en route. Current Biology 20:243–248.

Both, C., S. Bouwhuis, C. M. Lessells, and M. E. Visser. 2006. Climate change and population declines in a long-distance migratory bird. Nature 441:81–83.

Both, C., M. van Asch, R. G. Bijlsma, A. B. van den Burg, and M. E. Visser. 2009. Climate change and unequal phenological changes across four trophic levels: constraints or adaptations? Journal of Animal Ecology 78:73–83.

Bradley, N. L., A. C. Leopold, J. Ross, and W. Huffaker. 1999. Phenological changes reflect climate change in Wisconsin. Proceedings of the National Academy of Sciences of the USA 96:9701–9704.

Buler, J. J., F. R. Moore, and S. Woltmann. 2007. A multi-scale examination of stopover habitat use by birds. Ecology 88:1789–1802.

Carlisle, J. D., K. L. Olmstead, C. H. Richart, and D. L. Swanson. 2012. Food availability, foraging behavior, and diet of autumn migrant landbirds in the Boise foothills of southwestern Idaho. Condor 114:449–461.

Chernetsov, N. 2006. Habitat selection by nocturnal passerine migrants en route: mechanisms and results. Journal of Ornithology 147:185–191.

Chuine, I. 2010. Why does phenology drive species distribution? Philosophical Transactions of the Royal Society B 365:3149–3160.

Chuine, I., P. Yiou, N. Viovy, B. Seguin, V. Daux, and E. L. R. Ladurie. 2004. Historical phenology: grape ripening as a past climate indicator. Nature 432:289–290.

Coppack, T., and C. Both. 2002. Predicting life-cycle adaptation of migratory birds to global climate change. Ardea 90:369–378.

Cotton, P. A. 2003. Avian migration phenology and global climate change. Proceedings of the National Academy of Sciences of the USA 100:12219–12222.

Courter, J. R., R. J. Johnson, W. C. Bridges, and K. G. Hubbard. 2013. Assessing migration of Ruby-throated Hummingbirds (*Archilochus colubris*) at broad spatial and temporal scales. Auk 130:107–117.

Denny, E. G., K. L. Gerst, A. J. Miller-Rushing, G. L. Tierney, T. M. Crimmins, C. A. F. Enquist, P. Guertin, A. H. Rosemartin, M. D. Schwartz, K. A. Thomas, and J. F. Weltzin. 2014. Standardized phenology monitoring methods to track plant and animal activity for science and resource management applications. International Journal of Biometeorology 58:591–601.

Desante, D. F., and L. F. Baptista. 1989. Factors affecting the termination of breeding in Nuttall's White-crowned Sparrows. Wilson Bulletin 101:120–124.

Desante, D. F., K. M. Burton, J. F. Saracco, and B. L. Walker. 1995. Productivity indices and survival rate estimates from MAPS, a continent-wide programme of constant-effort mist-netting in North America. Journal of Applied Statistics 22:935–947.

Dickinson, J. L., B. Zuckerberg, and D. N. Bonter. 2010. Citizen science as an ecological research tool: challenges and benefits. Annual Review of Ecology, Evolution, and Systematics 41:149–172.

Ellwood, E. R., S. A. Temple, R. B. Primack, N. L. Bradley, and C. C. Davis. 2013. Record-breaking early flowering in the eastern United States. PloS One 8:e53788.

Elrod, M. L., N. E. Seavy, R. L. Cormier, and T. Gardali. 2011. Incidence of eccentric molt in first-year Wrentits increases with fledge date. Journal of Field Ornithology 82:325–332.

Enquist, C. A., J. L. Kellermann, K. L. Gerst, and A. J. Miller-Rushing. 2014. Phenology research for natural resource management in the United States. International Journal of Biometeorology 58:579–589.

Fink, D., W. M. Hochachka, B. Zuckerberg, D. W. Winkler, B. Shaby, M. A. Munson, G. Hooker, M. Riedewald, D. Sheldon, and S. Kelling. 2010. Spatiotemporal exploratory models for broad-scale survey data. Ecological Applications 20:2131–2147.

Forrest, J., and A. J. Miller-Rushing. 2010. Toward a synthetic understanding of the role of phenology in ecology and evolution. Philosophical Transactions of the Royal Society B 365:3101–3112.

Freeman, S. N., D. G. Noble, S. E. Newson, and S. R. Baillie. 2007. Modeling population changes using data from different surveys: the common birds census and the breeding bird survey. Bird Study 54:61–72.

Gillies, C. S., and C. C. St Clair. 2010. Functional responses in habitat selection by tropical birds moving through fragmented forest. Journal of Applied Ecology 47:182–190.

Goodman, R. E., G. Lebuhn, N. E. Seavy, T. Gardali, and J. D. Bluso-Demers. 2012. Avian body size changes and climate change: warming or increasing variability? Global Change Biology 18:63–73.

Gordo, O., and H. Doi. 2012. Spring phenology delays in an insular subtropical songbird: is response to climate change constrained by population size? Journal of Ornithology 153:355–366.

Gordo, O., and J. J. Sanz. 2010. Impact of climate change on plant phenology in Mediterranean ecosystems. Global Change Biology 16:1082–1106.

Greenberg, R., A. Kozlenko, M. Etterson, and T. Dietsch. 2008. Patterns of density, diversity, and the distribution of migratory strategies in the Russian boreal forest avifauna. Journal of Biogeography 35:2049–2060.

Howell, S. N. G., and T. Gardali. 2003. Phenology, sex ratios, and population trends of *Selasphorus* hummingbirds in central coastal California. Journal of Field Ornithology 74:17–25.

Hurlbert, A. H., and Z. F. Liang. 2012. Spatiotemporal variation in avian migration phenology: citizen science reveals effects of climate change. PloS One 7:e31662.

Jenni, L., and M. Kery. 2003. Timing of autumn bird migration under climate change: advances in long-distance migrants, delays in short-distance migrants. Proceedings of the Royal Society of London B 270:1467–1471.

Jeong, S. J., C. H. Ho, H. J. Gim, and M. E. Brown. 2011. Phenology shifts at start vs. end of growing season in temperate vegetation over the Northern Hemisphere for the period 1982–2008. Global Change Biology 17:2385–2399.

Jones, T., and W. Cresswell. 2010. The phenology mismatch hypothesis: are declines of migrant birds linked to uneven global climate change? Journal of Animal Ecology 79:98–108.

Kaheil, Y. H., M. K. Gill, M. McKee, L. A. Bastidas, and E. Rosero. 2008. Downscaling and assimilation of surface soil moisture using ground truth measurements. IEEE Transactions on Geoscience and Remote Sensing 46:1375–1384.

Kang, S. Y., S. W. Running, J. H. Lim, M. S. Zhao, C. R. Park, and R. Loehman. 2003. A regional phenology model for detecting onset of greenness in temperate mixed forests, Korea: an application of MODIS leaf area index. Remote Sensing of Environment 86:232–242.

Kelly, J. F., and R. L. Hutto. 2005. An east–west comparison of migration in North American wood warblers. Condor 107:197–211.

Lawler, J. J. 2009. Climate change adaptation strategies for resource management and conservation planning. Year in Ecology and Conservation Biology 2009 1162:79–98.

MacMynowski, D. P., and T. L. Root. 2007. Climate and the complexity of migratory phenology: sexes, migratory distance, and arrival distributions. International Journal of Biometeorology 51:361–373.

MacMynowski, D. P., T. L. Root, G. Ballard, and G. R. Geupel. 2007. Changes in spring arrival of Nearctic–Neotropical migrants attributed to multiscalar climate. Global Change Biology 13:2239–2251.

Marra, P. P., C. M. Francis, R. S. Mulvihill, and F. R. Moore. 2005. The influence of climate on the timing and rate of spring bird migration. Oecologia 142:307–315.

Marra, P. P., and R. L. Holberton. 1998. Corticosterone levels as indicators of habitat quality: effects of habitat segregation in a migratory bird during the nonbreeding season. Oecologia 116:284–292.

McGrath, L. J., C. van Riper, and J. J. Fontaine. 2009. Flower power: tree flowering phenology as a settlement cue for migrating birds. Journal of Animal Ecology 78:22–30.

McKellar, A. E., P. P. Marra, S. J. Hannon, C. E. Studds, and L. M. Ratcliffe. 2013. Winter rainfall predicts phenology in widely separated populations of a migrant songbird. Oecologia 172:595–605.

McKinney, A. M., P. J. CaraDonna, D. W. Inouye, B. Barr, C. D. Bertelsen, and N. M. Waser. 2012. Asynchronous changes in phenology of migrating Broad-tailed Hummingbirds and their early-season nectar resources. Ecology 93:1987–1993.

McWilliams, S. R., and W. H. Karasov. 2001. Phenotypic flexibility in digestive system structure and function in migratory birds and its ecological significance. Comparative Biochemistry and Physiology A 128:579–593.

Menzel, A. 2005. A 500 year pheno-climatological view on the 2003 heatwave in Europe assessed by grape harvest dates. Meteorologische Zeitschrift 14:75–77.

Menzel, A., T. H. Sparks, N. Estrella, E. Koch, A. Aasa, R. Ahas, and A. N. A. Zust. 2006. European phenological response to climate change matches the warming pattern. Global Change Biology 12:1969–1976.

Miller-Rushing, A. J., T. T. Høye, D. W. Inouye, and E. Post. 2010. The effects of phenological mismatches on demography. Philosophical Transactions of the Royal Society B 365:3177–3186.

Miller-Rushing, A. J., T. L. Lloyd-Evans, R. B. Primack, R. B. and P. Satzinger. 2008. Bird migration times, climate change, and changing population sizes. Global Change Biology 14:1959–1972.

Miller-Rushing, A. J., and R. B. Primack. 2008. Global warming and flowering times in Thoreau's concord: A community perspective. Ecology 89:332–341.

Miller-Rushing, A., R. Primack, and R. Bonney. 2012. The history of public participation in ecological research. Frontiers in Ecology and the Environment 10:285–290.

Møller, A. P. 2001. Heritability of arrival date in a migratory bird. Proceedings of the Royal Society of London B 268:203–206.

Møller A. P., D. Rubolini, and E. Lehikoinen. 2008. Populations of migratory bird species that did not show a phenological response to climate change are declining. Proceedings of the National Academy of Sciences of the USA 105:16195–16200.

Nemeth, Z., and F. R. Moore. 2007. Unfamiliar stopover sites and the value of social information during migration. Journal of Ornithology 148:S369–S376.

Norris, D. R., and P. P. Marra. 2007. Seasonal interactions, habitat quality, and population dynamics in migratory birds. Condor 109:535–547.

Parmesan, C. 2007. Influences of species, latitudes and methodologies on estimates of phenological response to global warming. Global Change Biology 13:1860–1872.

Pau, S., E. M. Wolkovich, B. I. Cook, T. J. Davies, N. J. B. Kraft, K. Bolmgren, J. L. Betancourt, and E. E. Cleland. 2011. Predicting phenology by integrating ecology, evolution and climate science. Global Change Biology 17:3633–3643.

Paxton, K. L., C. Van Riper, and C. O'Brien. 2008. Movement patterns and stopover ecology of Wilson's Warblers during spring migration of the lower Colorado River in southwestern Arizona. Condor 110:672–681.

Porzig, E. L., K. E. Dybala, T. Gardali, G. Ballard, G. R. Geupel, and J. A. Wiens. 2011. Forty-five years and counting: Reflections from the Palomarin Field Station on the contribution of long-term monitoring and recommendation for the future. Condor 113:713–723.

R Core Team. 2014. R: A language and environment for statistical computing. R Foundation for Statistical Computing, Vienna, Austria.

Richardson, A. D., T. F. Keenan, M. Migliavacca, Y. Ryu, O. Sonnentag, and M. Toomey. 2013. Climate change, phenology, and phenological control of vegetation feedbacks to the climate system. Agricultural and Forest Meteorology 169:156–173.

Rosemartin, A. H., T. M. Crimmins, C. A. Enquist, K. L. Gerst, J. L. Kellermann, E. E. Posthumus, P. Guertin, and J. F. Weltzin. 2013. Organizing phenological data resources to inform natural resource conservation. Biological Conservation 173:90–97.

Russell, R. W., F. L. Carpenter, M. A. Hixon, and D. C. Paton. 1994. The impact of variation in stopover habitat quality on migrant Rufous Hummingbirds. Conservation Biology 8:483–490.

Saino, N., R. Ambrosini, D. Rubolini, J. von Hardenberg, A. Provenzale, K. Huppop, O. Huppop, A. Lehikoinen, E. Lehikoinen, K. Rainio, M. Romano, and L. Sokolov. 2011. Climate warming, ecological mismatch at arrival and population decline in migratory birds. Proceedings of the Royal Society of London B 278:835–842.

Sandberg, R., and F. R. Moore. 1996. Fat stores and arrival on the breeding grounds: Reproductive consequences for passerine migrants. Oikos 77:577–581.

Sauer, J. R., B. G. Peterjohn, and W. A. Link. 1994. Observer difference in the North American breeding bird survey. Auk 111:50–62.

Schwartz, M. D. 1994. Monitoring global change with phenology: the case of the spring green wave. International Journal of Biometeorology 38:18–22.

Schwartz, M. D. 1999. Advancing to full bloom: planning phenological research for the 21st century. International Journal of Biometeorology 42:113–118.

Schwartz, M. D., J. L. Betancourt, and J. F. Weltzin. 2012. From Caprio's lilacs to the USA National Phenology Network. Frontiers in Ecology and the Environment 10:324–327.

Seewagen, C. L., C. G. Guglielmo, and Y. E. Morbey. 2013. Stopover refueling rate underlies protandry and seasonal variation in migration timing of songbirds. Behavioral Ecology 24:634–642.

Sherry, T. W. and R. T. Holmes. 1996. Winter habitat quality, population limitation, and conservation of Neotropical–Nearctic migrant birds. Ecology 77:36–48.

Sillett, T. S., and R. T. Holmes. 2002. Variation in survivorship of a migratory songbird throughout its annual cycle. Journal of Animal Ecology 71:296–308.

Silvertown, J. 2009. A new dawn for citizen science. Trends in Ecology and Evolution 24:467–471.

Singer, M. C., and C. Parmesan. 2010. Phenological asynchrony between herbivorous insects and their hosts: signal of climate change or pre-existing adaptive strategy? Philosophical Transactions of the Royal Society B 365:3161–3176.

Smith, R. J., and F. R. Moore. 2005. Arrival timing and seasonal reproductive performance in a long-distance migratory landbird. Behavioral Ecology and Sociobiology 57:231–239.

Smith, S. B., K. H. McPherson, J. M. Backer, B. J. Pierce, D. W. Podlesak, and S. R. McWilliams. 2007. Fruit quality and consumption by songbirds during autumn migration. Wilson Journal of Ornithology 119:419–428.

Sullivan, B. L., C. L. Wood, M. J. Iliff, R. E. Bonney, D. Fink, and S. Kelling. 2009. eBird: a citizen-based bird observation network in the biological sciences. Biological Conservation 142:2282–2292.

Taylor, P. D., S. A. Mackenzie, B. G. Thurber, A. M. Calvert, A. M. Mills, L. P. McGuire, and C. G. Guglielmo. 2011. Landscape movements of migratory birds and bats reveal an expanded scale of stopover. PloS One 6:e27054.

Tottrup, A. P., K. Rainio, T. Coppack, E. Lehikoinen, C. Rahbek, and K. Thorup. 2010. Local temperature fine-tunes the timing of spring migration in birds. Integrative and Comparative Biology 50:293–304.

van Asch, M., and M. E. Visser. 2007. Phenology of forest caterpillars and their host trees: the importance of synchrony. Annual Review of Entomology 52:37–55.

Vegvari, Z., V. Bokony, Z. Barta, and G. Kovacs. 2010. Life history predicts advancement of avian spring migration in response to climate change. Global Change Biology 16:1–11.

Visser, M. E., and C. Both. 2005. Shifts in phenology due to global climate change: the need for a yardstick. Proceedings of the Royal Society B 272:2561–2569.

Visser, M. E., L. te Marvelde, and M. E. Lof. 2012. Adaptive phenological mismatches of birds and their food in a warming world. Journal of Ornithology 153:S75–S84.

Walther, G. R. 2010. Community and ecosystem responses to recent climate change. Philosophical Transactions of the Royal Society B 365:2019–2024.

White, M. A., K. M. de Beurs, K. Didan, D. W. Inouye, A. D. Richardson, O. P. Jensen, J. O'Keefe, G. Zhang, R. R. Nemani, W. J. D. van Leeuwen, J. F. Brown, A. de Wit, M. Schaepman, X. M. Lin, M. Dettinger, A. S. Bailey, J. Kimball, M., D. Schwartz, D. D. Baldocchi, J. T. Lee, and W. K. Lauenroth. 2009. Intercomparison, interpretation, and assessment of spring phenology in North America estimated from remote sensing for 1982–2006. Global Change Biology 15:2335–2359.

Wilson, S., S. L. LaDeau, A. P. Tøttrup, and P. P. Marra. 2011. Range-wide effects of breeding- and nonbreeding-season climate on the abundance of a Neotropical migrant songbird. Ecology 92:1789–1798.

Yang, L. H., and V. H. W. Rudolf. 2010. Phenology, ontogeny and the effects of climate change on the timing of species interactions. Ecology Letters 13:1–10.

PART TWO

Migratory Connectivity

Spring Resource Phenology and Timing of Songbird Migration across the Gulf of Mexico[*]

Emily B. Cohen, Zoltán Németh, Theodore J. Zenzal, Jr., Kristina L. Paxton, Robert Diehl, Eben H. Paxton, and Frank R. Moore

Abstract. Migratory songbirds are advancing their arrival to breeding areas in response to climatic warming at temperate latitudes. Less is understood about the impacts of climate changes outside the breeding period. Every spring, millions of migrating songbirds that overwinter in the Caribbean and Central and South America stop to rest and refuel in the first available habitats after crossing the Gulf of Mexico (hereafter, the Gulf). We used capture data from a long-term banding station (20 years: 1993 to 2012) located on the northern coast of the Gulf to assess the passage timing of 17 species making northward migrations into eastern North America. We further assessed spring resource phenology as measured by normalized difference vegetation index (NDVI) on nonbreeding ranges and en route. We tested the hypotheses that spring passage timing has advanced during the past two decades and that annual variability in passage timing into eastern North America is related to spring resource phenology on stationary nonbreeding ranges and during passage south of the Gulf. Further, we assessed whether annual variability in resource phenology south of the Gulf was a good indicator of the conditions that migrants encountered upon first landfall in eastern North America. We found no trend in migration timing for species that migrate from South America and annual variability in their passage timing was unrelated to environmental conditions in nonbreeding ranges or en route. Species that migrate from Central America and the Caribbean delayed arrival by 2 to 3 days over the 20-year period and arrived later during years when conditions were dryer in nonbreeding ranges and passage areas south of the Gulf. Further, year-to-year variability in spring resource phenology in nonbreeding ranges and passage areas south of the Gulf were not good indicators of resource phenology upon arrival in eastern North America. Therefore, despite the fact that many migrant species have been arriving increasingly earlier to breeding grounds, the passage timing of 17 species into eastern North America has either not changed or is slightly later, due to drying spring conditions in Central America and the Caribbean. Our results suggest that Nearctic–Neotropical migratory birds adjust the rate of migration primarily within eastern North America and, in light of warmer temperatures in the temperate zone and earlier arrival timing to breeding ranges, species that overwinter in Central America and the Caribbean may be increasing the speed of migration within eastern North America.

Key Words: climate change, Nearctic–Neotropical migratory songbirds, resource phenology, timing of migration.

---

[*] Cohen, E. B., Z. Németh, T. J. Zenzal, Jr., K. L. Paxton, R. Diehl, E. H. Paxton, and F. R. Moore. 2015. Spring resource phenology and timing of songbird migration across the Gulf of Mexico. Pp. 63–82 in E. M. Wood and J. L. Kellermann (editors), Phenological synchrony and bird migration: changing climate and seasonal resources in North America. Studies in Avian Biology (no. 47), CRC Press, Boca Raton, FL.

igratory animals are particularly vulnerable to global climate change. A changing climate can alter the ecological conditions in any of the geographically disparate areas migrants depend upon, including breeding or nonbreeding stationary areas or stopover sites where migrants rest and refuel along the migratory route (Bairlein and Winkel 2001). Climate change in any of these areas may influence survival, reproductive success, or ecological cues that migrants use to optimize migration timing (Gienapp 2012). Many migratory songbird species have advanced their arrival to breeding areas in response to warmer spring temperatures (Jonzén et al. 2006, Parmesan 2006). However, in some cases, the rate of phenological change in prey is faster than the rate at which migrants are responding, which creates a phenological mismatch between the timing of migration and breeding in birds and the annual peak abundance of their prey (Both et al. 2006). Arrival timing has fitness consequences (Smith and Moore 2005, Møller et al. 2009, Gienapp and Bregnballe 2012), and trophic mismatches can lead to population declines (Both et al. 2006). Thus, it is essential that we understand how phenology, which is influenced by climate, influences migratory birds as well as the mechanisms by which migratory birds respond to climate change.

The onset of migration is initially determined by the interplay of an endogenous program and predictive cues (photoperiod) and then fine-tuned to local environmental conditions by supplementary cues such as temperature, food abundance, and social environment (Møller 2001, Pulido 2007, Wingfield 2008, Robson and Barriocanal 2011, Studds and Marra 2011, Newton 2012, Ramenofsky et al. 2012). An interactive mechanism appears to provide enough flexibility in the timing of migratory departures from overwintering areas to enable migrants to adjust their schedules appropriately and to arrive at the breeding grounds earlier in warmer springs and later in colder ones (Both et al. 2005, McKellar et al. 2012, Tryjanowski et al. 2013). Environmental conditions such as temperature have been correlated between the breeding grounds and along the migratory route providing migrants with a reliable cue to time their migration (Both et al. 2005, Smith et al. 2009). Moreover, rainfall and temperature, and consequently food abundance and body condition, at the overwintering grounds

influence timing of departure from overwintering areas (Saino et al. 2004, Gordo et al. 2005, Saino et al. 2007, Studds and Marra 2011), as well as arrival at the breeding grounds (McKellar et al. 2012), suggesting an important role for phenotypic plasticity in adjusting migration timing. On the other hand, recent analyses indicate that without genetic microevolution, phenotypic plasticity alone may not have been able to maintain the rate of change we have seen in migration timing during the past decades (Jonzén et al. 2006, Van Buskirk et al. 2012).

The consequences of climate change for migratory bird populations have been measured largely in relation to time and condition upon arrival to breeding grounds (Both et al. 2004, Lehikoinen and Sparks 2010). However, climate change may also differentially impact the availability of resources on stationary nonbreeding ranges and passage areas along the migration route which could influence a migrant's ability to deposit fuel loads necessary to migrate successfully, with consequences for timely arrival on the breeding grounds, not to mention survival during migration (Bairlein and Hüppop 2004). These effects may be magnified when migrants are faced with major geographic barriers. For example, coastal areas of the southeastern United States provide critical stopover habitat for migrant songbirds recovering from the energetic demands of the 18- to 24-hour nonstop flight across the Gulf (Moore 2000, Buler et al. 2007). Moreover, because the majority of species and individuals in the Nearctic–Neotropical migration system breed to the north of the Gulf Coast and overwinter to the south, this region hosts a disproportionately large number of migrant songbirds in fall and spring. Despite the high volume of migrants passing through the region, our project is the first study to determine if passage timing into eastern North America from the northern coast of the Gulf is influenced by resource phenology south of the Gulf.

Resource phenology in passage areas, as measured by NDVI, has been correlated with the speed of migration, timing of arrival at the breeding grounds (Tøttrup et al. 2008, 2012), and egg-laying date (Both et al. 2006). However, these studies have been primarily conducted in the European-African migration system, while studies within the Nearctic–Neotropical migration system are lacking. We used long-term migration banding data from the northern coast of the Gulf to test if Nearctic–Neotropical migrant passage timing has advanced over the past 20 years and if passage timing is related to resource phenology on stationary nonbreeding ranges or in passage areas south of the Gulf. To better understand the relationship between passage timing and resource phenology, we assessed whether spring resource phenology has advanced on stationary nonbreeding ranges or passage areas. In addition, we assessed if spring resource phenology south of the Gulf, on nonbreeding ranges and passage areas, is a good indicator of resource phenology north of the Gulf.

METHODS

Passage Timing

We used mist-netting capture data of spring migrants from the northern coast of the Gulf of Mexico to evaluate timing of passage (Figure 4.1). The study site at Johnson's Bayou (29°45′N, 93°37′W), is located about 1.3 km inland and is part of a chenier system that runs parallel with the coast. Southwestern Louisiana is dominated by open grassy marsh and wet prairie with forest occurring on narrow and elongated relic beach ridges called cheniers (Moore 1999, Barrow et al. 2000). Coastal woodlands often concentrate migratory birds because they are essentially habitat islands bordered by the Gulf to the south and marshes, agricultural fields, and other open landscapes to the north that are unsuitable for forest-dwelling songbirds (Gauthreaux 1971, 1972). Moreover, the volume of migratory passage across southwestern Louisiana and southeastern Texas is higher than other parts of the northern coast of the Gulf (Gauthreaux 1999), which raises the importance of cheniers as valuable stopover sites for songbirds during migratory passage (Moore 1999).

Migrants were captured using mist nets (12 × 2.6 m or 6 × 2.6 m, 30 mm mesh) from 28 March to 6 May from 1993 to 1996 and 1998 to 2012 (Table 4.1). Nets were opened daily between 7:00 and 16:00 CST, except in the case

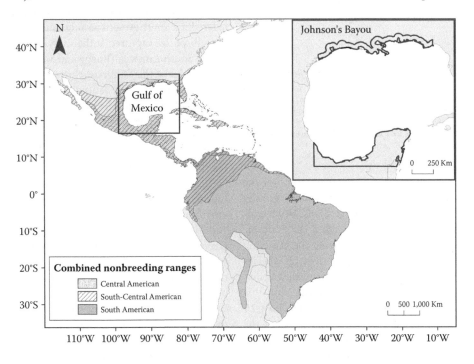

Figure 4.1. The combined nonbreeding ranges of the Central (Gray Catbird, Hooded Warbler, Indigo Bunting, Ovenbird, Painted Bunting, Worm-eating Warbler, Wood Thrush), South-Central (Kentucky Warbler, Northern Waterthrush, Rose-breasted Grosbeak, Tennessee Warbler, Black-and-white Warbler), and South American (Red-eyed Vireo, Scarlet Tanager, Swainson's Thrush, Veery, and Grey-cheeked Thrush) migrants captured during 19 spring migrations (1993–1996 and 1998–2012). Inset map shows high-density spring passage areas north and south of the Gulf of Mexico where environmental variables were measured (see text for information about how areas were selected) and Johnson's Bayou, Louisiana, where passage timing was measured.

TABLE 4.1
Capture effort at Johnson's Bayou from 28 March to 6 May (1993–1996, 1998–2012).

Year	Net hours	Total nets	Capture rate
1993	3593	12	28.69
1994	3277	12	21.06
1995	3608	12	17.74
1996	6005	24	10.74
1998	7107	26	26.82
1999	8440	25	10.23
2000	8648	25	23.61
2001	6474	24	21.24
2002	8209	24	9.25
2003	8000	24	19.11
2004	8017	24	20.18
2005	9068	24	15.54
2006	7700	21	13.64
2007	6898	20	15.51
2008	7388	21	13.14
2009	5721	21	18.27
2010	8088	22	26.19
2011	6329	22	20.40
2012	7862	24	19.92

NOTES: Both 6- and 12-m-long mist nets were used. Nets were arranged alone or as double long (length of up to 24 m) or double high (height of up to 6 m), all of which are counted here as one net. Net hours correct the number and size of nets for a measure of effort (1 net hour = 12 m of net open for 1 hour). The capture rate is the total number of individuals of the 17 species of migrants per 100 net hours.

of rain or high winds, and arranged to cover the 3.25-ha wooded study site from near the ground to higher in the canopy (up to 6 m). Upon capture, migrants were banded with a USGS aluminum leg band and released. During the sampling period, over 40,000 migrants of over 125 species were captured at the site (http://www.usm.edu/migratory-bird-research/Archive.html). For this analysis, we included 17 species with at least 25 captures per year during at least 10 years of the study. By definition, all spring migrants included were intercontinental migrants. We captured migrants on the northern coast of the Gulf and we did not include overwintering birds. Of the species included, the stationary nonbreeding range of one species overlaps the study site (Gray Catbird, *Dumetella carolinensis*), but none are known to overwinter in the chenier (Lowery and Tucker 1960,

Wiedenfeld et al. 2000). The breeding ranges of six species include the Gulf Coast region (Hooded Warbler, *Setophaga citrine*; Indigo Bunting, *Passerina cyanea*; Painted Bunting, *Passerina ciris*; Kentucky Warbler, *Geothlypis formosa*; Worm-eating Warbler, *Helmitheros vermivorum*; and Wood Thrush, *Hylocichla mustelina*), but only the Painted Bunting is known to breed in the chenier (Wiedenfeld and Swan 2000). Initial captures only were used to determine passage timing. While all 17 species were long-distance migrants, they varied in the proximity of their nonbreeding range to eastern North America. Therefore, we categorized them into three groups based on the extent of their nonbreeding range: South American migrants (nonbreeding ranges in South America), South-Central American migrants (nonbreeding ranges in South America, Central America, the Caribbean, and the southern United States), or Central American migrants (nonbreeding ranges in Central America, the Caribbean, and the southern United States); Table 4.2 and Figure 4.1). The nonbreeding range of the Swainson's Thrush (*Catharus ustulatus*) includes Central and South America but we considered it a South American migrant here because the subspecies captured at the study site overwinters in South America (Ruegg and Smith 2002).

Individuals that migrate through this region likely breed across the latitudinal extent of their breeding ranges (Langin et al. 2009), but for many species, the timing of passage varies with breeding destination with early migrants breeding farther south or farther north depending on the species (Langin et al. 2009, FRM and ZN, unpubl. data). Earlier and later individuals, in relation to breeding destination, may differ in their response to environmental change (Jenni and Schaub 2003), and different populations can move through the same location over the span of the migratory season (Paxton et al. 2007). Therefore, we included multiple measures of passage timing. We measured early, median, and late passage timing for each species as the ordinal day of the year that the 5th, 50th, and 95th individuals were captured each year (Table 4.2). We refer to these measures as "early," "peak," and "late" passage phases of migration. However, upon exploration of the data we determined that our sampling period, 28 March to 6 May, was biased for many species toward capturing earlier but not later migrants (Appendix 4.1). The maximum value for the day of the year that the 95th individual was captured

TABLE 4.2

Passage timing of all species with at least 25 captures per year during at least 10 years of the study (28 March to 6 May 1993–1996 and 1998–2012) at long-term banding station on the northern coast of the Gulf of Mexico (Figure 4.1).

Species	Code	Group	Mean captures	5Th mean	Min, max	50Th mean	Min, max	95Th mean	Min, max
Wood Thrush, *Hylocichla mustelina*	WOTH	Central	76	96	89, 103	110	98, 117	121	115, 126
Gray Catbird, *Dumetella carolinensis*	GRCA	Central	222	107	101, 116	117	112, 122	124	123, 127
Ovenbird, *Seiurus aurocapilla*	OVEN	Central	63	103	93, 114	116	109, 122	125	123, 126
Worm-eating Warbler, *Helmitheros vermivorum*	WEWA	Central	33	92	87, 99	105	99, 110	118	109, 125
Hooded Warbler, *Setophaga citrine*	HOWA	Central	89	90	87, 95	102	91, 110	118	112, 125
Indigo Bunting, *Passerina cyanea*	INBU	Central	171	100	95, 108	114	110, 119	124	121, 126
Painted Bunting, *Passerina ciris*	PABU	Central	46	102	96, 106	115	110, 118	124	122, 127
Northern Waterthrush, *Parkesia noveboracensis*	NOWA	South-Central	74	100	89, 107	115	108, 121	125	125, 126
Black-and-white Warbler, *Mniotilta varia*	BAWW	South-Central	46	92	87, 108	109	95, 123	123	115, 126
Tennessee Warbler, *Oreothlypis peregrina*	TEWA	South-Central	82	103	90, 111	113	106, 119	123	116, 127
Kentucky Warbler, *Geothlypis formosa*	KEWA	South-Central	55	93	87, 100	107	99, 116	120	108, 126
Rose-breasted Grosbeak, *Pheucticus ludovicianus*	RBGR	South-Central	29	106	99, 115	117	111, 123	124	114, 126
Red-eyed Vireo, *Vireo olivaceus*	REVI	South	105	97	87, 107	113	96, 124	123	115, 126
Veery, *Catharus fuscescens*	VEER	South	31	110	100, 117	120	111, 124	125	121, 126
Grey-cheeked Thrush, *Catharus minimus*	GCTH	South	32	109	103, 115	118	113, 123	125	122, 127
Swainson's Thrush, *Catharus ustulatus*	SWTH	South	63	104	96, 116	118	110, 125	125	122, 127
Scarlet Tanager, *Piranga olivacea*	SCTA	South	27	104	98, 110	116	109, 124	124	121, 127

NOTES: The common name, scientific name, and four-letter alpha abbreviation (Pyle and DeSante 2012) of the 17 species of intercontinental migrants grouped based on the extent of their stationary nonbreeding ranges in South America (South); South America, Central America, and the Caribbean (South-Central); and Central America and the Caribbean (Central). The number of migrants captured varied by species and time of year. We present the mean, minimum, and maximum number of individuals captured during spring migration and the ordinal day of the year when 5%, 50%, and 95% of each species were captured (ordinal day 100 = 10 April).

was within 1 day of the last day of banding for all but one species so we excluded the late phase of migration from analyses (Table 4.2). We did not include first passage date because it may be confounded by annual variability in capture numbers (Miller-Rushing et al. 2008b), although we only included the more abundant species at this site.

Resource Phenology

We measured the annual variability in NDVI as a proxy for primary productivity of vegetation. NDVI is a measure of vegetation vigor and density reflective of availability of food resources for insectivorous migratory birds (Pettorelli et al. 2005). It is assessed from remotely sensed imagery with higher NDVI values associated with increased primary productivity (Tucker et al. 2005). We measured NDVI in the stationary nonbreeding range of each species (http://www.natureserve.org/) and at high-density stopover areas north and south of the Gulf (Figure 4.1). Field studies reveal that intercontinental migrating songbirds often concentrate south of the Gulf in Veracruz and Yucatan, Mexico, and Belize (Bayly and Gómez 2011, Shaw and Winker 2011), though little work has been done in this region in spring so it is uncertain to what extent spring migrants deposit substantial fat stores during stopover south of the Gulf (Bayly and Gómez 2011; Bayly et al. 2012, 2013). After crossing the Gulf, billions of migrants concentrate in hardwood forests from Florida to Texas. Weather radar and field studies reveal that intercontinental migrating songbirds stop in high densities in coastal habitats after crossing the Gulf in spring (Gauthreaux 1971, Barrow et al. 2000, Gauthreaux et al. 2006, Buler et al. 2007, Buler and Moore 2011). However, radar studies indicate that while many migrants fly beyond immediate coastal habitats to more contiguous, inland forests after crossing the Gulf, fewer make initial land-fall beyond 50 km from the coast, the northern extent of our sampling site (Gauthreaux 1971, Buler and Moore 2011, Figure 4.1). Little is known about where migrants stop over in spring prior to crossing the Gulf. Therefore, we ran a second set of analyses with a southern passage area similar in size and shape to the northern stopover area, 50 km along the northern coast of the Yucatan. The magnitude of the effects differed but the patterns were largely the same when we included the alternative southern stopover area. Therefore, we acknowledge that more information is needed about the distribution of spring migrants south of the Gulf but our results are somewhat robust to the extent of the southern passage area included.

We obtained NDVI values from the Global Inventory Modeling Mapping Studies (GIMMS, http://glcf.umiacs.umd.edu/data/gimms/) data set derived from imagery obtained from the advanced very high resolution radiometer (AVHRR) sensor onboard the National Oceanic and Atmospheric Agency (NOAA) satellites (1° resolution). The GIMMS data set is corrected for calibration, view geometry, cloud cover, and other effects not related to vegetation change (Pinzon et al. 2005, Tucker et al. 2005) and therefore is ideal for looking at changes in NDVI values over time. However, the GIMMS collection is available only through 2006, with no other series of images that are directly comparable, so we limited our analysis of phenology and NDVI values from 1993 to 2006. We took an annual average of mean NDVI from three biweekly composites for nonbreeding ranges (1 February to 15 March) and passage areas (16 March to 30 April).

As additional measures of annual environmental conditions, we used mean minimum spring temperature and cumulative growing degree days in the northern and southern passage areas (Figure 4.1). We obtained these data from the Oak Ridge National Laboratory Daymet data server for 16 March to 30 April from 1993 to 2006 (Thornton et al. 2012). Daymet delivers daily modeled min and max temperature data at 1 km² resolution. The data are spatially comprehensive across North America and our sampling regions were large, so we subsampled temperature within each region based on a 55-km interval grid. Our criteria resulted in 26 and 63 sample locations for the northern and southern passage areas, respectively. We calculated cumulative growing degree days starting 1 January of each year with a base temperature of 50°F. We averaged the daily minimum temperature for each location. These values were in turn averaged across all locations and days to generate mean estimates for each of the three biweekly periods for each passage area. Last, we obtained a measure of spring temperature for the area surrounding the Johnson's Bayou study site from 19 temperature stations in southwest Louisiana (http://www.srcc.lsu.edu/ranking_tool.html). We found local temperature was strongly correlated with regional temperature (Pearson's product-moment r = 0.74, t = 3.62, n = 13, P = 0.004). Therefore, we included only the regional measure of temperature in analyses.

ANALYSES

We assessed if passage timing advanced for Central, South-Central, and South American migrants with linear mixed-effects models. Comparisons were made across species, included as a random intercept, with year as the fixed effect. Groups of migrants (South, South-Central, and Central American migrants) and phases of migration (5th and 50th passage days) were analyzed separately. We also tested for differences in timing between migrant groups with the group (South, South-Central, and Central American migrants) as the fixed effect, and species and year as uncorrelated random intercepts.

We evaluated if resource phenology advanced to be greener and warmer during spring from 1993 to 2006. We used linear models for NDVI and minimum temperature in the passage areas. We tested for associations between annual NDVI, minimum temperature, and cumulative growing degree days within each passage area with

Pearson's product moment correlation coefficient. We also tested for NDVI changes in the nonbreeding ranges of South, South-Central, and Central American migrant groups separately. Species was included as a random intercept because the nonbreeding ranges were unique to each species.

To determine if resource phenology farther south could be a good indicator of resource phenology farther north, we tested for relationships between environmental variables in the geographically disparate nonbreeding ranges and passage areas. We used linear models to determine if resource phenology (NDVI, minimum temperature, and cumulative growing degree days) in the southern passage area was related to resource phenology in the northern passage area and mixed-effect models, with species as a random intercept, to determine if resource phenology (NDVI) on nonbreeding ranges was related to resource phenology in the southern or northern passage areas. We also looked for a relationship between NDVI in the nonbreeding ranges of South, South-Central, and Central American migrants with a mixed-effects model with the group as a fixed effect and species as a random intercept.

Last, to determine if passage timing to the northern coast of the Gulf is related to environmental conditions either on the nonbreeding ranges (NDVI) and/or in the southern passage areas (NDVI and temperature), we developed seven *a priori* candidate models to examine the influence of environmental factors south of the Gulf on passage timing to the northern coast of the Gulf:

1. "NDVI" model, including NDVI in passage areas and nonbreeding ranges

2. "Passage area" model, including NDVI, mean minimum temperature, and cumulative growing degree days in the southern passage area

3. "Nonbreeding range NDVI" model, including NDVI measured in nonbreeding ranges

4. "Passage NDVI" model, including NDVI in the southern passage area

5. "Passage temperature" model, including mean minimum temperature in the southern passage area

6. "Passage degree days" model, including cumulative growing degree days in the southern passage area

7. "Full" model, including southern passage area mean minimum temperature, NDVI and cumulative growing degree days, and nonbreeding range NDVI.

Mixed-effects models included the species as a random variable. For comparison, we included an intercept-only (null) model. Because the southern passage area and Nonbreeding range NDVI of Central American migrants were strongly related ($\beta = 1.42 \pm 0.15$, n = 7, 91, t = 12.30, P < 0.001), we excluded models that included southern passage area NDVI from comparison for Central American migrants. Therefore, for Central American migrants we included four competing models: (3) Nonbreeding range NDVI and (5) passage temperature (6) passage degree days and altered (7) full model to exclude passage area mean minimum temperature (described previously). Although some Nonbreeding ranges of South-Central American migrants also overlapped the southern passage area, there was no relationship between NDVI in those areas ($\beta = 0.14 \pm 0.12$, n = 5, 65, t = 1.11, P = 0.27). We selected among competing models using Akaike's information criteria adjusted for small sample sizes (AIC_c, aictab function in package AICcmodavg; Burnham and Anderson 2002, Mazerolle and Mazerolle 2013). Models with the lowest AIC_c are the most parsimonious of the candidate models and models ≤ 2 Δi ($\Delta i = AIC_ci - $ minimum AIC_c) were considered equally plausible. When more than one model was supported, we performed model averaging for estimates with unconditional SE and 95% confidence intervals (modavg function in package AICcmodavg; Burnham and Anderson 2002, Mazerolle and Mazerolle 2013). Confidence intervals that do not include zero indicate a strong effect of the variable.

Analyses were conducted in R (R Development Core Team 2012). Passage days, NDVI, and minimum temperature were standardized to mean zero. For mixed-effects models, random effects were uncorrelated with random intercepts (lmer function in R package lme4; Bates et al. 2008, Zuur et al. 2009). For the fixed effects in the mixed-effects models, we used a Markov chain Monte Carlo (MCMC) approach to compute P-values and 95% highest posterior density (HPD) credible intervals for fixed effects on the basis of an MCMC sample with 10,000 simulations (pvals. fnc function in package languageR; Baayen 2009).

For each model, goodness of fit was confirmed by visual inspection of plots of residuals against predicted values. Means ± SE are reported unless stated otherwise.

RESULTS

Passage Timing

Passage timing across the Gulf varied annually but we did not find evidence for advancement in passage timing to the northern coast of the Gulf from 1993 to 2012 (Figure 4.2, Appendix 4.1). Spring passage timing varied by species and among phases of migration but neither South, South-Central, or Central American migrants were arriving earlier (Figure 4.2). In fact, Central American migrant early passage timing shifted 2.41 days later ($\beta = 0.02 \pm 0.01$, n = 7 species, 133 observations, t = 2.24, P = 0.03) and peak passage timing

shifted 2.91 days later ($\beta = 0.02 \pm 0.01$, n = 7, 133, t = 2.77, P < 0.01). South-Central American migrant passage timing did not change for early ($\beta = 0.01 \pm 0.01$, n = 5, 95, t = 1.13, P = 0.26) or peak migrating individuals ($\beta = 0.01 \pm 0.02$, n = 5, 95, t = 0.77, P = 0.45), nor was there a change in passage timing of South American migrants during early ($\beta = 0.02 \pm 0.01$, n = 5, 95, t = 1.46, P = 0.15) or peak migration ($\beta = 0.006 \pm 0.02$, n = 5, 95, t = 0.35, P = 0.73). Early and peak South American migrants arrived 6.37 and 5.66 days later than early and peak Central American migrants (early: $\beta = 0.85 \pm 0.46$, n = 17, 323, t = 1.86, P = 0.06; peak: $\beta = 0.84 \pm 0.42$, n = 17, 323, t = 2.04, P = 0.04), while South-Central American migrants did not differ in their timing from Central American migrants (early: $\beta = 0.04 \pm 0.46$, n = 17, 323, t = 0.10, P = 0.92; peak: $\beta = 0.13 \pm 0.42$, n = 17, 323, t = 0.32, P = 0.75; Figure 4.3).

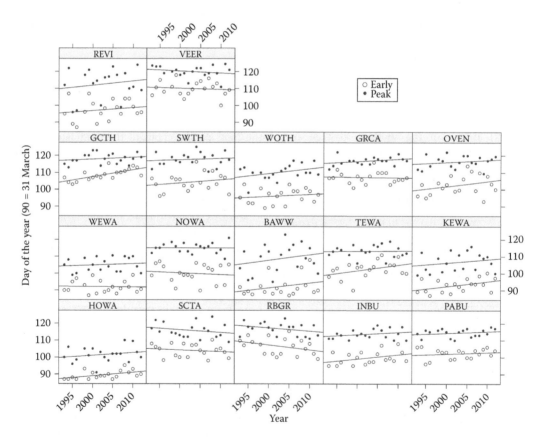

Figure 4.2. Passage timing of 17 migrant species of landbirds commonly captured on the north coast of the Gulf of Mexico (see Figure 4.1 for capture location) during early (day of the year that the 5% of individuals of each species was captured) and peak (day of the year that the 50% of individuals of each species was captured) migration from 1993 to 1996 and 1998 to 2012.

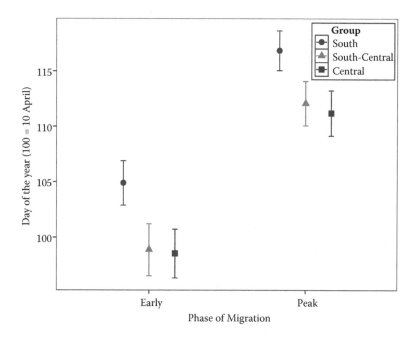

Figure 4.3. Mean passage timing (± SE) of Central, South-Central, and South American migrants (see Table 4.2 for species) during early (ordinal day of the year that 5% of individuals of each species was captured) and peak (ordinal day of the year that the 50% of individuals of each species was captured) migration across the Gulf of Mexico (see Figure 4.1 for capture location) of species captured during spring migration on the north coast of the Gulf of Mexico from 1993 to 1996 and 1998 to 2012.

Resource Phenology

We did not find evidence that spring greening of vegetation (NDVI) increased during the spring period in passage areas or on nonbreeding ranges from 1993 to 2006. Rather, spring NDVI declined in the southern passage area ($\beta = -0.17 \pm 0.05$, n = 13, t = −3.69, P = 0.004) and did not advance or decline in the northern passage area but was considerably variable year to year ($\beta = 0.01 \pm 0.07$, n = 13, t = 0.15, P = 0.89; Figure 4.4). NDVI also declined in the nonbreeding ranges of Central American migrants ($\beta = -0.10 \pm 0.01$, n = 7, 91, t = −12.37, P < 0.001) but did not change in the nonbreeding ranges of either South-Central ($\beta = -0.02 \pm 0.01$, n = 5, 65, t = −1.31, P = 0.20) or South American ($\beta = 0.009 \pm 0.02$, n = 5, 65, t = 0.59, P = 0.56; Figure 4.5) migrants. Nor did we find evidence that mean minimum temperature (northern: $\beta = 0.05 \pm 0.07$, n = 13, t = 0.72, P = 0.49, southern: $\beta = -0.07 \pm 0.07$, n = 13, t = −0.89, P = 0.39) or cumulative growing degree days increased in either passage area (northern: $\beta = 7.11 \pm 4.67$, n = 13, t = 1.52, P = 0.16, southern: $\beta = -2.60 \pm 3.07$, n = 13, t = −0.85, P = 0.42).

Southern resource phenology was not a good indicator of northern resource phenology. NDVI in nonbreeding ranges was not related to resource phenology in passage areas. For Central American migrants, spring NDVI on nonbreeding ranges was not related to NDVI on the northern coast of the Gulf ($\beta = -0.07 \pm 0.10$, n = 7, 91, t = −0.67, P = 0.50). Similarly, for South-Central and South American migrants, there was no relationship between NDVI in nonbreeding ranges and the southern passage area (south: $\beta = 0.005 \pm 0.10$, n = 5, 65, t = 0.04, P = 0.97; south-central: $\beta = 0.14 \pm 0.12$, n = 5, 65, t = 1.11, P = 0.27) or the northern passage area (south: $\beta = -0.12 \pm 0.12$, n = 5, 65, t = −1.01, P = 0.32; south-central: $\beta = -0.14 \pm 0.12$, n = 5, 65, t = −0.94, P = 0.35). We excluded nonbreeding ranges of Central American migrants and the southern passage area from these analyses because they strongly overlapped ($\beta = 1.42 \pm 0.15$, n = 7, 91, t = 12.30, P < 0.001). Between migrant groups, NDVI was higher in the nonbreeding ranges of Central American migrants as compared to South-Central migrants ($\beta = -1.15 \pm 0.49$, n = 17, 221, t = −2.35, P = 0.02), but there was no difference in nonbreeding range NDVI

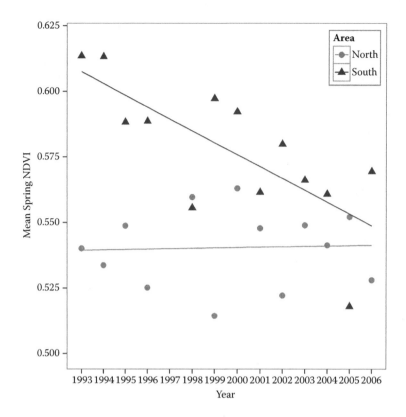

Figure 4.4. Mean spring NDVI values from 1993 to 2006 at high-density passage areas north and south of the Gulf of Mexico (see Figure 4.1 for passage areas). Annual NDVI values for each area are a mean of three spring composites during peak passage timing (16–31 March, 1–15 April, and 16–30 April). Higher NDVI values indicate increased greening. Mean spring NDVI declined in the southern passage region ($\beta = -0.17 \pm 0.05$, n = 13, t = −3.69, P = 0.004), but did not change in the northern passage region ($\beta = 0.01 \pm 0.07$, n = 13, t = 0.15, P = 0.89).

between Central and South American migrants ($\beta = -0.45 \pm 0.49$, n = 17, 221, t = −0.92, P = 0.36; Figure 4.5). For passage areas, NDVI values in the south were not related to NDVI values in the north ($\beta = -0.38 \pm 0.28$, n = 13, t = −1.40, P = 0.19). Nor were minimum temperature ($\beta = 0.59 \pm 0.24$, n = 13, t = 2.44, P = 0.03) or cumulative growing degree days ($\beta = 0.10 \pm 0.49$, n = 13, t = 0.20, P = 0.85) in the southern passage area related to the northern passage area. NDVI was also not strongly correlated with the other resource phenology variables within the northern (NDVI and minimum temperature, r = −0.39, t = −1.41, n = 13, P = 0.19; NDVI and cumulative growing degree days, r = −0.12, t = −0.38, n = 13, P = 0.71) or southern (NDVI and minimum temperature, r = 0.20, t = 0.68, n = 13, P = 0.51; NDVI and cumulative growing degree days, r = −0.05, t = −0.18, n = 13, P = 0.86) passage areas.

Effects of Resource Phenology on Passage Timing

For both early and peak migrating Central American migrants there was support for the influence of nonbreeding range NDVI on passage timing and no support for the influence of temperature in the passage area (Table 4.3). Central American and peak migrating South-Central American migrants passed through the northern coast of the Gulf later in years when NDVI values on nonbreeding ranges and the southern passage region were lower (Table 4.4). Conversely, for South-Central American migrants, the influence of environmental variables on passage timing differed for early and peak migrating birds. There was little support for the influence of passage area NDVI on the timing of early South-Central American migrants while there was support for the influence of both nonbreeding range and

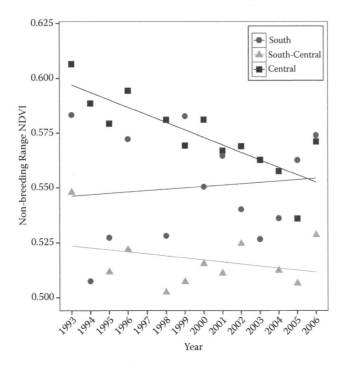

Figure 4.5. Mean spring NDVI values from 1993 to 2006 in the nonbreeding ranges of Central, South-Central, and South American migrants (see Table 4.2 for species). Annual NDVI values are a mean of three spring composites (1–15 February, 16–30 February, and 1–15 March). Higher NDVI values indicate increased greening. Mean nonbreeding range NDVI for Central American migrants declined (b = −0.17 ± 0.05, n = 13, t = −3.69, P = 0.004) but did not change for South-Central American migrants (β = −0.02 ± 0.01, n = 5, 65, t = −1.31, P = 0.20) or South American migrants (β = 0.009 ± 0.02, n = 5, 65, t = 0.59, P = 0.56).

passage area NDVI on the timing of peak South-Central American migrants (Table 4.3). For early and peak migrating South American birds, there was little evidence for the influence of NDVI but some evidence for the influence of cumulative growing degree days on the timing of early migrating South American birds (Table 4.3). There was little evidence that the explanatory variables we included in candidate models had an effect on the passage timing of early migrating South-Central American migrants and peak migrating South American migrants. The null, intercept-only model was the top supported model for these groups (Table 4.3) and the 95% confidence intervals of variable estimates overlap zero (Table 4.4).

DISCUSSION

Migratory animals face the unique challenge of optimizing the timing of migration to match environmental conditions at distant locations. In North America, migratory birds are arriving earlier to their temperate breeding grounds with warming spring temperatures (Miller-Rushing et al. 2008a, Knudsen et al. 2011, Hurlbert and Liang 2012). We found annual variability in spring passage timing across the Gulf. In contrast to more northern latitudes, we did not find evidence of advancement in spring greening or temperature along the southern coast of eastern North America. Further, in spite of earlier arrival dates to breeding areas, passage timing of 17 common eastern migrant species into eastern North America either has not changed or has been delayed by a few days from 1993 to 2012. The observed lack of change in passage timing could be related to the fact that we found no relationship between spring resource phenology north and south of the Gulf—in passage areas or stationary nonbreeding ranges—indicating that migrants may not have reliable cues to the resource phenology they will encounter after crossing the Gulf into eastern North America. Therefore, Nearctic–Neotropical migrants may make the non-top flight across the

TABLE 4.3

Results of model selection from seven candidate models examining the influence of environmental variables on migration passage timing.

Model	K	AIC_c	ΔAIC_c	w_i	Cumulative w_i
South American migrants, early					
Passage degree days	4	163.09	0.00	0.33	0.33
Null	3	163.84	0.76	0.23	0.56
Passage NDVI	4	164.95	1.86	0.13	0.69
Passage area	6	165.59	2.50	0.09	0.78
Passage temperature	4	165.98	2.90	0.08	0.86
Nonbreeding NDVI	4	166.10	3.01	0.07	0.93
NDVI	5	167.28	4.19	0.04	0.97
Full	7	168.09	5.00	0.03	1.00
South American migrants, peak					
Null	3	186.29	0.00	0.28	0.28
Passage temperature	4	186.88	0.58	0.21	0.49
Passage NDVI	4	187.02	0.73	0.19	0.68
Passage degree days	4	188.38	2.09	0.10	0.78
Nonbreeding NDVI	4	188.53	2.24	0.09	0.87
NDVI	5	189.33	3.04	0.06	0.93
Passage area	6	189.59	3.29	0.05	0.98
Full	7	191.97	5.67	0.02	1.00
South-Central migrants, early					
Null	3	142.10	0.00	0.36	0.36
Passage degree days	4	143.52	1.42	0.18	0.53
Passage NDVI	4	143.82	1.71	0.15	0.69
Passage temperature	4	144.23	2.13	0.12	0.81
Nonbreeding NDVI	4	144.35	2.25	0.12	0.93
NDVI	5	146.16	4.06	0.05	0.97
Passage area	6	147.79	5.69	0.02	0.99
Full	7	150.30	8.19	0.01	1.00
South-Central migrants, peak					
NDVI	5	170.08	0.00	0.34	0.34
Nonbreeding NDVI	4	170.62	0.53	0.26	0.60
Passage NDVI	4	170.89	0.81	0.23	0.83
Null	3	173.25	3.17	0.07	0.90
Full	7	174.94	4.86	0.03	0.93
Passage degree days	4	175.37	5.29	0.02	0.96
Passage temperature	4	175.52	5.44	0.02	0.98
Passage area	6	175.53	5.45	0.02	1.00

TABLE 4.3 (continued)
Results of model selection from seven candidate models examining the influence of environmental variables on migration passage timing.

Model	K	AIC_c	ΔAIC_c	w_i	Cumulative w_i
Central American migrants, early					
Nonbreeding NDVI	4	179.44	0.00	0.57	0.57
Null	3	181.71	2.27	0.18	0.76
Full	6	182.96	3.52	0.10	0.85
Passage degree days	4	183.36	3.92	0.08	0.94
Passage temperature	4	183.80	4.36	0.06	1.00
Central American migrants, peak					
Nonbreeding NDVI	4	167.16	0.00	0.51	0.51
Full	6	167.35	0.19	0.47	0.98
Null	3	174.07	6.91	0.02	0.99
Passage temperature	4	175.89	8.73	0.01	1.00
Passage degree days	4	183.36	16.19	0.00	1.00

NOTES: K = number of parameters, ΔAIC_c = AIC_c – minimum AIC, w_a = Akaike weight represents support for each model for 17 species of South, South-Central, and Central American migrant species (Table 4.2) captured during early (day of the year when 5% of each species was captured) and peak (day of the year when 50% of each species was captured) spring migration 1993–1996 and 1998–2006. Passage area and nonbreeding ranges of Central American migrants overlapped, so models with passage area NDVI were excluded for this group.

TABLE 4.4
Model averaged parameter estimates and unconditional 95% confidence intervals (CI) from the supported models (≤2 ΔAIC_c) for the relationship between environmental variables and passage timing of early and peak migrating birds from South, South-Central, and Central America.

Migrant group[a]	Variable[b]	Estimate	95% CI
South, early	Passage NDVI	-0.10 ± 0.09	$-0.29, 0.08$
	Passage degree days	0 ± 0	$0, 0.01$
South, peak	Passage NDVI	-0.15 ± 0.12	$-0.38, 0.08$
	Passage temp	-0.15 ± 0.12	$-0.39, 0.08$
South-Central, early	Passage NDVI	-0.06 ± 0.08	$-0.21, 0.09$
South-Central, peak	Winter NDVI	-0.37 ± 0.19	$-0.73, 0$
	Passage NDVI	-0.19 ± 0.10	$-0.39, 0.01$
Central, early	Winter NDVI	-0.21 ± 0.10	$-0.43, -0.02$
Central, peak	Winter NDVI	-0.30 ± 0.10	$-0.49, -0.11$

NOTES: The peak and early passage is the date that 5 and 50% of individuals of each species were captured each spring (1993–1996 and 1998–2006) on the northern coast of the Gulf of Mexico (see Figure 4.1 for study site).

[a] See Table 4.2 for species.
[b] NDVI, temperature, and cumulative growing degree days.

Gulf once they have sufficient fuel stores and then fine-tune the progress of migration after they arrive in eastern North America where southern environmental conditions may be better indicators of what they will encounter on breeding grounds (Marra et al. 2005, Fraser et al. 2013). Annual variability in spring temperatures in southeastern North America has been correlated with spring temperatures in northeastern North America (Marra et al. 2005). Migrants are capable of increasing the speed of migration as they encounter warmer spring temperatures (Marra et al. 2005, Tøttrup et al. 2008, Balbontín et al. 2009, Hurlbert and Liang 2012). In fact, the rate of migration between the site in this study and two sites in northeastern North America declined by almost 1 day for every 1°C decrease in mean spring temperature (Marra et al. 2005).

However, increasing spring temperatures in eastern North America may not be the only climatic determinants of migration speed. Drought and, consequently, scarce food resources at stopover and overwintering sites in the tropics can also delay the arrival of long-distance migrants (Ahola et al. 2004, Studds and Marra 2011, McKellar et al. 2012, Tøttrup et al. 2012, González-Prieto and Hobson 2013). For example, Central American migrants may have been constrained in their passage timing across the Gulf by the increasingly dry conditions in Central America and the Caribbean (Heartsill-Scalley et al. 2007, Slingo 2011, Studds and Marra 2011). Central American migrants arrived 2 to 3 days later than they did 20 years earlier and arrived later when NDVI values were reduced in nonbreeding ranges and in passage areas south of the Gulf. Moreover, the unpredictable nature of tropical droughts and inhospitable weather conditions further magnify the uncertainties migrants face in timing their migrations thousands of kilometers away from their destinations (González-Prieto and Hobson 2013). Thus, by increasing the speed of migration in response to reliable environmental cues within North America, late migrants may be able to catch up with the northward advancing spring warming (Marra et al. 2005, Smith et al. 2009, Hurlbert and Liang 2012).

Phenological responses to environmental conditions may vary with migratory distance (Berthold 1996). In general, species that migrated from Central America and the Caribbean arrived later during years when conditions were drier south of the Gulf and species that migrated from South America arrived more consistently across years and may have been less influenced by environmental conditions on nonbreeding ranges or along the migratory route. Consistent arrival dates support the idea that longer distance migrants have a "stricter" endogenous program, influenced less by environmental cues en route, ensuring that they get to the breeding grounds on time (Gienapp et al. 2007, Knudsen et al. 2011). Environmental conditions did not influence departure date from South America or the rate of migration for Purple Martins (*Progne subis*; Fraser et al. 2013). On the other hand, shorter distance migrants may exhibit more flexibility in use of environmental cues, which is attributed to their success in adjusting migration timing to warming spring temperatures. The stronger response of Central American migrants to warming trends would be adaptively significant when breeding in less seasonal environments (Mills 2005, Swanson and Palmer 2009, Tøttrup et al. 2010, Hurlbert and Liang 2012). However, the Wood Thrush, which was one of the shorter distance migrants in this study, has also been shown to have high within-individual repeatability of spring departure from overwintering areas, supporting the idea that phenotypic plasticity has little role in timing spring migration in this species (Stanley et al. 2012).

All of the migrants included in this study were intercontinental migrants typically classified as long-distance migrants; however one explanation for the difference in the variability in timing between longer and shorter distance migrants has been that shorter distance migrants overwinter closer to their breeding grounds and, therefore, have increased access to information about annual variability in resource phenology at stopover sites or on breeding grounds. We did not find evidence that the shorter distance migrants in this study, overwintering in Central America and the Caribbean, benefited from increased information about breeding ground resource phenology, as compared to those overwintering in South America. However, long-distance migrants should also be constrained in their timing by drying conditions en route but we found little support for that pattern (Tøttrup et al. 2012). Unfortunately, little is known about where South American migrants stop to refuel prior to crossing the Gulf during vernal migration. South American migrants do stop south of the Gulf during vernal migration (Leyva et al. 2009, Shaw and Winker 2011, Fraser et al. 2013), but it is possible that this area is of little importance for refueling or maintaining energy stores of South American migrants, suggesting that most refueling could happen before they leave the South American continent and after they cross the Gulf (Bayly et al. 2013). We did not quantify the condition of migrants upon arrival across the Gulf, so it is possible that long-distance migrants adjust their condition in favor of arriving on time to the southern coast of eastern North America. For example, migrants departing the Amazon may migrate northward consistently, without adjusting flight speed or stopover duration. During a year with little spring rainfall, they would likely arrive on the northern coast of the Gulf in poorer condition (Paxton et al. 2014).

Finer scale geographic information about migratory movements and local resources are necessary to understand fully if, when, and how migrants adjust the timing of migration in relation to environmental conditions. NDVI is variable regionally, so it is unlikely that measuring across an entire nonbreeding range will reflect the extent of variability in resource phenology encountered by the migrants we captured (Ichii et al. 2002). For example, here we could not evaluate the degree to which environmental conditions in passage areas versus nonbreeding ranges influence passage timing of Central American migrants because the two areas overlapped considerably and annual NDVI values were strongly related. However, Wood Thrush from all of their nonbreeding range stop over during spring migration in the passage area we measured on the northern Yucatan peninsula (E. McKinnon, pers. comm.). Therefore, we could not assess the relative influence of overwinter and passage area resource phenology for this species due to the lack of information about migratory connectivity from overwintering to stopover sites, even though individuals and populations of this species likely experience differences in resources.

Further, we found a difference in the effects of environmental conditions on passage timing between early and peak South-Central American migrants. Without an understanding of migratory connectivity it is not possible to differentiate if this result is due to variability in the influence of resource phenology on birds from different parts of the nonbreeding ranges or variability in the resources encountered. Nevertheless, even with individuals grouped by continent, which likely contain extensive variability in resource availability, we found evidence for the influence of nonbreeding range and passage area resource phenology on timing of migrants caught at our study site. Migrants caught at Johnson's Bayou breed throughout eastern North America, so it is possible that individuals passing through this site are also from throughout the stationary nonbreeding ranges (Langin et al. 2009). Further, our methods were conservative to avoid bias in the sample of migrants and resources in that we did not use the date of the first capture of the year as a measure of passage timing. Also we included only species that were consistently captured in high numbers to avoid confounding annual variability in passage timing with differences in abundance (Miller-Rushing et al. 2008b).

Our results suggest that some species are experiencing opposing trends in climate change in the geographically disparate areas they depend upon. Studies of the impact of climate change on migratory birds have focused largely on the potential for mismatches between arrival timing and food abundance in breeding areas. However, global climate change is also affecting migrants in the tropics where they spend the majority of the year. Dry conditions in passage areas and nonbreeding ranges south of the Gulf could also mean that Central American migrants arrive in eastern North America in poorer condition (González-Prieto and Hobson 2013). Another consequence of the later arrival timing of Central American migrants could be increasing density-dependent competition with South American migrants, which has not changed in their passage timing. Therefore, in order to arrive optimally to breeding grounds, migrants from Central America and the Caribbean arriving later and in poorer condition may need to migrate faster through eastern North America while refueling at stopover sites with greater competition. The widespread warming and decreased precipitation in Central America and the Caribbean are projected to intensify and continue (Neelin et al. 2006), emphasizing the importance and lack of information about how migrants use interior North American stopover sites (Cohen et al. 2012), and how characteristics of sites influence fuel deposition rates and stopover duration (Cohen et al. 2014). For example, the extent to which density affects the ability of migrants to refuel at stopover sites remains poorly understood (Moore and Yong 1991, Kelly et al. 2002), limiting our ability to understand the effects of rapidly changing North American landscapes (Cohen et al. 2014).

ACKNOWLEDGMENTS

We thank the banding crews who helped with data collection, the student workers who assisted with data management, and Wylie Barrow for use of banding data from 1993 to 1995. Research was supported by the National Science Foundation (DEB 0554754, IBN 0078189, IOS 844703, GK-12 0947944).

APPENDIX 4.1

Mean Number of Individuals Captured at Johnson's Bayou, Louisiana, on Each Day of Banding from 1993 to 1996 and 1998 to 2012 (28 March to 6 May is Day of the Year 87 to 127)

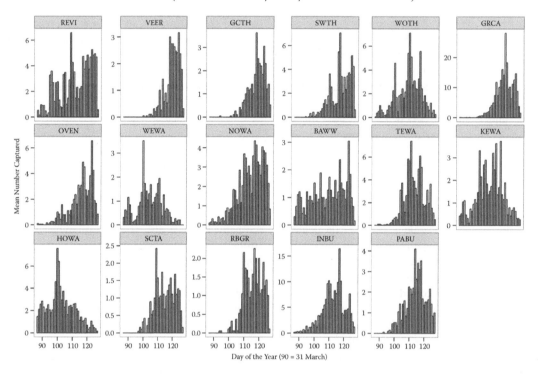

LITERATURE CITED

Ahola, M., T. Laaksonen, K. Sippola, T. Eeva, K. Rainio, and E. Lehikoinen. 2004. Variation in climate warming along the migration route uncouples arrival and breeding dates. Global Change Biology 10:1610–1617.

Baayen, H. R. [online]. 2009. LanguageR. Version 1.4. <http://CRAN.R-Project.org/package=languageR>

Bairlein, F., and O. Hüppop. 2004. Migratory fuelling and global climate change. Advances in Ecological Research 35:33–47.

Bairlein, F., and W. Winkel. 2001. Birds and climate change. Pp. 278–282 in J. L. Lozan, H. Graßl, and P. Hupfer (editors), Climate of the 21st century: changes and risks. Wissenschaftliche Auswertungen, Hamburg, Germany.

Balbontín, J., A. P. Møller, I. G. Hermosell, A. Marzal, M. Reviriego, and F. De Lope. 2009. Individual responses in spring arrival date to ecological conditions during winter and migration in a migratory bird. Journal of Animal Ecology 78:981–989.

Barrow, W. C., C. C. Chen, R. B. Hamilton, K. Ouchley, and T. J. Spengler. 2000. Disruption and restoration of en route habitat, a case study: the Chenier Plain. Studies in Avian Biology 20:71–87.

Bates, D., M. Maechler, and B. Dai. [online]. 2008. The lme4 package. <ftp://ftp.uni-bayreuth.de/pub/math/statlib/R/CRAN/doc/packages/lme4.pdf>

Bayly, N. J., and C. Gómez. 2011. Comparison of autumn and spring migration strategies of Neotropical migratory landbirds in northeast Belize. Journal of Field Ornithology 82:117–131.

Bayly, N. J., C. Gómez, and K. A. Hobson. 2013. Energy reserves stored by migrating Gray-cheeked Thrushes *Catharus minimus* at a spring stopover site in northern Colombia are sufficient for a long-distance flight to North America. Ibis 155:271–283.

Bayly, N. J., C. Gómez, K. A. Hobson, A. M. González, and K. V. Rosenberg. 2012. Fall migration of the Veery (*Catharus fuscescens*) in northern Colombia: determining the energetic importance of a stopover site. Auk 129:449–459.

Berthold, P. 1996. Control of bird migration, Chapman and Hall, London, UK.

Both, C., A. V. Artemyev, B. Blaauw, R. J. Cowie, A. J. Dekhuijzen, T. Eeva, A. Enemar, L. Gustafsson, E. V. Ivankina, and A. Järvinen. 2004. Large–scale geographical variation confirms that climate change causes birds to lay earlier. Proceedings of the Royal Society of London B 271:1657–1662.

Both, C., R. G. Bijlsma, and M. E Visser. 2005. Climatic effects on timing of spring migration and breeding in a long-distance migrant, the Pied Flycatcher *Ficedula hypoleuca*. Journal of Avian Biology 36:368–373.

Both, C., S. Bouwhuis, C. M. Lessells, and M. E. Visser. 2006. Climate change and population declines in a long-distance migratory bird. Nature 441:81–83.

Buler, J. J., and F. R. Moore. 2011. Migrant–habitat relationships during stopover along an ecological barrier: extrinsic constraints and conservation implications. Journal of Ornithology 152:101–112.

Buler, J. J., F. R. Moore, and S. Woltmann. 2007. A multi-scale examination of stopover habitat use by birds. Ecology 88:1789–1802.

Burnham, K. P., and D. R. Anderson. 2002. Model selection and multi-model inference: a practical information-theoretic approach. Springer, New York, NY.

Cohen, E. B., F. R. Moore, and R. A. Fischer. 2012. Experimental evidence for the interplay of exogenous and endogenous factors on the movement ecology of a migrating songbird. PLoS One 7:e41818.

Cohen, E. B., S. M. Pearson, and F. R. Moore. 2014. Effects of landscape composition and configuration on migrating songbirds: inference from an individual-based model. Ecological Applications 24:169–180.

Fraser, K. C., C. Silverio, P. Kramer, N. Mickle, R. Aeppli, and B. J. Stutchbury. 2013. A trans-hemispheric migratory songbird does not advance spring schedules or increase migration rate in response to record-setting temperatures at breeding sites. PloS One 8:e64587.

Gauthreaux, S. A., Jr. 1971. A radar and direct visual study of passerine spring migration in southern Louisiana. Auk 88:343–365.

Gauthreaux, S. A., Jr. 1972. Behavioral responses of migrating birds to daylight and darkness: a radar and direct visual study. Wilson Bulletin 84:136–148.

Gauthreaux, S. A., Jr. 1999. Neotropical migrants and the Gulf of Mexico: The view from aloft. P. 27 in K. P. Able (editor), A gathering of angels: migrating birds and their ecology. Cornell University Press, York, NY.

Gauthreaux, S. A., Jr., C. G. Belser, and C. M. Welch. 2006. Atmospheric trajectories and spring bird migration across the Gulf of Mexico. Journal of Ornithology 147:317–325.

Gienapp, P. 2012. Migration. Pp. 80–92 in U. Candolin and B. B. Wong (editors), Behavioural responses to a changing world: mechanisms and consequences. Oxford University Press, Oxford, UK.

Gienapp, P., and T. Bregnballe. 2012. Fitness consequences of timing of migration and breeding in Cormorants. PLoS One 7:e46165.

Gienapp, P., R. Leimu, and J. Merila. 2007. Responses to climate change in avian migration time—microevolution versus phenotypic plasticity. Climate Research 35:25–35.

González-Prieto, A. M., and K. A. Hobson. 2013. Environmental conditions on wintering grounds and during migration influence spring nutritional condition and arrival phenology of Neotropical migrants at a northern stopover site. Journal of Ornithology 154:1067–1078.

Gordo, O., L. Brotons, X. Ferrer, and P. Comas. 2005. Do changes in climate patterns in wintering areas affect the timing of the spring arrival of trans-Saharan migrant birds? Global Change Biology 11:12–21.

Heartsill-Scalley, T., F. N. Scatena, C. Estrada, W. H. McDowell, and A. E. Lugo. 2007. Disturbance and long-term patterns of rainfall and throughfall nutrient fluxes in a subtropical wet forest in Puerto Rico. Journal of Hydrology 333:472–485.

Hurlbert, A. H., and Z. Liang. 2012. Spatiotemporal variation in avian migration phenology: citizen science reveals effects of climate change. PLoS One 7:e31662.

Ichii, K., A. Kawabata, and Y. Yamaguchi. 2002. Global correlation analysis for NDVI and climatic variables and NDVI trends: 1982–1990. International Journal of Remote Sensing 23:3873–3878.

Jenni, L., and M. Schaub. 2003. Behavioural and physiological reactions to environmental variation in bird migration: a review. Pp. 155–171 in P. Berthold, E. Gwinner, and E. Sonnenschein (editors), Avian migration. Springer-Verlag, Berlin, Germany.

Jonzén, N., A. Lindén, T. Ergon, E. Knudsen, J. O. Vik, D. Rubolini, D. Piacentini, C. Brinch, F. Spina, and L. Karlsson. 2006. Rapid advance of spring arrival dates in long-distance migratory birds. Science 312:1959–1961.

Kelly, J. F., L. S. DeLay, D. M. Finch, and C. Blem. 2002. Density-dependent mass gain by Wilson's Warblers during stopover. Auk 119:210–213.

Knudsen, E., A. Lindén, C. Both, N. Jonzén, F. Pulido, N. Saino, W. J. Sutherland, L. A. Bach, T. Coppack, and T. Ergon. 2011. Challenging claims in the study of migratory birds and climate change. Biological Reviews 86:928–946.

Langin, K. M., P. P. Marra, Z. Németh, F. R. Moore, T. Kurt Kyser, and L. M. Ratcliffe. 2009. Breeding latitude and timing of spring migration in songbirds crossing the Gulf of Mexico. Journal of Avian Biology 40:309–316.

Lehikoinen, E., and T. H. Sparks. 2010. Effects of climate change on birds, Oxford University Press, Oxford, UK.

Leyva, E. M., E. R. Inzunza, O. C. Carretero, J. L. Barr, E. P. Rojas, I. C. Dominguez, G. R. Lara, R. R. Mesa, A. G. Miranda, and N. F. Dominguez. 2009. Dynamics of passerine migration in Veracruz, Mexico. Proceedings of the Fourth International Partners in Flight Conference: Tundra to Tropics 62–70.

Lowery, G. H., and R. E. Tucker. 1960. Louisiana birds. Louisiana Wildlife and Fisheries Commission, Baton Rouge, LA.

Marra, P. P., C. M. Francis, R. S. Mulvihill, and F. R. Moore. 2005. The influence of climate on the timing and rate of spring bird migration. Oecologia 142:307–315.

Mazerolle, M. J., and M. M. J. Mazerolle. [online]. 2013. Package AICcmodavg. <http://cran.rproject. org/web/packages/AICcmodavg/AICcmodavg.pdf>

McKellar, A. E., P. P. Marra, S. J. Hannon, C. E. Studds, and L. M. Ratcliffe. 2012. Winter rainfall predicts phenology in widely separated populations of a migrant songbird. Oecologia 172:1–11.

Miller-Rushing, A. J., T. L. Lloyd-Evans, R. B. Primack, and P. Satzinger. 2008a. Bird migration times, climate change, and changing population sizes. Global Change Biology 14:1959–1972.

Miller-Rushing, A. J., R. B. Primack, and R. Stymeist. 2008b. Interpreting variation in bird migration times as observed by volunteers. Auk 125:565–573.

Mills, A. M. 2005. Changes in the timing of spring and autumn migration in North American migrant passerines during a period of global warming. Ibis 147:259–269.

Møller, A. P. 2001. Heritability of arrival date in a migratory bird. Proceedings of the Royal Society of London B 268:203–206.

Møller, A. P., J. Balbontín, J. J. Cuervo, I. G. Hermosell, and F. De Lope. 2009. Individual differences in protandry, sexual selection, and fitness. Behavioral Ecology 20:433–440.

Moore, F. R. 1999. Cheniers of Louisiana and the stopover ecology of migrant landbirds in K. P. Able (editor), A gathering of angels: ecology and conservation of migrating birds. Cornell University Press, Ithaca, NY.

Moore, F. R. 2000. Stopover ecology of Nearctic– Neotropical landbird migrants: habitat relations and conservation implications. Studies in Avian Biology 20.

Moore, F. R., and W. Yong. 1991. Evidence of food-based competition among passerine migrants during stopover. Behavioral Ecology and Sociobiology 28:85–90.

Neelin, J. D., M. Münnich, H. Su, J. E. Meyerson, and C. E. Holloway. 2006. Tropical drying trends in global warming models and observations. Proceedings of the National Academy of Sciences of the USA 103:6110–6115.

Newton, I. 2012. Obligate and facultative migration in birds: ecological aspects. Journal of Ornithology 153:171–180.

Parmesan, C. 2006. Ecological and evolutionary responses to recent climate change. Annual Review of Ecology, Evolution, and Systematics 37:637–669.

Paxton, K. L., C. Van Riper, T. C. Theimer, and E. H. Paxton. 2007. Spatial and temporal migration patterns of Wilson's Warbler (*Wilsonia pusilla*) in the southwest as revealed by stable isotopes. Auk 124:162–175.

Paxton, K.L., E.B. Cohen, E.H. Paxton, Z. Németh, and F.R. Moore. 2014. El Niño-Southern Oscillation is linked to decreased energetic condition in long-distance migrants. PLoS One 9:e95383.

Pettorelli, N., J. O. Vik, A. Mysterud, J.-M. Gaillard, C. J. Tucker, and N. C. Stenseth. 2005. Using the satellite-derived NDVI to assess ecological responses to environmental change. Trends in Ecology and Evolution 20:503–510.

Pinzon, J., M. E. Brown, and C. J. Tucker. 2005. EMD correction of orbital drift artifacts in satellite data stream. Pp. 167–83 in N. E. Huang and S. S. Shen (editors), Hilbert-Huang transform and its applications. World Scientific, Singapore.

Pulido, F. 2007. Phenotypic changes in spring arrival: evolution, phenotypic plasticity, effects of weather and condition. Climate Research 35:5–23.

Pyle, P., and D. DeSante. [online]. 2012. List of North American birds and alpha codes according to American Ornithologists' Union taxonomy through the 53rd AOU Supplement. The Institute for Bird Populations, Point Reyes Station, CA. <http://www.birdpop. org/alphacodes.htm>

R Development Core Team. [online]. 2012. R: A language and environment for statistical computing. R Foundation for Statistical Computing, Vienna, Austria. <http://www.R-project.org/>

Ramenofsky, M., J. M. Cornelius, and B. Helm. 2012. Physiological and behavioral responses of migrants to environmental cues. Journal of Ornithology 153:181–191.

Robson, D., and C. Barriocanal. 2011. Ecological conditions in wintering and passage areas as determinants of timing of spring migration in trans-Saharan migratory birds. Journal of Animal Ecology 80:320–331.

Ruegg, K. C., and T. B. Smith. 2002. Not as the crow flies: a historical explanation for circuitous migration in Swainson's Thrush (Catharus ustulatus). Proceedings of the Royal Society of London B 269:1375–1381.

Saino, N., D. Rubolini, N. Jonzén, T. Ergon, A. Montemaggiori, N. C. Stenseth, and F. Spina. 2007. Temperature and rainfall anomalies in Africa predict timing of spring migration in trans-Saharan migratory birds. Climate Research 35:123–134.

Saino, N., T. Szép, M. Romano, D. Rubolini, F. Spina, and A. P. Møller. 2004. Ecological conditions during winter predict arrival date at the breeding quarters in a trans-Saharan migratory bird. Ecology Letters 7:21–25.

Shaw, D. W., and K. Winker. 2011. Spring stopover and refueling among migrant passerines in the Sierra De Los Tuxtlas, Veracruz, Mexico. Wilson Journal of Ornithology 123:575–587.

Slingo, J. [online]. 2011. Climate: observations, projections and impacts. Reference Mexico, Met Office Hadley Centre, Devon, UK. <http://www.met-office.gov.uk/climate-change/policy-relevant/obs-projections-impacts>

Smith, R. J., S. E. Mabey, and F. R. Moore. 2009. Spring passage and arrival patterns of American Redstarts in Michigan's eastern Upper Peninsula. Wilson Journal of Ornithology 121:290–297.

Smith, R. J., and F. R. Moore. 2005. Arrival timing and seasonal reproductive performance in a long-distance migratory landbird. Behavioral Ecology and Sociobiology 57:231–239.

Stanley, C. Q., M. MacPherson, K. C. Fraser, E. A. McKinnon, and B. J. M. Stutchbury. 2012. Repeat tracking of individual songbirds reveals consistent migration timing but flexibility in route. PLoS One 7:e40688.

Studds, C. E., and P. P. Marra. 2011. Rainfall-induced changes in food availability modify the spring departure programme of a migratory bird. Proceedings of the Royal Society of London B 278:3437–3443.

Swanson, D. L., and J. S. Palmer. 2009. Spring migration phenology of birds in the Northern Prairie region is correlated with local climate change. Journal of Field Ornithology 80:351–363.

Thornton, P. E., M. M. Thornton, B. W. Mayer, N. Wilhelmi, Y. Wei, and R. B. Cook. [online]. 2012. DAYMET: daily surface weather on a 1 km grid for North America, 1980–2008. Oak Ridge National Laboratory Distributed Active Archive Center, Oak Ridge, TN. <http://doi:10.3334/ORNLDAAC/Daymet_V2>

Tøttrup, A. P., R. H. G. Klaassen, M. W. Kristensen, R. Strandberg, Y. Vardanis, A. Lindstrom, C. Rahbek, T. Alerstam, and K. Thorup. 2012. Drought in Africa caused delayed arrival of European Songbirds. Science 338:1307–1307.

Tøttrup, A. P., K. Rainio, T. Coppack, E. Lehikoinen, C. Rahbek, and K. Thorup. 2010. Local temperature fine-tunes the timing of spring migration in birds. Integrative and Comparative Biology 50:293–304.

Tøttrup, A. P., K. Thorup, K. Rainio, R. Yosef, E. Lehikoinen, and C. Rahbek. 2008. Avian migrants adjust migration in response to environmental conditions en route. Biology Letters 4:685–688.

Tryjanowski, P., T. H. Sparks, S. Kuźniak, P. Czechowski, and L. Jerzak. 2013. Bird migration advances more strongly in urban environments. PLoS One 8:e63482.

Tucker, C. J., J. E. Pinzon, M. E. Brown, D. A. Slayback, E. W. Pak, R. Mahoney, E. F. Vermote, and N. El Saleous. 2005. An extended AVHRR 8-km NDVI dataset compatible with MODIS and SPOT vegetation NDVI data. International Journal of Remote Sensing 26:4485–4498.

Van Buskirk, J., R. S. Mulvihill, and R. C. Leberman. 2012. Phenotypic plasticity alone cannot explain climate-induced change in avian migration timing. Ecology and Evolution 2:2430–2437.

Wiedenfeld, D. A., M. M. Swan, and Louisiana Sea Grant College Program. 2000. Louisiana breeding bird atlas. Louisiana Wildlife and Fisheries Commission. Louisiana State University Press, Baton Rouge, LA.

Wingfield, J. C. 2008. Organization of vertebrate annual cycles: implications for control mechanisms. Philosophical Transactions of the Royal Society B 363:425–441.

Zuur, A. F., E. N. Ieno, N. Walker, A. A. Saveliev, and G. M. Smith. 2009. Mixed effects models and extensions in ecology with R. Springer, New York, NY.

CHAPTER FIVE

Climate on Wintering Grounds Drives Spring Arrival of Short-Distance Migrants to the Upper Midwest*

Benjamin Zuckerberg, Eric J. Ross, Karine Princé, and David N. Bonter

Abstract. Modern climate change has resulted in an advancement of spring phenology throughout the Northern Hemisphere, and studies suggest that many birds are arriving earlier to their breeding grounds in response to these changing spring conditions. Past studies on spring arrival of migrating birds often rely on long-term records from a single banding station and focus on identifying climatic cues at species' breeding grounds to explain variation in arrival times. As a result, the role of climate variability on wintering grounds and its associated impacts on spring migration in birds remain unresolved. Using data from Project FeederWatch, a continentalwide citizen science program focused on wintering birds, we studied the variation in spring arrival for five short-distance, temperate migrants from 1990 to 2011. We hypothesize that short-distance migrants respond strongly to climate conditions on their wintering grounds, which then play a critical role in driving spring arrival to breeding regions. To test this hypothesis, we quantified spring arrival to the upper midwestern United States by calculating first the median arrival dates and explored how interannual variation in winter climate conditions altered these arrival times. For American Robins (*Turdus migratorius*), we found annual variation in temperature on wintering grounds was strongly related to first-arrival date, but not a strong predictor of median arrival dates. For other species, such as Red-winged Blackbirds (*Agelaius phoeniceus*), first-arrival dates were most influenced by precipitation during winter and spring months. These findings suggest that climate on wintering grounds plays an important role in the migration phenology of temperate migrants, but that these effects will likely be variable across species and different subpopulations. Our findings highlight the impact winter climate variability can have on migration phenology and demonstrate the importance of considering the heterogeneous nature of climate change.

Key Words: citizen science, climate change, first arrival, migration, phenology, Project FeederWatch, resident birds.

* Zuckerberg, B., E. J. Ross, K. Princé, and D. N. Bonter. 2015. Climate on wintering grounds drives spring arrival of short-distance migrants to the Upper Midwest. Pp. 83–94 in E. M. Wood and J. L. Kellermann (editors), Phenological synchrony and bird migration: changing climate and seasonal resources in North America. Studies in Avian Biology (no. 47), CRC Press, Boca Raton, FL.

A prominent result of climate change has been a shift toward earlier phenological events across the globe (Walther et al. 2002, Edwards and Richardson 2004, Steltzer and Post 2009). In many regions of the world, less intense freezing events and warmer spring conditions have led to an advance of springtime phenological events over the past few decades (Parmesan 2006, Schwartz and Hanes 2009). Bird migration phenology is thought to have strong ties to climate variability, and as spring conditions warm, a growing body of evidence demonstrates that many birds are migrating and nesting earlier than before (Both et al. 2004, Jonzén et al. 2007, Végvári et al. 2010). Even in the earliest studies of avian migration, temperature was found to be an important factor influencing migration onset (Moffat et al. 1929, Richardson 1978, Harmata 1980), and more recent studies based on long-term banding station data suggest that a wide range of climatic factors play a strong role in migration events (Gordo 2007, Kerlinger 2009). In a recent review on the impacts of climate change on avian migration, Gordo (2007) found that, although studies have explored temperature and precipitation cues, modes of climate variability (e.g., North Atlantic oscillation or El Niño southern oscillation), and combinations of wind speed and direction, most of these studies focus on climatic conditions on breeding grounds or nearby migratory passage areas.

There is little doubt that climate variability often plays an important role in the migration phenology of many bird species, but further research is needed to disentangle the role of climate at different stages of species' life cycles. Studies based on banding station data have historically focused on how weather conditions affect spring arrival to breeding grounds (Jonzén et al. 2007, Knudsen et al. 2011), and although there is strong evidence that this is the case, there is still a great deal of variation in the strength of these relationships between species and even between different populations of the same species—suggesting that shifts in migration may also be driven by broader changes in conditions at wintering grounds (Jonzén et al. 2007, Swanson and Palmer 2009).

Short-distance migrants of North America are considered to be more plastic in their response to climate variability compared to long-distance, Neotropical migrants whose migratory cues are more strongly influenced by photoperiod (Newton

2008). In support of European studies (Both and te Marvelde 2007, Rubolini et al. 2007, Végvári et al. 2010), Swanson and Palmer (2009) found that of 44 migratory species in South Dakota and Minnesota, the 10 species that had uniform advances in spring arrival were all short-distance migrants that had first-arrival dates in early spring such as before 10 April. Migratory plasticity of early spring, short-distance migrants is likely due to this group of species having climatic tolerances necessary to survive variable winter and spring conditions and ride the edge of tolerable climatic conditions (Hurlbert and Liang 2012).

The goal of this study was to test the hypothesis that early spring, short-distance migrants respond to favorable climate conditions on their wintering grounds by migrating earlier. To test this hypothesis, we focused on over two decades of shifts in spring migration phenology across the Upper Midwest (Figure 5.1), using data from Project FeederWatch, a broad-scale citizen science database.

METHODS

Project FeederWatch and Study Species

Project FeederWatch is a winter-long repeated survey of birds that visit supplemental feeding stations throughout the United States and Canada (for program details, see Wells et al. 1998). More than 10,000 FeederWatch sites are located across the United States and Canada annually, with as many as 22 counts submitted from each site per winter. From early November to late April, participants record the maximum number of each species seen at a supplemental feeding station during 2-day periods. Two-day observational periods are defined as single count and are each separated by a minimum of 5 calendar days. For our analyses, we summarized FeederWatch counts as detection/nondetection data for each species. Participants reported the geographic location (latitude and longitude) of their feeding station(s) using online mapping tools, geographic positioning systems, or address information. When only address information was provided, we used geocoding to obtain precise spatial coordinates.

We analyzed arrival dates for five early spring migrants: American Robins (*Turdus migratorius*), Song Sparrows (*Melospiza melodia*), Red-winged Blackbirds (*Agelaius phoeniceus*), Common Grackles (*Quiscalus*

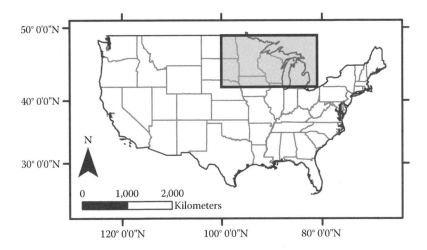

Figure 5.1. The Upper Midwest study region used to determine spring arrival for our study species. Project FeederWatch sites in the study region included the states shown, primarily Michigan, Minnesota, and Wisconsin, as well as small segments of southern Manitoba and Ontario (not shown).

quiscula), and Brown-headed Cowbirds (*Molothrus ater*). These five species were selected because they are relatively common (>10% of sites in this study reporting a sighting at least once during the winter season), are well sampled by broad-scale monitoring programs, and have a clearly delineated early spring migration peak in the Upper Midwest.

Study Region

To document arrival to a breeding region, we selected FeederWatch sampling sites within the upper midwestern United States (Figure 5.1). Although several of the bird species are detected within this study region during the winter months, their numbers are low (<10% of active sites) and the majority of their winter range is south of this region.

Estimating First-Arrival Dates

We restricted our analyses to data collected from 1990 to 2011 (1989 was the first year of records from the program and had relatively low participation). We described species- and year-specific estimates of arrival using generalized additive models (GAM; La Sorte et al. 2013). GAMs are particularly useful as an adaptive method that adjusts automatically to nonlinear relationships between a predictor and response variable (Wood 2006). In the GAM model, we fitted the response variable as the proportion of FeederWatch sites (0–1)

within the breeding region where a given species occurred on a specific date. Although the number of FeederWatch sites submitting observations for any particular date was variable, we excluded count periods that did not have data from a minimum of 23 independent FeederWatch sites (lower quartile of sampling intensity during the winter). In the GAM model, Julian date was modeled as a sum of smooth functions using penalized regression splines plus a conventional parametric component. For this model, we specified a standard "premigratory" period that consisted of a 10-day window for each year. The premigratory period was meant to capture the baseline proportion of occupied sites during the winter season within the breeding region and represented the last 10 days of February.

Next, we fit a GAM over this 10-day period and quantified the mean predicted proportion of sites (p_{winter}) that a species was present. We used the upper limit of the 95% confidence interval of p_{winter} as a premigratory threshold. Last, we fitted a GAM model across the full range observations from the end of the premigratory period to the termination of the FeederWatch season (1 March–1 April). Using this approach, we estimated first arrival (FA) and median arrival (MA) dates as the Julian date at which the minimum and median proportion of sites exceeded the upper limit of the premigratory threshold ($p_{winter} + 2.5 \times SE$), respectively (Figure 5.2). Our estimation of both metrics allowed us to capture different aspects of species' migratory timing for each FeederWatch

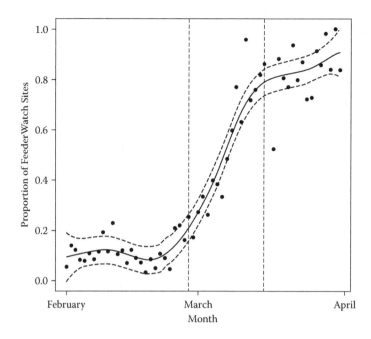

Figure 5.2. Proportion of sites within our study region that were occupied by American Robins in a single migratory season. Occupancy is relatively low during the winter months and quickly rises during spring migration in early March. The solid line represents the GAM for this year along with its 95% confidence interval. The first vertical dashed line represents the calculated FA date followed by the second representing the MA date.

season (1990–2011). FA dates are traditionally the metric used in phenology studies to characterize arrival on the breeding grounds, while MA dates capture the peak of the population migration and have been shown to reduce the bias associated with first-arrival dates, which can be sensitive to changes in cohort size of migrants (Tryjanowski et al. 2005, Miller-Rushing et al. 2008).

Climate Conditions on Wintering Grounds

We analyzed data from the NOAA National Climatic Data Center (NCDC; www.ncdc.noaa. gov). The NCDC provides climate summaries for the real-time analysis of monthly temperature and precipitation data across the contiguous United States. Climate observations are adjusted to account for confounding factors such as instrument changes, station relocation, observer practice changes, and urbanization.

The NCDC has identified nine climatic regions within the contiguous United States (Karl and Koss 1984). We analyzed climate data from the central US climate region, which consists of Illinois, Indiana, Ohio, West Virginia, Missouri, Tennessee, and Kentucky. We chose the central US region to capture generalized winter climate conditions for

our study species because avian migration routes tend to follow a north–south axis (Newton 2008, La Sorte et al. 2013). We wanted to capture climate variability on wintering grounds directly south of the breeding study region with the strongest potential influence on arrival date. For the central region, we calculated mean winter temperature (°C, WTemp; January and February) and mean winter precipitation (cm, WPpt; January and February) as 2-month composites and mean spring temperature (°C, STemp; March) and mean spring precipitation (cm, SPpt; March) as a 1-month composite. We did not include February as a spring metric because the majority of our study species usually do not arrive to the Upper Midwest until the beginning of March (Temple et al. 1997). Because we were using monthly aggregates, a higher proportion of February would be included in the winter composites or the time window preceding spring arrival and therefore was not included with March as representing spring climatic conditions. Additionally, the majority of FeederWatch observations begin to diminish in early April; therefore, we felt justified in focusing solely on March to quantify spring climatic conditions for capturing variation in FA and MA dates.

Statistical Analysis

For each year and species, we calculated FA and MA dates. We constructed generalized linear models (GLMs) using the arrival metrics (FA and MA) as the response variable and the suite of winter climate variables as predictors. After checking for multicollinearity in the predictor variables, we developed an *a priori* set of 14 candidate approximating models with different combinations of climate variables. We also included year as a continuous predictor variable to test for temporal trends in arrival dates. We used an information-theoretic approach to assess the relative support of the models. For each model, we calculated the log likelihood value, number of model parameters (K), sample-size adjusted AIC_c, ΔAIC_c, and Akaike weights (w_i; Burnham and Anderson 2002). When there was more than one competing model ($\Delta AIC_c < 2$), we used model averaging to calculate parameter estimates (β) and unconditional standard errors. To calculate relative variable importance for each predictor, we summed the Akaike weights over all models including the explanatory variable.

We performed all modeling with R (version 3.0.0; R Development Core Team 2013) and used package MuMin for model selection analysis (Barton 2013). Except when otherwise stated, values presented are parameter estimates ± unconditional SE.

RESULTS

The number of FeederWatch sites contributing data within the breeding season region ranged from 497 in 1991 to 1,064 in 2008 ($\bar{x} = 810 \pm 46$). We found interannual differences in FA dates across the study period that ranged from 22 days for Common Grackles to 26 days for American Robins. The ranges reflect strong variability in arrival dates and a potential shift of up to ~3 weeks during any given year across the 22 years evaluated. Despite interannual variability in FA dates, we did not find evidence of any strong temporal trends in FA or MA. Year was not an important predictor of arrival dates except for Common Grackle, which had weak model support when year was specified as a predictor (Table 5.1).

Over 21 years, mean climate conditions in the overwintering region varied considerably, but did not demonstrate any strong trends over time.

We found that winter temperature ranged from $-2.94°C$ to $4.17°C$ ($\bar{x} = 0.53°C$; $\beta = -0.065 \pm 0.05$), winter precipitation from 99.06 to 210.06 mm ($\bar{x} = 151.57$ mm; $\beta = -0.052 \pm 0.13$), spring temperature from $2.83°C$ to $10.06°C$ ($\bar{x} = 6.44°C$; $\beta = -0.030 \pm 0.06$), and spring precipitation from 52.32 to 157.48 mm ($\bar{x} = 94.63$; $\beta = 0.004 \pm 0.09$).

The spring arrival of American Robins into the Upper Midwest breeding region was most strongly influenced by winter and spring temperatures. Of the five models with the strongest weight of support ($\Delta AIC_c < 2$), winter temperature had the highest relative variable importance (0.83), followed by spring temperature (0.82), winter precipitation (0.17), and spring precipitation (0.15, Table 5.1). Earlier first-arrival dates of robins into the Upper Midwest were more common in years characterized by milder winter (-1.27 ± 0.69) and spring conditions (-1.32 ± 0.74; Figure 5.3). We found similar patterns with dates of MA; however, MA models had higher uncertainty (seven models with $\Delta AIC_c < 2$), and MA arrival–climate relationships were of lower magnitude for all climate factors, including winter (-0.63 ± 0.35) and spring (0.65 ± 0.38) temperatures (Table 5.2).

We found that the relationship between Song Sparrow arrival and climate parameters was more variable and resulted in higher model selection uncertainty (Table 5.2). Spring precipitation had the highest relative variable importance (0.58), followed by winter (0.56) and spring (0.42) temperatures (Table 5.1). We found that Song Sparrow FA dates were generally earlier during springs characterized by higher precipitation (-0.91 ± 0.77, Figure 5.3). The effects of temperature were less strong, but demonstrated a negative response where the arrival of Song Sparrows was generally earlier during warmer winters and springs (Table 5.1). We found similar patterns when analyzing MA dates; however, winter temperature had a slightly higher relative variable importance (0.57; Table 5.2).

Similar to Song Sparrows, Red-winged Blackbirds were most strongly influenced by spring precipitation. Three models had strong support ($\Delta AIC_c < 2$), but spring precipitation had the highest relative variable importance (0.78), followed by winter temperature (0.51, Table 5.1). We found that the FA dates of blackbirds were earlier during years of wet spring conditions (-0.08 ± 0.05), and warmer winters (-0.86 ± 0.68, Figure 5.3). Analysis of MA dates showed similar results;

TABLE 5.1
Results from model selection for the relationship between first-arrival (FA) dates and climatic conditions on the wintering grounds.

Species	Year	WTemp	WPpt	STemp	SPpt	K	AIC$_c$	ΔAIC$_c$	w_i
American Robin		−1.26		−1.38		4	144.56	0	0.21
		−1.16				3	145.77	1.21	0.11
				−1.27		3	145.84	1.28	0.11
		−1.50	0.06	−1.25		5	145.86	1.30	0.11
		−1.20		−1.33	−0.06	5	146.07	1.51	0.10
Song Sparrow					−0.08	3	145.58	0	0.17
		−1.06				3	145.66	0.08	0.16
		−0.99			−0.07	4	146.02	0.44	0.13
		−1.13		−0.98		4	146.62	1.04	0.10
				−0.89		3	146.84	1.26	0.09
				−0.83	−0.07	4	147.14	1.57	0.08
		−1.05		−0.92	−0.07	5	147.38	1.80	0.07
Red-winged Blackbird					−0.09	3	143.58	0	0.30
		−0.82			−0.08	4	144.65	1.08	0.18
		−0.91				3	145.19	1.61	0.14
Common Grackle		−1.02				3	143.13	0	0.19
	0.29					3	143.33	0.20	0.17
					−0.07	3	143.51	0.38	0.15
		−0.95			−0.06	4	143.91	0.78	0.13
			0.04			3	144.92	1.79	0.08
Brown-headed Cowbird			0.12		−0.08	4	137.12	0	0.32
		−1.15	0.13	−0.42	−0.07	6	138.00	0.87	0.21
			0.12			3	138.56	1.43	0.16
		−1.24	0.14	−0.48		5	138.83	1.70	0.14

NOTES: Climate conditions on the wintering grounds included winter temperature (WTemp), winter precipitation (WPpt), spring temperature (STemp), and spring precipitation (SPpt). We report model coefficients associated with each predictor, number of model parameters (K), sample-size adjusted Akaike Information Criteria (AIC$_c$), model differences (ΔAIC$_c$), and model weight of evidence (w_i). We present only models with a ΔAIC$_c$ < 2.

however, the parameter estimates were reduced (Table 5.2).

Similarly to American Robins, first-arrival dates of Common Grackles were most strongly influenced by winter temperature. We found that five models had strong support, but winter temperature had the highest relative variable importance (0.44), followed by spring precipitation (0.39), and weak relationships with year (0.24) and winter precipitation (0.11, Table 5.1). The FA dates of grackles were earlier during warmer winters (−0.98 ± 0.66) and wet spring conditions (−0.07 ± 0.05, Figure 5.3). The MA dates of Common Grackles showed similar patterns but

with a slightly stronger influence of spring precipitation (Table 5.2).

The arrival of Brown-headed Cowbirds into the Upper Midwest was strongly influenced by all wintering ground climate conditions. We found that the global climate model had support (w_i = 0.25) and the only other competing model had both spring and winter precipitation (w_i = 0.39, Table 5.1). Variation in winter and spring precipitation was a strong predictor of first arrival, but with different associated impacts. We found a strong effect that FA dates for cowbirds were earlier during wetter springs (−0.08 ± 0.04) and drier winters (0.12 ± 0.04, Figure 5.3; Table 5.1). Winter and

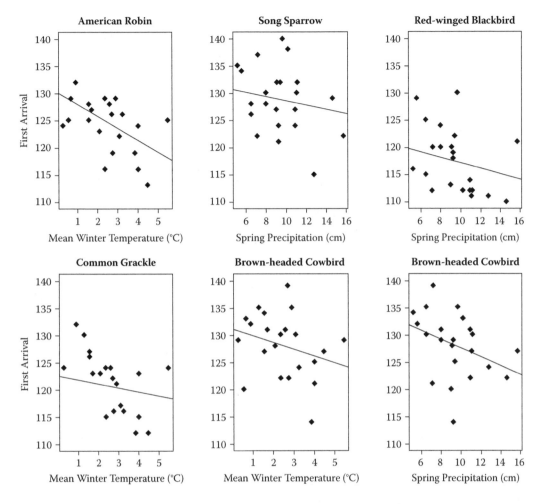

Figure 5.3. Top candidate climate variables from our model selection for first arrival dates for each study species. We focus on NCDC climate data from the wintering grounds in the central United States. We present the relationship between annual first arrival dates and the most strongly supported climate variable for each species.

spring temperatures played a much weaker role in explaining the arrival of Brown-headed Cowbirds into the Upper Midwest; however, earlier first arrival continued to be associated with milder winters (−1.18 ± 0.56, Table 5.1). We found similar patterns for MA; however, the magnitude of the relationships was less pronounced (Table 5.2).

DISCUSSION

Using data generated from a long-term and international citizen science program, we found that the interannual variation in spring migration for short-distance migrants entering the Upper Midwest was strongly influenced by wintering ground climate conditions for at least five common species. Knudsen et al. (2011) concluded

that there has been no consistent pattern of short-distance migrants responding more strongly to favorable climate conditions than long-distance migrants. In the case of spring arrival, our results support the hypothesis that the spring phenology of short-distance migrants is influenced by interannual variation in temperature and precipitation on their wintering grounds, and that conditions associated with milder winter temperatures or increased spring precipitation translate into earlier arrival of these species to the Upper Midwest region. Our findings add to the growing body of research exploring the interaction and importance of climate on nonbreeding grounds and its influence on migration phenology (Gordo et al. 2005, Both et al. 2006, Stirnemann et al. 2012, Tøttrup et al. 2012, McKellar et al. 2013).

TABLE 5.2

Model selection results and parameter estimates comparing interannual variation in median arrival (MA) dates of five temperate migrants and climatic conditions on the wintering grounds.

Species	Year	WTemp	WPpt	STemp	SPpt	K	AIC$_c$	ΔAIC$_c$	w_i
American Robin		−0.64		−0.68		4	114.99	0	0.19
		−0.60				3	115.91	0.93	0.12
		−0.61		−0.65	−0.03	5	116.18	1.20	0.11
				−0.62		3	116.28	1.29	0.10
		−0.76	0.03	−0.62		5	116.66	1.67	0.08
		−0.56			−0.03	4	116.79	1.80	0.08
					−0.04	3	116.88	1.89	0.08
Song Sparrow		−0.56				3	115.66	0	0.18
					−0.04	3	115.73	0.08	0.17
		−0.52			−0.04	4	115.95	0.29	0.15
		−0.59		−0.46		4	116.95	1.29	0.09
				−0.41		3	117.34	1.68	0.08
				−0.38	−0.04	4	117.60	1.94	0.07
Red-winged Blackbird					−0.04	3	112.98	0	0.27
		−0.47			−0.04	4	113.45	0.47	0.21
		−0.51				3	114.08	1.10	0.15
Common Grackle					−0.04	3	113.52	0	0.18
		−0.50				3	113.56	0.04	0.17
	0.13					3	114.09	0.57	0.13
		−0.46			−0.03	4	114.12	0.60	0.13
			0.02			3	115.16	1.64	0.08
Brown-headed Cowbird			0.05		−0.04	4	106.22	0	0.31
		−0.57	0.06	−0.22	−0.03	6	106.92	0.71	0.22
			0.05			3	107.76	1.54	0.14
		−0.62	0.06	−0.25		5	107.90	1.69	0.13

NOTES: Climate conditions on the wintering grounds included winter temperature (WTemp), winter precipitation (WPpt), spring temperature (STemp), and spring precipitation (SPpt). We report model coefficients associated with each predictor, number of model parameters (K), sample-size adjusted Akaike information criteria (AIC$_c$), model differences (ΔAIC$_c$), and model weight of evidence (w_i). We present only models with a ΔAIC$_c$ < 2.

Of the climate conditions that we considered, it is clear that winter temperature was a consistent predictor of arrival dates. Across all five species, 14 of the 24 models for FA dates included winter temperature as an important climate variable influencing arrival to the Upper Midwest. Each species demonstrated the pattern of arriving earlier following a milder winter, resulting in an estimated 1.5 day shift per 1°C increase in mean winter temperature. Hurlbert and Liang (2012) used data from eBird as another citizen science database to explore spatiotemporal variation in

mean arrival dates versus mean minimum spring temperature for 18 common species in North America over the past decade. The authors found median arrival dates shifted 0.8 days earlier for every 1°C of warming of spring temperature. Our finding of 1.5 day shift per 1°C provides stronger support for the role of winter temperature, as opposed to spring temperature, in migration timing. Several studies have reported a stronger relationship between arrival dates and winter weather on the wintering grounds as opposed to spring conditions on the breeding grounds (Gordo and

Sanz 2008, Balbontín et al. 2009, McKellar et al. 2013). However, these studies have focused on long-distance migrants and Neotropical wintering sites; few studies have focused on the role of winter climate on migration phenology in temperate migrants.

We found that precipitation played a secondary, but important, role in the timing of migration, especially for Song Sparrows, Red-winged Blackbirds, and Brown-headed Cowbirds. For Neotropical–Nearctic migrants, drought is an important climate factor that can delay migrant birds at any stage of their migratory journey (Studds and Marra 2011, Rockwell et al. 2012, Tøttrup et al. 2012). It is assumed that years of high precipitation on Neotropical wintering grounds lead to greater primary productivity, increased condition of overwintering birds, and earlier migration onset (Studds and Marra 2011, Rockwell et al. 2012). However, we found limited influence of drought on migration phenology. In fact, Brown-headed Cowbird arrival was delayed during times of high winter precipitation, which seems to contradict the response found in long-distance migrants, where drought caused significant delays in arrival (Tøttrup et al. 2012). Short-distance migrants such as cowbirds, however, may be taking advantage of increased productivity on their wintering grounds to delay their migration onset to their breeding grounds, or increased winter precipitation on temperate wintering grounds could be associated with inclement weather conditions, such as snow cover, thus delaying their migration. Future research incorporating more detailed exploration of climate variability, snow cover and primary productivity on temperate wintering grounds would allow for a better exploration of the influence of winter precipitation on earlier arrival.

The implications of shifting migration phenology highlight the potential effects of phenological mismatches between nesting/migration phenology and peak prey availability. Birds often rely on pulses of insects and plant productivity during early spring and thus appear to be shifting both their migratory and nesting phenologies in response to changing climate (Both et al. 2006, Knudsen et al. 2011, Lof et al. 2012). A growing conservation concern is that of the avian species that are shifting their spring arrival and nesting dates forward; they are not advancing at the same rate as the peak in prey phenology,

leading to a phenological mismatch (Both et al. 2004). Phenological mismatches can have varied effects on fitness (Visser et al. 1998, Møller et al. 2008). Several studies have found that, for long-distance migrants, a failure to advance in their spring phenology could lead to rapid population declines (Both et al. 2006, Møller et al. 2008). Recently, however, studies have shown that the effect of these mismatches can be alleviated by having high food availability regardless of seasonal peaks in insect densities (Dunn et al. 2011), and that although seasonal differences in phenology can impact individual fitness, the long-term population-level demographic consequences may not be felt due to relaxed competition (Reed et al. 2013). Altogether, this evidence suggests that there is potential for selective pressures to drive changes in migration and nesting phenology for a wide array of species, including those that are not declining (Visser et al. 2011, Lof et al. 2012).

Identifying where and when climate conditions are important to migration phenology is important to understanding population dynamics, especially in the presence of climate change. Changes in temperature and precipitation due to climate change are not uniform throughout the year; therefore, identifying the impact of climate across full-life cycle phenology will be critical if we hope to make any predictions as to how migration phenology of avian species might change in the future (Houghton et al. 2001). Consequently, it is critical that future assessments of climate variability on bird migration phenology consider both spring and winter climate conditions within a single analytical framework. With the emergence of broad-scale citizen science databases, comprehensive data are now available for bird distributions at breeding and wintering grounds (short-distance migrants), which can be used to assess population trends (Robbins et al. 1989, Link and Sauer 2007), track disease (Dhondt and Tessaglia 1998), and investigate shifts in geographic ranges (Zuckerberg et al. 2011, La Sorte and Jetz 2012). Most recently these data sets have shown their value in being able to capture broad-scale changes in migration phenology (Hurlbert and Liang 2012, Jones et al. 2012). With the continuation of citizen scientists collecting data over increasingly longer periods and multiple geographic regions, the potential to explore phenology across broad scales has never been greater.

ACKNOWLEDGMENTS

We are grateful for the comments, support, and patience of the editors of this special issue on phenological synchrony, J. L. Kellermann and E. M. Wood. We thank W. Hochachka and F. La Sorte for early discussions on the use of citizen science databases in capturing migration phenology.

LITERATURE CITED

Balbontín, J., A. P. Møller, I. G. Hermosell, A. Marzal, M. Reviriego, and F. de Lope. 2009. Individual responses in spring arrival date to ecological conditions during winter and migration in a migratory bird. Journal of Animal Ecology 78:981–989.

Barton, K. 2013. MuMIn: Multi-model inference. R package version 1.9.5. http://CRAN.R-project.org/package=MuMI.

Both, C., A. V. Artemyev, B. Blaauw, R. J. Cowie, A. J. Dekhuijzen, T. Eeva, A. Enemar, L. Gustafsson, E. V. Ivankina, A. Järvinen, N. B. Metcalfe, N. E. I. Nyholm, J. Potti, P. Ravussin, J. J. Sanz, B. Silverin, F. M. Slater, L. V. Sokolov, J. Török, W. Winkel, J. Wright, H. Zang, and M. E. Visser. 2004. Large-scale geographical variation confirms that climate change causes birds to lay earlier. Proceedings of the Royal Society of London B 271:1657–1662.

Both, C., J. J. Sanz, A. V. Artemyev, B. Blaauw, R. J. Cowie, A. J. Dekhuizen, A. Enemar, A. Järvinen, N. E. I. Nyholm, J. Potti, P. Ravussin, B. Silverin, F. M. Slater, L. V. Sokolov, M. E. Visser, W. Winkel, J. Wright, and H. Zang. 2006. Pied Flycatchers Ficedula hypoleuca traveling from Africa to breed in Europe: differential effects of winter and migration conditions on breeding date. Ardea 94:511–525.

Both, C., and L. te Marvelde. 2007. Climate change and timing of avian breeding and migration throughout Europe. Climate Research 35:93–105.

Burnham, K. P., and D. R. Anderson. 2002. Model selection and multi-model inference: a practical information-theoretic approach. Springer, New York, NY.

Dhondt, A., and L. Tessaglia. 1998. Epidemic mycoplasmal conjunctivitis in House Finches from eastern North America. Journal of Wildlife Diseases 34:265–280.

Dunn, P. O., D. W. Winkler, L. A. Whittingham, S. J. Hannon, and R. J. Robertson. 2011. A test of the mismatch hypothesis: How is timing of reproduction related to food abundance in an aerial insectivore? Ecology 92:450–641.

Edwards, M., and A. J. Richardson. 2004. Impact of climate change on marine pelagic phenology and trophic mismatch. Nature 430:881–884.

Gordo, O. 2007. Why are bird migration dates shifting? A review of weather and climate effects on avian migratory phenology. Climate Research 35:37–58.

Gordo, O., L. Brotons, X. Ferrer, and P. Comas. 2005. Do changes in climate patterns in wintering areas affect the timing of the spring arrival of trans-Saharan migrant birds? Global Change Biology 11:12–21.

Gordo, O., and J. J. Sanz. 2008. The relative importance of conditions in wintering and passage areas on spring arrival dates: the case of long-distance Iberian migrants. Journal of Ornithology 149:199–210.

Harmata, W. 1980. Phenological dynamics of arrivals and departures on migratory birds in Cracov, Poland, and the environments in the 19th and 20th century. International Journal of Biometeorology 24:137–140.

Houghton, J. T., Y. Ding, D. J. Griggs, M. Noguer, P. J. van der Linden, X. Dai, K. Maskell, and C. A. Johnson (editorss.). 2001. IPCC, 2001: climate change 2001: the scientific basis. Contribution of Working Group I to the Third Assessment Report of the Intergovernmental Panel on Climate Change. Cambridge University Press, Cambridge, UK and New York, NY.

Hurlbert, A. H., and Z. Liang. 2012. Spatiotemporal variation in avian migration phenology: citizen science reveals effects of climate change. PloS One 7:e31662.

Jones, G. M., B. Zuckerberg, and A. T. Paulios. 2012. The early bird gets earlier: a phenological shift in migration timing of the American Robin (Turdus migratorius) in the state of Wisconsin. Passenger Pigeon 74:131–142.

Jonzén, N., T. Ergon, A. Lindén, and N. Stenseth. 2007. Bird migration and climate: the general picture and beyond. Climate Research 35:177–180.

Karl, T., and W. J. Koss. 1984. Regional and national monthly, seasonal, and annual temperature weighted by area, 1895–1983. National Climatic Data Center.

Kerlinger, P. 2009. How birds migrate. Stackpole Books, Mechanicsburg, PA.

Knudsen, E., A. Lindén, C. Both, N. Jonzén, F. Pulido, N. Saino, W. J. Sutherland, L. A. Bach, T. Coppack, T. Ergon, P. Gienapp, J. A. Gill, O. Gordo, A. Hedenström, E. Lehikoinen, P. P. Marra, A. P. Møller,

A. L. K. Nilsson, G. Péron, E. Ranta, D. Rubolini, T. H Sparks, F. Spina, C. E. Studds, S. A. Sæther, P. Tryjanowski, and N. C. Stenseth. 2011. Challenging claims in the study of migratory birds and climate change. Biological Reviews 86:928–946.

La Sorte, F. A., D. Fink, W. M. Hochachka, J. P. DeLong, and S. Kelling. 2013. Population-level scaling of avian migration speed with body size and distance for powered fliers. Ecology 94:1839–1847.

La Sorte, F. A., and W. Jetz. 2012. Tracking of climatic niche boundaries under recent climate change. Journal of Animal Ecology 81:914–925.

Link, W. A., and J. R. Sauer. 2007. Estimating population change from count data: application to the North American Breeding Bird Survey. Ecological Applications 8:258–268.

Lof, M. E., T. E. Reed, J. M. McNamara, and M. E. Visser. 2012. Timing in a fluctuating environment: environmental variability and asymmetric fitness curves can lead to adaptively mismatched avian reproduction. Proceedings of the Royal Society of London B 279:3161–3169.

McKellar, A. E., P. P. Marra, S. J. Hannon, C. E. Studds, and L. M. Ratcliffe. 2013. Winter rainfall predicts phenology in widely separated populations of a migrant songbird. Oecologia 172:595–605.

Miller-Rushing, A. J., T. L. Lloyd-Evans, R. B. Primack, and P. Satzinger. 2008. Bird migration times, climate change, and changing population sizes. Global Change Biology 14:1959–1972.

Moffat, C. B., J. P. Brunker, and A. W. Stelfox. 1929. Phenological report for 1928. Irish Naturalists' Journal 2:132–146.

Møller, A. P., D. Rubolini, and E. Lehikoinen. 2008. Populations of migratory bird species that did not show a phenological response to climate change are declining. Proceedings of the National Academy of Sciences of the USA 105:16195–16200.

Newton, I. 2008. The migration ecology of birds. Academic Press, London, UK.

Parmesan, C. 2006. Ecological and evolutionary responses to recent climate change. Annual Review of Ecology, Evolution, and Systematics 37:637–669.

Reed, T. E., S. Jenouvrier, and M. E. Visser. 2013. Phenological mismatch strongly affects individual fitness but not population demography in a woodland passerine. Journal of Animal Ecology 82:131–144.

Richardson, J. W. 1978. Timing and amount of bird migration in relation to weather: a review. Oikos 30:224–272.

Robbins, C. S., J. R. Sauer, R. S. Greenberg, and S. Droege. 1989. Population declines in North American birds that migrate to the Neotropics. Proceedings of the National Academy of Sciences of the USA 86:7658–7662.

Rockwell, S. M., C. I. Bocetti, and P. P. Marra. 2012. Carry-over effects of winter climate on spring arrival date and reproductive success in an endangered migratory bird, Kirtland's Warbler (Setophaga kirtlandii). Auk 129:744–752.

Rubolini, D., A. P. Møller, K. Raino, and E. Lehikoinen. 2007. Intraspecific consistency and geographic variability in temporal trends of spring migration phenology among European bird species. Climate Research 35:135–146.

Schwartz, M. D., and J. M. Hanes. 2009. Continental-scale phenology: warming and chilling. International Journal of Climatology 30:1595–1598.

Steltzer, H., and E. Post. 2009. Seasons and life cycles. Science 324:886–887.

Stirnemann, R. L., J. O'Halloran, M. Ridgway, and A. Donnelly. 2012. Temperature-related increases in grass growth and greater competition for food drive earlier migrational departure of wintering Whooper Swans. Ibis 154:542–553.

Studds, C. E., and P. P. Marra. 2011. Rainfall-induced changes in food availability modify the spring departure programme of a migratory bird. Proceedings of the Royal Society of London B 278:3437–3443.

Swanson, D. L., and J. S. Palmer. 2009. Spring migration phenology of birds in the Northern Prairie region is correlated with local climate change. Journal of Field Ornithology 80:351–363.

Temple, S. A., J. R. Cary, and R. E. Rolley. 1997. Wisconsin birds: a seasonal and geographical guide. University of Wisconsin Press, Madison, WI.

Tøttrup, A. P., R. H. G. Klaassen, M. W. Kristensen, R. Strandberg, Y. Vardanis, Å. Lindström, C. Rahbek, T. Alerstam, and K. Thorup. 2012. Drought in Africa caused delayed arrival of European songbirds. Science 338:1307.

Tryjanowski, P., S. Kuźniak, and T. H. Sparks. 2005. What affects the magnitude of change in first-arrival dates of migrant birds? Journal of Ornithology 146:200–205.

Végvári, Z., V. Bakóny, Z. Barta, and G. Kovács. 2010. Life history predicts advancement of avian spring migration in response to climate change. Global Change Biology 16:1–11.

Visser, M. E., L. te Marvelde, and M. E. Lof. 2011. Adaptive phenological mismatches of birds and their food in a warming world. Journal of Ornithology 153:75–84.

Visser, M. E., A. J. van Noordwijk, J. M. Tinbergen, and C. M. Lessells. 1998. Warmer springs lead to mistimed reproduction in Great Tits (*Parus major*). Proceedings of the Royal Society of London B 265:1867–1870.

Walther, G., E. Post, P. Convey, A. Menzel, C. Parmesank, T. J. C. Beebee, J. Fromentin, O. Hoegh-Guldberg, and F. Bairlein. 2002. Ecological responses to recent climate change. Nature 416:389–395.

Wells, J. V., K. V. Rosenberg, and E. H. Dunn. 1998. Feeder counts as indicators of spatial and temporal variation in winter abundance of resident birds. Journal of Field Ornithology 69:577–586.

Wood, S. N. 2006. Generalized additive models: an introduction with R. Chapman & Hall/CRC Press, Boca Raton, FL.

Zuckerberg, B., D. N. Bonter, W. M. Hochachka, W. D. Koenig, A. T. DeGaetano, and J. L. Dickinson. 2011. Climatic constraints on wintering bird distributions are modified by urbanization and weather. Journal of Animal Ecology 80:403–413.

Spring Migration

Phenological Asynchrony between Migrant Songbirds and Food Resources during Early Springs*

INITIATION OF A TROPHIC CASCADE AT A STOPOVER SITE

Paul K. Strode

Abstract. This chapter compares the phenology of forest trees, caterpillars, and Neotropical–Nearctic migrant songbirds during three successive spring seasons at Trelease Woods, a 24-ha deciduous forest fragment near Urbana, Illinois, in the United States, and describes a trophic cascade in years of phenological asynchrony. I show that the phenological patterns of budbreak and leaf emergence and the occurrence of canopy caterpillars varied significantly among springs, in accordance with variation in temperature accumulation. The timing of migration of migrant songbirds through the stopover, however, did not vary among years at the community or the species levels. As a result, activity by the three trophic levels was synchronized in 2002, but unsynchronized in 2001 and 2003. In the synchronized year, mean leaf area lost to folivory was 9.3% in both bur oak (*Quercus macrocarpa*) and northern red oak (*Q. rubra*), but increased significantly to 20.5% in the year with least synchrony (2003). In the two years when black walnut (*Juglans nigra*) was included in the study, folivory increased from 1.8% in 2002 to 7.8% in 2003. Increases in folivory were likely an indirect result of trophic level asynchrony and are evidence of a trophic cascade. My findings raise concerns that more frequent early springs generated by climate change, with resulting trophic asynchrony, may affect the productivity and future composition of temperate deciduous forests, as well as the fitness of many Neotropical–Nearctic birds.

Key Words: black walnut, caterpillars, climate change, folivory, *Juglans*, migrant birds, migration, oak, phenology, *Quercus*, stopover, vireo, wood-warbler.

When Hairston et al. (1960) introduced the green-world hypothesis, they argued that producer biomass is high because predators limit herbivore populations. Top-down trophic cascades with indirect effects of predators on plants via herbivores are now the focus of numerous studies of indirect effects in communities (Pace et al. 1999, Borer et al. 2006). In many systems, trophic cascades can occur by way of phenological variance among trophic levels (Hunter

* Strode, P. K. 2015. Phenological asynchrony between migrant songbirds and food resources during early springs: initiation of a trophic cascade at a stopover site. Pp. 97–116 in E. M. Wood and J. L. Kellermann (editors), Phenological synchrony and bird migration: changing climate and seasonal resources in North America. Studies in Avian Biology (no. 47), CRC Press, Boca Raton, FL.

2001, Rodríguez-Casteñeda 2013)—especially in trophic cascades in which species occupying different positions have narrow temporal windows for synchrony (Stenseth and Mysterud 2002, van Asch and Visser 2007). One such system involves migrant songbirds, their folivorous insect prey, and trees at stopover habitats in the spring.

Wood-warblers (F. Parulidae) and vireos (F. Vireonidae) are Neotropical–Nearctic migrant songbirds and are potentially vulnerable to trophic asynchrony at stopover habitats in the spring. Trophic asynchrony may arise because the birds are strongly dependent on photoperiod to time their northward movements from the tropics (Hagan et al. 1991, Breuner and Wingfield 2000, Dawson et al. 2001), while the availability of their herbivore prey (Lepidoptera) and the prey's host plants (forest trees) is linked to spring temperatures (Lysyk 1989, Hunter and Lechowicz 1992b, Lawrence et al. 1997, Ivashov et al. 2002, van Asch and Visser 2007). The timing of arrival at stopover habitats for many insectivorous songbirds has varied little over the last century (Wilson et al. 2000, Strode 2003). In early or late springs, it may be predicted that these birds could miss the peak in their food resources at stopover habitats. It also can be predicted that enough asynchrony could initiate a trophic cascade and result in increased folivory to the trees and decreased fitness for the birds.

In temperate deciduous forests, spring-feeding caterpillars emerge and feed when the nitrogen and water contents of young leaves are high (Hunter and Lechowicz 1992a). In Britain, Dury et al. (1998) found that the organic nitrogen content of leaves of pedunculate oak (*Quercus robur*) declined by 60% during the spring warm-up and leaf growth phase. Similarly, Hunter and Lechowicz (1992a) found that leaf nitrogen in the young leaves of nine species of deciduous hardwood trees in the eastern United States had declined to half of their initial levels (percentage of dry weight) after the date of a predictable temperature threshold (225°C–250°C day accumulation above a 5°C threshold).

Many spring-feeding insect herbivores have evolved spring emergence phenologies that correspond to the temperature-driven emergence of their host plant food sources (Lysyk 1989, Hunter and Lechowicz 1992b, Lawrence et al. 1997, Ivashov et al. 2002). Phenophase synchrony has been particularly well documented

for economically important insect species, especially Lepidoptera (Buse and Good 1996, Parry et al. 1998, Bryant et al. 2002). Insect herbivores in general are most dense in the spring in many forest systems (Feeny 1970, Reichle et al. 1973). Caterpillars, in particular, are at their highest densities in May in temperate deciduous forests (Kendeigh 1979), and insectivorous migrant songbirds typically arrive in these forests to rest and feed during migration.

Caterpillars are the main prey of wood-warblers and vireos during spring migration, and they account for almost 100% of the diet of some species of migratory birds (Graber et al. 1983, 1985). Previous studies have shown that forest insectivorous birds are significant population regulators of caterpillars (Holmes et al. 1979, Moore and Yong 1991), and that caterpillar-consuming birds can have both weak and strong top-down effects on forest tree growth (Marquis and Whelan 1994, Lichtenberg and Lichtenberg 2002). Indeed, predation on caterpillars by birds can reduce caterpillar abundance in saplings of forest trees by up to 50% (Marquis and Whelan 1994). Therefore, wood-warblers and vireos are likely key players in the population dynamics of spring-feeding caterpillars. However, the impact of insectivorous migrant songbirds on temperate forest ecosystems may depend on the local density of folivorous insects and the phenology of songbird migration as it relates to the phenology of their insect prey and the host plants.

Most studies to date on the leaf–caterpillar–songbird tritrophic system used understory saplings as the foraging substrate for birds and insects; only recently have researchers looked for effects in the forest canopy (Barber and Marquis 2009, Böhm et al. 2011). Most published studies also focus on a single common plant species within the study area. For example, Holmes et al. (1979) studied effects on striped maple (*Acer pensylvanicum*); Marquis and Whelan (1994), Lichtenberg and Lichtenberg (2002), and Barber and Marquis (2009) on white oak (*Quercus alba*); and Böhm et al. (2011) on pedunculate oak. Indeed, studies that focus on the understory may underestimate the impact of insectivorous birds on temperate forest ecosystems because they do not account for leaf–caterpillar–songbird interactions in the forest canopy (MacArthur 1958), where food resource levels can elicit numerical and functional responses by insectivorous birds (Morse 1978, Haney 1999) and where most of the primary productivity occurs. Studies that focus on a single tree species as the foraging substrate may miss large impacts on tree species where insect herbivore abundance is higher and thus attracts more bird predation (Buler et al. 2007). While migrant songbirds occupy stopover habitats only briefly during the growing season (Moore et al. 1995), they occur at much higher densities than summer residents (Kendeigh 1982, Finch and Yong 2000, Rodewald and Abrams 2002), and can depress the abundance of insect prey during their short stay (Moore and Yong 1991, Barber and Marquis 2009).

A detailed study of plant–herbivore–predator relationships is especially timely in light of the accumulating evidence that climate change is uncoupling the timing of some predator species from the lower levels of the trophic system on which they depend (Harrington et al. 1999, Parmesan 2006). In this observational study, I tested the hypothesis that the phenology of budbreak, leaf expansion, and caterpillar development is linked to accumulating spring temperatures and that the spring phenology of wood-warbler and vireo migration is unlinked from these environmental cues. I predicted that the timing of wood-warbler and vireo presence at the stopover will be independent of tree and caterpillar phenology across years, whereas the phenology of trees and caterpillars will be synchronized and will vary among springs with different temperature patterns. I also tested the hypothesis that reduction of leaf area lost to folivory by spring-feeding caterpillars is indirectly mediated by migrant songbirds and depends on phenological synchrony between trophic levels. The hypothesis allows for the prediction of a trophic cascade: defoliation should be increased in years of greater phenological asynchrony among different trophic levels.

METHODS

Study Area

I conducted this study from 2001 to 2003 in Trelease Woods, a rectangular (400 m E–W by 600 m N–S) 24-ha deciduous forest fragment located 3 km northeast of Urbana, Illinois (40°09′ N, 88°10′ W). Before European settlement of east central Illinois, Trelease Woods was part of a 2,500-ha forest tract called the "Big Grove" (Miceli et al. 1977). The Trelease Woods forest

fragment has been managed as a research forest by the University of Illinois since 1917 and has been protected from human impacts. Prior to 1950, the relative basal area of trees ≥7 cm dbh (diameter at breast height) was 45% mixed hardwoods, 25% American elm (Ulmus americana), and 30% sugar maple (Acer saccharum; Boggess 1964). While diseases have reduced American elm in importance, Trelease Woods is still described as a mixed hardwood forest that maintains some characteristics of old-growth forest, including a number of large, mature oaks (Pelz and Rolfe 1977, J. Edgington, unpubl. data). The forest has no slope and is dominated by 20 tree species with a maximum canopy height of 20 to >25 m (Augspurger and Bartlett 2003). The landscape in this region is dominated by rowcrop agriculture (Illinois Department of Agriculture 2002). The Trelease Woods forest fragment is heavily used by migrant songbirds in the spring (Kendeigh 1982, Diehl 2003, Strode 2009).

Study Trees and Sampling Transects

For my first objective, I test the hypothesis that the phenology of budbreak, leaf expansion, and caterpillar development is linked to accumulating spring temperatures and the spring phenology of wood-warbler and vireo migration is unlinked from this environmental cue. The trees selected for this study occur along established bird census and tree phenology routes on both the edge and interior of the Trelease Woods forest patch. The routes include four linear 400-m transects of forest edge habitat representing each aspect (N, W, S, E) and five linear 300-m transects of forest interior habitat. The study tree species were bur oak (Quercus macrocarpa), northern red oak (Q. rubra), and black walnut (Juglans nigra). I selected oak and walnut trees because of their known status as preferred foraging substrates by wood-warblers and vireos in the eastern US deciduous forest in general (Holmes and Robinson 1981) and in east central Illinois in particular (Graber and Graber 1983). All three species are economically valuable forest hardwood trees (Cogliastro et al. 1990) that may indirectly benefit from predation by birds on leaf-eating caterpillars (Graber and Graber 1983, Marquis and Whelan 1994).

A total of 209 study trees were surveyed along transect lines and all were mature canopy trees of reproductive size (≥33 cm dbh) with an average canopy height of 25 m (bur oak, n = 89, northern red oak, n = 21; black walnut, n = 99). Study trees comprised 16.3% of the total basal area of trees on transect lines (J. Edgington, unpubl. data). Of my sample, 28 bur oak, 22 northern red oak, and 27 black walnut trees were already part of a forest tree phenology study that has been ongoing in Trelease Woods since 1993 (37%, n = 77 of 209; Augspurger and Bartlett 2003).

Spring Temperature Patterns

I generated a pattern each year of spring temperature accumulation that used degree-days (DD) above 5°C (DD_5) that had accumulated since Julian day (JD) 32 (1 February) to predict the peak in number of canopy caterpillars in any given spring. I chose a base mean daily temperature of 5°C to generate the spring temperature pattern because temperatures above 5°C have been shown to drive spring development in temperate deciduous tree foliage and forest caterpillars (Hunter and Lechowicz 1992a, Parry et al. 1998). I chose JD 32 to begin the DD_5 accumulation because days with a mean temperature above 5°C in January (JD 1–31) in Urbana are uncommon (<7% of January days 1948–2003), and a consistent DD_5 accumulation in most years does not begin until after JD 31 (Illinois State Water Survey 2014). I also determined the day on which a 300 DD_5 accumulation in temperature was reached because the 300 DD_5 accumulation dates provide a thermal model for the beginning of the peak in spring caterpillar activity (Lysyk 1989, Strode 2003). I used the daily mean spring temperatures for Urbana during the 114-year period of 1889–2003 to generate a historical mean spring temperature pattern for comparison with the years of this study (National Climatic Data Center 2014).

Tree Budbreak and Leaf Growth Phenology

I recorded phenological data during the springs of 2001–2003 for each of the 77 oak and walnut study trees. I surveyed the live portion of each study tree crown with binoculars on a consistent day each week beginning before budbreak each year and ending at full leaf expansion. For detailed methods for recording tree leaf phenology data, see Augspurger and Bartlett (2003). Briefly, the date of budbreak (BB) for each individual study tree is defined as the first day on

which, through visual estimation, greater than or equal to one-third of a tree's buds have swollen so that leaf tissue can be seen between the bud scales. The date that leaf emergence (LE) begins is defined as the first day on which leaf tissue extends beyond the bud scales of more than or equal to one-third of a tree's buds. The date that leaf growth (LG) begins is defined as the day on which flattening and leaf form have occurred in more than or equal to one-third of a tree's leaves, while still miniature. The date of full leaf expansion (FE, the end of leaf growth) is defined as the first day on which more than two-thirds of a tree's leaves are deemed fully grown and no growth has occurred for two consecutive weeks of observations. The single phenological stage recorded for each individual included only the visible portion of the entire crown. Tree phenology can vary considerably within and among trees, but this study focuses on monitoring tree leaf-out among tree species to broadly understand how development varies with temperature and, in turn, how timing of leaf emergence affects caterpillar folivory and bird use of deciduous trees. Other tree phenophases such as flowering were not included in this study because only leaves were searched for caterpillars and because leaf area lost to folivory was used as evidence of a trophic cascade.

I made between- and among-year comparisons of each tree phenophase by considering the mean date by which the individuals of each species had completed each phase. I chose the completion of each phase because leaves decline in nutritional quality and acceptability to spring-feeding caterpillars during the period from the end of budbreak to the end of the leaf growth phase (Hunter and Lechowicz 1992a).

I calculated mean and standard deviation for the tree phenophases by converting the phase for each individual each week to a corresponding ordinal value similar to the five-point Likert scale (Likert 1932), in which before budbreak = 0, budbreak = 1, leaf emergence = 2, leaf growth = 3, and full leaf expansion = 4. The conversion allowed me to calculate the mean leaf phase for each tree species each week and determine the overall leaf phase (budbreak, leaf emergence, leaf growth, or full leaf expansion) during which the mean occurrence date occurred for the caterpillar community in the three study tree species in 2002 and 2003. I also calculated the cumulative proportion of individuals observed on each survey date of each study species that had completed each leaf phenophase to compare overall budbreak and leaf development phenological patterns among years. For example, if only 2 of the 28 bur oak study trees had initiated budbreak on a given survey day, this survey day would return a proportion of 0.07 (2/28). In the next survey, if five more individual trees had initiated budbreak, the second survey would return a cumulative proportion of $(2 + 5)/28 = 0.25$.

Caterpillar Phenology

I recorded caterpillar abundance in the springs of 2001–2003 on 2,400 canopy leaves evenly distributed among 20 edge trees. All 20 edge trees were also part of the leaf phenology study. I based my selection of each individual in the subsample on its accessibility with a Terex TB60 telescopic boom lift with a 20-m working height. The lift maximized the number of trees that could be sampled weekly; however, its size limited sampling to the canopies of edge trees only. Only five (n = 5) of the edge northern red oak trees were accessible by the lift so I chose an equal number (n = 5) of the most easily accessed bur oak trees for the bur oak sample. The mean dbh of the oak trees was more than twice the dbh of the 13 accessible black walnut trees, so I chose the 10 most easily accessed black walnut individuals as the black walnut sample to control for basal area. The sample trees were distributed among all four forest edges (N, W, S, E).

Edge trees may be exposed to more light and wind than the interior trees of a forest habitat, and different microclimates could influence the number of insects on each tree. However, microclimate gradients in light, temperature, and humidity can extend up to 700 m into a forest patch, but the Trelease Woods forest patch was only 400 m wide (Bergès et al. 2013). Thus, the microclimate of the edge canopy of Trelease Woods is likely indistinguishable from the microclimate of the canopy in the middle of the forest habitat 200 m away. Moreover, forest fragments of the size of Trelease Woods (24 ha) are essentially all edge with respect to many community interactions (Robinson et al. 1995, Ewers et al. 2007).

Methods for estimating caterpillar abundance and determining sampling effort were based in large part on a similar study of trophic cascades in bird–caterpillar–white oaks (Q. alba; Marquis and Whelan 1994). Sampling units were 30 leaves and total sampling effort per season was 2,700 leaves. Before budbreak in 2001, I located and tagged four branches in the upper half of the crown of each sample tree. I only selected branches that were easily accessed and similar in architecture. Beginning with leaf growth in 2001, but with budbreak in 2002 and 2003, I searched branches at least weekly for caterpillars. Using the terminal bud as a starting point, I searched all surfaces of only the first 30 buds, leaves, or a combination of the two. I also searched the length of branch that included those leaves. I recorded the length of each caterpillar to the nearest millimeter. I held all leaves with rolled sections up to the light to detect any dark caterpillar image visible through the thin, young foliage, and the length of the image was measured. Based on lengths, I estimated biomass using a length–mass regression across species ($y = 0.0314 + 0.0042x + 0.0005x^2$, r = 0.91, P (x) < 0.0001, P (x^2) < 0.0001; J. Graber, unpubl. data). Length is related to biomass of invertebrates, but the biomass calculations account for additional size differences among Lepidoptera larvae, thus providing a better indication of their quality as food resources for migratory birds.

In 2003, I conducted an additional study to determine the phenology of the end of the larval phase of spring-feeding forest caterpillars and capture the timing of when caterpillars are ready to pupate on the ground. Beginning soon after budbreak, I made daily counts of caterpillars descending from the trees to pupate in leaf litter that were caught in two drop traps beneath each of seven trees and represented species monitored in canopy populations of caterpillars (Varley et al. 1974, Hunter et al. 1997).

I generated three phenological patterns for caterpillars. The first pattern was the population density of caterpillars (caterpillars per 2,400 leaves) each week (Holmes et al. 1979, Martel and Mauffette 1997). The second pattern was the mean occurrence date for caterpillar observations each year. For the final phenological pattern, I calculated the cumulative proportion of all caterpillars counted for the spring season for each survey period. The method allowed me to compare the caterpillar phenological patterns to the leaf phase phenological patterns described earlier.

Migratory Songbird Phenology

I conducted daily foraging observations and tallies of foraging songbirds, except during inclement weather. I recorded all wood-warblers and Red-eyed Vireos (*Vireo olivaceus*—hereafter, migratory songbirds) detected on edge and interior transects. Beginning on JD 91 (1 April) each year, I began surveying birds using the Trelease Woods habitat and identified and recorded the number of migratory songbirds foraging in mid-canopy and canopy trees within a fixed transect width (20 m of the forest edge transect and within 20 m of either side of each interior transect). Thus, the sampling unit was the total number of migratory songbirds detected using oak and walnut trees per 9.4 ha per day. I located birds by sight and sound, and used binoculars to identify their species. I began each census 30 min after sunrise because the focal species of migratory songbirds are nocturnal migrants, ended their flights before sunrise, and then used the day's first light to find habitats for foraging and roosting (Wiedner et al. 1992, Moore and Aborn 2000). All transects were visited on each census day. Similarly to the leaf development and caterpillar phenology patterns described before, I generated two phenological patterns for migratory songbirds: mean occurrence date each year and the cumulative proportion of each species of migratory songbird counted in the study trees for each survey day.

I also analyzed spring survey records dating from 1903 to determine the mean date of the peak in combined wood-warbler and vireo species richness (WV peak) and to gain a historical perspective on the timing of wood-warbler and vireo migration through Urbana, Illinois. The WV peak is used here as a proxy for the peak in total numbers of wood-warblers and vireos at the stopover site during migration (Rotenberry and Chandler 1999). The date indicates when potential predation pressure by Neotropical–Nearctic migrant songbirds on forest caterpillars is likely to be greatest (Graber and Graber 1983). Use of the WV peak does not take into account early migrants like the Yellow-rumped Warbler (*Setophaga coronata*), when early migrants are feeding on arthropods other than caterpillars (Strode 2009). Bird survey records are stored in the library archives at the University of Illinois at Urbana-Champaign, and further details of how I collected and used these data can be found in Strode (2003).

Folivory

For my second objective, I tested the hypothesis that the reduction of leaf area lost to folivory by spring-feeding caterpillars is indirectly mediated by migrant songbirds, which depend on phenological synchrony between trophic levels. To test this hypothesis, I measured folivory by spring-feeding caterpillars in the same 20 edge trees where I studied for caterpillar abundance. I flagged two branches in the upper half of the crown of each tree before budbreak. I chose branches that were easily accessible by the boom lift and not used for the caterpillar abundance study. In 2001, my sampling did not include the black walnut study trees because an herbicide overspray from an adjacent agricultural field along the east-facing edge of Trelease Woods caused the leaves on most of the black walnut trees to grow abnormally. The abnormal growth in 2001 prevented me from using a leaf shape–leaf area allometric relationship for estimating folivory.

The wood-warbler and vireo migration season usually ended around JD 151 (31 May). Each year, I collected 15 leaves from each oak branch (n = 30 leaves per tree, n = 150 each for both oak species; Lowman 1985) and 30 leaflets from each black walnut branch (n = 600 for walnut, except during 2001). The larger sample size for black walnut leaflets was necessary to reflect the fact that the largest northern red oak and bur oak leaves are greater in leaf area (pooled $\bar{x}$ = 123 cm^2) than the largest black walnut leaflets ($\bar{x}$ = 18 cm^2; P. Strode, unpubl. data). For oak leaf selection on each branch, I removed the longest leaf, as determined by inspection, in each of 15 terminal leaf clusters (Feeny 1970). The longest leaves are assumed to have been the first to emerge in early spring and to have been exposed to folivory for the longest time (Reichle et al. 1973). It was necessary to sample some leaf clusters twice when there were fewer than 15 clusters on the branch. For black walnut leaf selection, I first selected the five longest leaves. With numbering beginning at the petiole of each leaf, I removed one leaflet at each position from five to nine where the leaflets on black walnut are largest. I collected leaves in the morning, placed them in plastic bags, and transported them to a laboratory for leaf area measurements made later that day. Plastic bags minimized water loss and any change in leaf size.

I recorded leaf length (oaks) and leaflet length and width (black walnut) to the nearest millimeter. I used a leaf area meter (Li-Cor area meter, model 3100, LI-COR, Inc., Lincoln, Nebraska) to determine the observed area of each leaf and leaflet (Lichtenberg and Lichtenberg 2002). On the same day that I collected leaves for folivory measurements, I also collected a series of undamaged leaves from the same branches that were used as substitutes for length measurements when too much area was missing from the study leaves for length to be determined. To predict the expected area (y) from the length (x) for the study leaves of bur oak and northern red oak, I used regression equations that were generated from length (independent variable) and area (response variable) measurements of 150 additional undamaged leaves of each species (bur oak: $r = 0.98$, $y = 0.2238x^{2.1545}$, $P < 0.001$; northern red oak: $r = 0.98$, $y = 0.1706x^{2.2549}$, $P < 0.001$; Linit et al. 1986). For black walnut, I generated a regression equation that predicted area (y, response variable) from the log product of length multiplied by width (x, independent variable) from measurements of 150 leaflets ($r = 0.97$, $y = 18.873 - 39.241x + 26.409x^2$, $P (x) < 0.0001$, $P (x^2) < 0.0001$). I collected all leaves and leaflets used to generate these allometric relationships at the end of the leaf growth stage in 2001 (bur oak and northern red oak) and 2002 (black walnut) and samples represented a nearly complete range of leaf sizes present in each tree.

Statistical Analysis

I calculated mean completion dates for each tree phase (budbreak, leaf emergence, leaf growth) for each of the three study tree species for each spring in the 3 years of the study. I considered the effects of the temperature pattern each spring on the phenology of each phase for each tree species separately, using paired t-tests (Sokal and Rohlf 1995). I also pooled tree species to generate an overall pattern.

I used the Kolmogorov-Smirnov (K-S) goodness-of-fit test (Zar 1999, Yin et al. 2005) to test the null hypotheses that the spring temperature patterns among years (generated by the DD_5 accumulation) were statistically indistinguishable. The K-S test returns a test statistic (D) that is a measure of the maximum difference between the cumulative proportions of two patterns. I compared each test statistic to a critical value for D (D_{crit}), based on each sample size and a rejection level of $\alpha \leq 0.05$.

I used mean occurrence dates during the spring sampling period each year for caterpillars and migratory songbirds to detect differences in phenology between years. Mean dates have been shown to be accurate estimators of phenological shifts (Moussus et al. 2010). Mean occurrence date calculations were weighted by the counts of caterpillars or migratory songbirds on each day a survey was taken. I used t-tests to analyze the mean occurrence dates for caterpillars between years (2002 and 2003) in the three tree species and the mean occurrence dates for migratory songbirds between paired years (2001–2003). I also separately analyzed mean occurrence dates for migratory songbird species where the total number of birds counted within a species was ≥10 individuals. In all cases, I tested the null hypothesis that the mean occurrence dates for all comparisons were statistically indistinguishable.

I used the leaf phase Likert score means to determine the leaf phase during which the number of spring-feeding caterpillars peaked in each of the three tree species in 2002 and 2003. I used one-way analysis of variance (ANOVA) to test the null hypothesis that the mean leaf phase during which caterpillars peaked did not vary among years.

Last, I used the response variable of proportion leaf area missing per leaf to analyze the levels of folivory in the canopy. I square root transformed the data to improve normality and then tested for differences within species among years and among species within years in leaf area lost to folivory with ANOVA. Differences between years were analyzed with paired t-tests. ANOVA and t-tests were considered significant at a rejection level of $\alpha \leq 0.05$.

RESULTS

Spring Temperature Patterns

The DD_5 accumulation patterns for the springs of 2001, 2002, and 2003 were distinctive (Figure 6.1). The 2002 pattern closely matched the average pattern generated by the 114-year mean, while the 2001 and 2003 patterns were earlier than the 114-year mean pattern for most of each spring season. In all 3 years, the DD_5 accumulations had

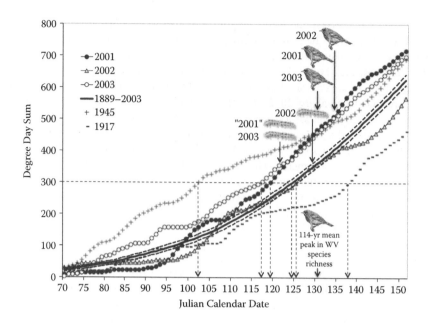

Figure 6.1. Spring temperature pattern as generated by accumulating degree-days (DD, base 5°C) for the three spring seasons of 2001–2003, the 114-year mean (dashed line—95% CI) for the years 1889–2003, and the earliest (1945) and latest (1917) springs at Urbana, Illinois. Horizontal dashed line indicates an accumulation of 300 DD$_5$. Caterpillar and bird images with years and vertical solid lines indicate mean occurrence dates during the spring seasons when data were recorded. Mean occurrence date for caterpillars (2002) was inferred from a thermal developmental model for caterpillars. Vertical dashed lines with arrows indicate the date by which 300 DD$_5$ had accumulated for each year and the 114-year mean.

begun by JD 73. From this date to the end of the migration season (JD 151), K-S tests showed that all three spring temperature patterns were significantly different from each other (2001 vs. 2003: D = 0.218, P = 0.049; 2001 vs. 2002: D = 0.256, P = 0.012; 2002 vs. 2003: D = 0.244, P = 0.020). Degree-days in 2003 accumulated most rapidly for most of the season, 2002 degree-days accumulated most slowly, and 2001 was intermediate. Overall, 2003 DD sums preceded 2002 sums by 5 to 14 days through the end of the spring season and ran 7 to 12 days ahead of the 114-year mean (Figure 6.1).

Tree Budbreak and Leaf Growth Phenology

Taken together, each phase of leaf development for bur oak, northern red oak, and black walnut reflected the observed differences in the temperature patterns (Table 6.1). Compared to mean completion dates for 2002, the budbreak phase of leaf development for the three tree species was earlier by 3 days in 2001 (t = −5.87, df = 76, P < 0.001; paired t-tests, Table 6.1) and 5 days in 2003 (t = 10.09, df = 76, P < 0.001). Compared to

2002, the leaf emergence phase was also earlier (5 and 7 days, respectively) in 2001 (t = -11.22, df = 76, P < 0.001) and 2003 (t = 12.99, df = 76, P < 0.001). In 2003, the mean completion date of the leaf growth phase was 4 days earlier than in 2001 (t = 10.36, df = 76, P < 0.001). The mean completion date of the leaf growth phase was the same in 2001 and 2002 (t = −0.52, df = 76, P = 0.60). For all three tree species, mean completion date for budbreak, leaf emergence, and leaf growth was significantly earlier (paired t-tests, df = 27, 21, and 26 for bur oak, northern red oak, and black walnut, respectively, P < 0.05) in 2001 and 2003 compared to 2002 (Table 6.1). However, in 2001, the earlier mean leaf growth completion date for black walnut was statistically indistinguishable from the 2002 date (t = 1.55, df = 26, P = 0.13).

Caterpillar Phenology

Caterpillar phenology reflected the annual differences among temperature patterns. In 2002, the total number of caterpillars on 2,400 canopy leaves of the three study tree species peaked at

TABLE 6.1

Summary of mean completion dates (Julian calendar day) for three phenophases of Bur Oak (Quercus macrocarpa), Northern Red Oak (Quercus rubra), and Black Walnut (Juglans nigra) during three spring seasons at Trelease Woods, Urbana, Illinois.

Species	Budbreak			Leaf emergence			Leaf growth		
	2001	2002	2003	2001	2002	2003	2001	2002	2003
All trees	111[a]	114[b]	**109[c]**	124[a]	129[b]	**122[c]**	149[a]	149[a]	**145[b]**
Bur oak	112[a]	114[b]	**108[c]**	123[a]	129[b]	**119[c]**	147[a]	145[b]	**139[c]**
Northern red oak	**111[a]**	115[b]	113[a]	**121[a]**	124[b]	122[a]	142[a]	144[b]	**141[a]**
Black walnut	111[a]	112[b]	**107[c]**	128[a]	132[b]	**125[c]**	156[a]	157[a]	**154[b]**

NOTES: Mean dates within phenophases and species with different superscript letters (a, b, c) are different at P = 0.05 (paired t-tests). Bold entries indicate earliest mean completion dates among years for each phase by species.

15 individuals on JD 131 (366 DD_5), with a mean occurrence date of JD 128 and an estimated mean biomass per caterpillar of 0.11 g ($\pm$0.005, n = 66). The timing of the 2002 mean occurrence date of caterpillars was 3 days later than the 114-year Urbana mean date (JD 125) of a 300 DD_5 accumulation (coefficient of variation (CV) = 0.06; Figure 6.1) when thermal models estimate the peak in abundance of forest caterpillars. In 2003, the total number of caterpillars peaked on JD 117 (302 DD_5) at 23 individuals, 14 days earlier than 2002, with a mean occurrence date of JD 122 and 0.13 ($\pm$0.005) g of estimated mean biomass per caterpillar (n = 95). Comparisons of mean occurrence dates between 2002 (JD 130) and 2003 (JD 125) were nonsignificant for caterpillars in northern red oak (t-test, t = 1.56, df = 27, P = 0.13) and between 2002 (JD 124) and 2003 (JD 121) for black walnut trees (t = 1.06, df = 48, P = 0.30). However, in bur oak trees, the mean occurrence date of JD 123 in 2003 came significantly earlier than the mean occurrence date of JD 129 for 2002 (t = 2.24, df = 44, P = 0.03). The 6-day difference in mean occurrence date between 2002 (JD 128) and 2003 (JD 122) for the entire caterpillar community sample was also significant (t = 2.99, df = 101, P = 0.003).

In 2002 and 2003, the peak of caterpillar activity, as determined by the mean occurrence date, occurred during the leaf emergence phase in all three tree species searched, but with considerable variability among tree species within years and between years within tree species. Analysis of variance revealed that the mean leaf phase during which caterpillar populations peaked varied by tree species in both 2002 ($F_{2,74}$ = 28.3, P < 0.001) and 2003 ($F_{2,74}$ = 17.1, P < 0.001). The t-tests showed that the mean leaf phase during which caterpillar populations peaked was significantly different between 2002 and 2003 for bur oak (t = 2.5, df = 54, P = 0.02), but not for northern red oak (t = 1.9, df = 35, P = 0.07) or black walnut (t = 0.46, df = 50, P = 0.64).

In 2003, caterpillars began appearing in the drop traps on JD 119. By JD 128, more than half (56%) of the total caterpillars captured by the end of songbird migration (JD 149) had been recorded (n = 49; mostly Hymenoptera, Geometridae, Noctuidae, and Tortricidae; P. K. Strode, unpubl. data). The last caterpillar was recorded on JD 139.

Migratory Songbird Phenology

I counted totals of 241 (27 species), 234 (25 species), and 428 (24 species) foliage-gleaning migratory songbirds, respectively, each migration season (2001–2003) in the three study tree species along the edge and in the interior of Trelease Woods. Five species of birds had adequate counts each year for analyses by species ($\geq$10; Table 6.2). The migration phenology each spring for the entire community of these birds, as determined by mean occurrence date, varied with year (ANOVA: $F_{2,898}$ = 30.2, P < 0.001). The effect of year on mean occurrence date for songbird migration phenology was due primarily to a significantly later mean occurrence date in 2001 (JD 134) compared to 2002 (JD 129, t = 5.9, df = 465, P < 0.001) and 2003 (JD 129, t = 7.7, df = 516, P < 0.001; Figure 6.1). Mean

TABLE 6.2
Summary and comparative statistics[a] for all neotropical–nearctic migrant songbirds and the five species with large enough sample sizes for analysis (n ≥ 10) counted in Bur Oak, Northern Red Oak, and Black Walnut Trees during the Spring seasons of 2001–2003 at Trelease Woods, Urbana, Illinois.

Species	n	First appearance date			Mean occurrence date			ANOVA F (df)	Student's t (df)		
		2001	2002	2003	2001	2002	2003	All years	2001 vs. 2002	2001 vs. 2003	2002 vs. 2003
All birds	903	117	106	**104**	134	**129**	129	**30.15 (2,898)**	**5.93 (465)**	**7.73 (516)**	0.74 (465)
American Redstart	80	131	130	**128**	140	137	**135**	**8.60 (2,77)**	1.63 (30)	**3.38 (25)**	1.93 (22)
Black-throated-Green Warbler	81	117	**112**	118	131	133	**128**	2.22 (2,78)	0.67 (23)	1.33 (40)	1.65 (17)
Palm Warbler	88	122	**109**	113	126	**124**	126	0.72 (2,85)	1.20 (31)	0.33 (65)	0.94 (33)
Red-eyed Vireo	63	131	127	**122**	140	**138**	139	0.60 (2,60)	1.03 (39)	0.79 (42)	0.27 (38)
Yellow-rumped Warbler	157	117	106	**104**	125	121	**119**	**5.09 (2,154)**	**2.80 (39)**	**4.55 (31)**	1.68 (132)

NOTES: First appearance and mean occurrence dates are Julian calendar days and bold entries are the earliest dates for each species among years. Individual species (with species code) were the American Redstart (AMRE), *Setophaga ruticilla*; Black-throated Green Warbler (BTGW), *Setophaga virens*; Palm Warbler (PAWA), *Setophaga palmarum*; Red-eyed Vireo (REVI), *Vireo olivaceus*; and Yellow-rumped Warbler (YRWA), *Setophaga coronata*. First appearance dates for "All birds" determined by YRWA arrival because YRWA were always first to arrive. First appearance dates in oak and walnut trees are not to be confused with first arrival dates at Trelease Woods. Bolded ANOVA and t-test results indicate significance at α ≤ 0.05.

[a] ANOVA and t-tests for mean occurrence dates for all years and between pairs of years.

occurrence dates for 2002 and 2003 were both JD 129 (t = 0.74, df = 465, P = 0.46).

Five species of migratory songbirds had an adequate number of detections for phenological comparisons (≥10, Table 6.2). The songbird species (with species code) were the American Redstart (AMRE, *S. ruticilla*), Black-throated Green Warbler (BTGW, *S. virens*), Palm Warbler (PAWA, *S. palmarum*), Red-eyed Vireo (REVI, *Vireo olivaceus*), and Yellow-rumped Warbler (YRWA, *S. coronata*). Only the American Redstart and the Yellow-rumped Warbler showed shifts in mean occurrence date in at least one year-to-year t-test comparison. The mean occurrence date for the American Redstart in 2003 (JD 135) was significantly earlier than in 2001 (JD 140, t = 3.38, df = 25, P = 0.002). The mean occurrence date for the Yellow-rumped Warbler in 2001 (JD 125) was significantly later than in 2002 (JD 121, t = 2.80, df = 39, P = 0.008) and 2003 (JD 119, t = 4.55, df = 31, P < 0.001). Taken as a whole, none of the mean occurrence dates reflected the annual differences between temperature patterns with any consistency (Table 6.2; Figure 6.1).

The peak in species richness for wood-warblers and Red-eyed Vireos (WV peak) occurred on

JD 133, 128, and 129 in 2001, 2002, and 2003, respectively. Analysis of the historical migration records from 1903 to 2003 placed the mean WV peak on JD 131 (11 May, n = 75 years, SD = 4 days, CV = 0.030; Figure 6.1). The date coincides within 6 days (JD 125) of the 114-year mean peak date (1889–2003) for caterpillar availability in the forest canopy, as predicted by the 300 DD$_5$ thermal model explained previously.

Trophic Level Synchrony

The general temporal patterns of leaf development, numbers of caterpillars, and numbers of birds each year were distinctive (Figure 6.2). All three trophic levels were synchronized in 2002 (Figure 6.2b) but asynchrony was detected between the first two trophic levels and the migratory songbirds in 2001 and 2003 (Figure 6.2a, c). In 2002, half of all study trees had completed the leaf emergence phase and half of all caterpillars had been recorded by JD 125; half of all migratory songbirds had arrived at the stopover by JD 128 (Figure 6.3b). In 2001, half of all study trees had completed the leaf emergence phase and, based on the 300 DD thermal developmental model,

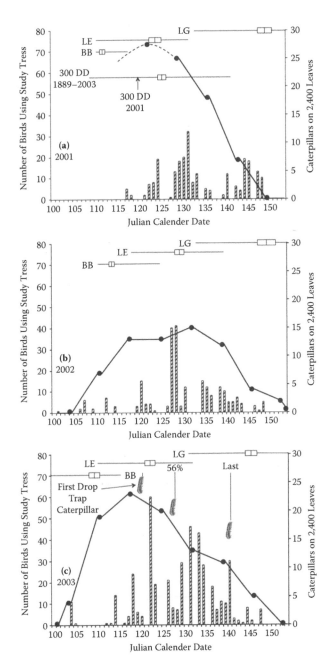

Figure 6.2. Summaries for three trophic phenophases for the spring seasons of (a) 2001, (b) 2002, and (c) 2003 at Trelease Woods, Urbana, Illinois. Hatched bars (primary y-axis) are the number of wood warblers and vireos observed using bur oak, northern red oak, and black walnut study trees in 9.4 ha of transect area; solid line (secondary y-axis) is the number of caterpillars observed on 2,400 edge canopy leaves of five bur oak, five northern red oak, and 10 black walnut trees. Dashed line in (a) indicates the peak in caterpillar activity predicted by a degree-day model (see "Methods"). In (a), 300 DD box and whisker plot indicates the mean, 95% CI, and range for the date of a 300 degree-day accumulation for the years of 1889–2003; upward arrow indicates the Julian date (119) of a 300 DD accumulation in 2001. In (c), suspended caterpillar images indicate the dates on which proportions of 0.07 (first caterpillar), 0.56, and 1.0 (last caterpillar) of the total caterpillars collected in drop traps were recorded. Box and whisker plots in (a)–(c) indicate mean completion date, 95% CI, and range for bud break (BB), leaf emergence (LE), and leaf growth (LG) phases of the study trees (see "Methods" for sample sizes). The single-day high abundance of birds on JD 122 in 2003 was driven by the arrival of Yellow-rumped Warblers (*Setophaga coronata*) and is not representative of seasonal peaks in wood-warbler and vireo migration.

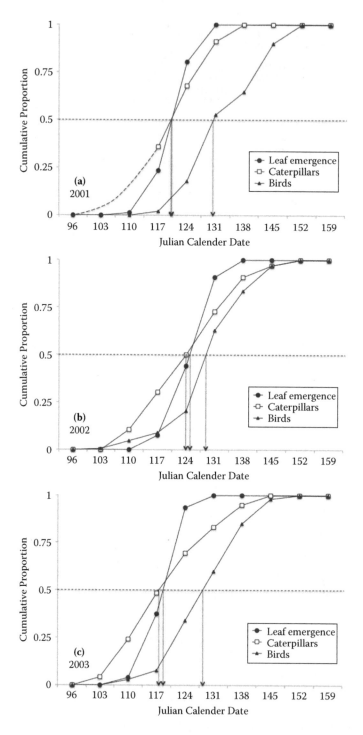

Figure 6.3. Timing of each trophic level during three springs (a–c) at Trelease Woods, Urbana, Illinois. Seasonal variation is shown for the cumulative proportion of bur oak, northern red oak, and black walnut trees (n = 77) that had completed the leaf emergence phase (leaf emergence), the number of caterpillars observed on 2,400 leaves in the canopies of a sample of the study trees (n = 20 trees, caterpillars), and the number of migratory songbirds observed in the study trees (birds). Vertical lines with arrows point to the dates on which 0.5 of each trophic level event had occurred. The curved dashed line in (a) is based on a thermal developmental model for caterpillars.

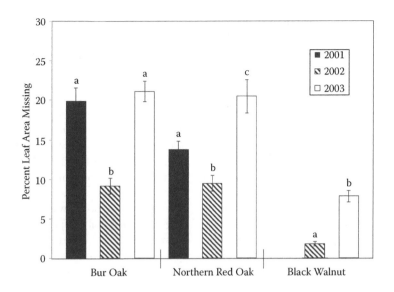

Figure 6.4. Mean percent area lost to folivory (±1 SE) on individual leaves in bur oak, northern red oak, and black walnut at the end of the three spring seasons at Trelease Woods, Urbana, Illinois. ANOVA indicates differences exist among years within species and among species within years (P < 0.05). Different small letters indicate significant differences within species between years as determined by post-hoc t-tests (P < 0.05).

half of all caterpillars would likely have been recorded by JD 120. However, the date by which half of all migratory songbirds had arrived at the stopover was 11 days later on JD 131 (Figure 6.3a). In 2003, half of all study trees had completed the leaf emergence phase and half of all caterpillars had been recorded by JD 118, while the date by which half of all migratory songbirds had arrived at the stopover was also 11 days later on JD 129 (Figure 6.3c).

Folivory

Folivory was the lowest for all three tree study species in 2002, the year with least asynchrony (Figure 6.4). Analysis of variance showed that the mean percentage leaf area lost to folivory varied significantly among years for bur oak ($F_{2,447}$ = 31.89, P < 0.001) and northern red oak ($F_{2,447}$ = 10.78, P < 0.001). Folivory in black walnut was significantly greater in 2003 (7.8%) than in 2002 (1.8%) according to a paired t-test (t = 7.78, df = 812, P < 0.001). Within year and among species comparisons also showed variation for 2002 ($F_{2,897}$ = 77.22, P < 0.001) and 2003 ($F_{2,897}$ = 68.19, P < 0.001), the 2 years in which black walnut was included in the folivory study. Folivory levels in bur oak were statistically indistinguishable for 2001 and 2003 (19.9% and 21.1%,

respectively; paired t-test, t = 0.56, df = 298, P = 0.58). However, these folivory levels were an increase of >100% from levels quantified in 2002 (9.1%) and both differences from 2002 were significant (2001: t = 5.45, df = 298, P < 0.001; 2003: t = 7.27, df = 298, P < 0.001). The 3-year pattern of folivory in northern red oak matched the pattern of bur oak. While folivory in northern red oak in 2001 (13.7%) was less than that of bur oak, folivory levels in northern red oak nearly matched those of bur oak in 2002 (9.5%) and 2003 (20.5%). All three levels of folivory in northern red oak were statistically distinct from each other as revealed by paired t-tests (2001 vs. 2002: t = 2.78, df = 298, P < 0.001; 2001 vs. 2003: t = 2.63, df = 298, P < 0.001; 2002 vs. 2003: t = 4.28, df = 298, P < 0.001).

DISCUSSION

The results from my 3-year study support the hypothesis that the communitywide timing and pattern of wood-warbler and vireo spring migration were unlinked from variable spring temperatures at a migratory stopover site in the midwestern United States. However, the timing of the caterpillar food resources upon which these birds depend for refueling at the stopover site does track seasonal variation in spring temperature.

Indeed, the number of caterpillars in the forest canopy of the three study tree species (bur oak, northern red oak, and black walnut) was predicted by accumulating spring temperatures, with the spring peak (mean occurrence date) for available caterpillars occurring earlier in the early springs of 2001 and 2003. Timing of budbreak, leaf emergence, and leaf growth in the study tree species also varied with the temperature pattern and all phases were completed earlier in either 2001 or 2003. Caterpillar activity was generally linked to the leaf emergence phase of forest tree phenology. Yet, because the arrival of Neotropical migrants did not vary between years, the top-down predation pressure of the migratory songbirds was less in 2001 and 2003 than in 2002. As a result, these findings suggest that a trophic cascade occurs in warm years when tree and caterpillar phenology becomes decoupled from migratory songbird stopover habitat use.

The most likely hypothesis that explains the lack of a behavioral response to variation in temperature at this temperate site by the migrant songbird community is that most of these species are long-distance migrants that are physiologically linked to day length, which controls their early spring departure from the tropics (Hagan et al. 1991, Breuner and Wingfield 2000, Dawson et al. 2001). The 6-day difference between the mean WV peak date at Urbana for 1903–2003 and the 114-year predicted mean peak of available caterpillars is notable. Selection for the photoperiodic response mechanism may have been driven in large part by the overlap of the birds' northward movements with the "green wave" of spring foliage development and the availability of food resources (Schwartz 1998). Overlap may have occurred historically in enough years to maintain fitness. The small coefficient of variation in the date of the WV peak (0.03) compared to that of the 114-year mean date of 300 DD_5 (0.06) in this study suggests that many of these species are traveling through the Illinois stopover location at similar times year after year, as well as farther north (Strode 2003). Songbirds must arrive at northern breeding grounds with time to select and defend optimal breeding territory and produce young (Coppack and Both 2002, Goodenough et al. 2010). However, photoperiod may be less of a determinant for migration timing in years of low food availability in late winter on the non-breeding grounds. For example, Studds and Marra

(2011) showed that drought years produced low food resources in March in the wintering grounds and that American Redstarts delayed the start of northward spring migration 3–5 days as a result.

The patterns for the three trophic levels (trees, caterpillars, birds) show an overlap in 2002 among the leaf emergence phase, the activity peak and mean occurrence date for caterpillars, and the peak in abundance of foliage-gleaning birds and a mismatch in 2001 and 2003. As a likely result of trophic synchrony in 2002, mean levels of folivory on leaves of bur oaks and northern red oaks (<10% for both) was within the 2%–15% range reported for temperate broad-leaved forest systems (Mattson and Addy 1975, Schowalter et al. 1986, Coley and Barone 1996) and close to the level of 10.6% (3-year range: 7.4%–12.4%) reported by Bray (1964) from two oak stands in southern Ontario, Canada. The level of 10.6% was recorded in the presence of actively foraging foliage-gleaning birds (Bray 1964). Mean folivory in black walnut (1.8%) in 2002 was comparable to levels reported in other temperate forests (Gosz et al. 1972), but was considerably less than in oaks. Lower leaf consumption in black walnut may have been influenced by the presence of a secondary metabolite (juglone) in the leaves, which is toxic to some species of Lepidoptera (Piskorski et al. 2011).

Leaf area loss can be expected to continue in the forest canopy after the end of the spring season as summer-feeding folivorous insects become active. Yet, most of the leaf damage in the canopy of temperate deciduous forests occurs in the spring when caterpillars are most numerous and leaf consumption rates are highest on young leaves (leaf area per unit time; Reichle et al. 1973). In addition, spring-feeding caterpillars consume disproportionately more leaf tissue when feeding on buds and small leaves than on fully grown leaves, which decreases the photosynthetic potential of individual leaves (Schowalter et al. 1986, Strong et al. 2000).

In 2002–2003, peak caterpillar activity was synchronized with forest tree leaf emergence. The link was likely a product of the narrow developmental window of opportunity for spring-feeding caterpillars before leaves decline in nutritional quality during the leaf growth phase (Hunter and Lechowicz 1992a, Hunter 1993, Lawrence et al. 1997, Dury et al. 1998). The phenologies of the lower two trophic levels, however, were significantly advanced in 2003 versus 2002, and both events in 2003 occurred well in advance of the mean occurrence date and main peak in abundance of foliage-gleaning birds. The drop in number of caterpillars in the canopy in 2003 before the main pulse of migrant songbirds may have been caused by two additional factors:

1. Early migrants like Yellow-rumped Warblers may have consumed the caterpillars and depressed their number.

2. Caterpillars that had their development advanced by the early spring warmth and available food were completing the larval stage early and descending from the trees to pupate in the soil while birds were still arriving.

Indeed, the drop-trap data indicated that more than half (56%) of all caterpillars collected in the drop traps in 2003 had been collected before the main pulse of bird migration (Figure 6.2). The advances in the timing of tree leaf-out and caterpillar activity were statistically significant, but whether or not these shifts—with the lack of a response by the birds—were biologically significant enough to generate a trophic cascade is the more important ecological question.

Leaf area consumed by herbivorous insects was significantly greater in 2003 compared to 2002 for all three tree species that were monitored (Figure 6.4), a difference that provides support for the hypothesis that asynchrony between trophic levels can elicit a trophic cascade. Peak numbers of caterpillars on 2,400 canopy leaves ($n = 15$) and the total caterpillars counted in the season ($n = 66$) were lower in 2002 compared to 2003 ($n = 23$ and 95, respectively). The mean biomass per caterpillar was 15% lower in 2002 (0.11 g) than in 2003 (0.13 g), so the observed 100% increase in folivory in 2003 was most likely a consequence of a trophic asynchrony between the caterpillars and their predators. The density of caterpillar-eating birds may have been too low during the peak in caterpillar activity to depress their number effectively and thus reduce their consumptive impact in the canopy. A similar effect on the trophic cascade has been generated empirically in a deciduous forest of Pyrenean oak (Quercus pyrenaica) in Spain. Sanz (2001) experimentally manipulated the density of foliage-gleaning insectivorous birds in two similar forest plots and, in the plot with a lower bird density, he showed a significant increase in caterpillar density with an increase in leaf area lost to folivory.

The peak in caterpillar activity for 2001 predicted by the caterpillar DD_5 model also shows asynchrony between the birds and their food resources (Figures 6.2a–6.3a). As a possible consequence, the level of folivory in bur oak was significantly higher in 2001 than in 2002 and similar to the 2003 level, and the significant increase in folivory in northern red oak in 2001 compared to 2002 was consistent with that of the bur oak.

Reviews and meta-analyses have shown that trophic cascades often reduce damage to plants and consequences for forest ecosystems as a result of trophic asynchrony can be considerable (Halaj and Wise 2001, Mäntylä et al. 2011, this study). Low levels of defoliation (~10%) have little effect on forest tree growth (Marquis and Whelan 1994) or fitness (Schowalter et al. 1986, Hochwender et al. 2003), and most of the nitrogen freed from leaves consumed by defoliating insects ends up back in the soil through production of frass (Lovett et al. 2002). However, high levels of defoliation (>15%), especially on young leaves, can dramatically reduce tree fitness by way of significant increases in fruit abortion (Schowalter et al. 1986, Hochwender et al. 2003) and subsequent regeneration failure of some species. Oak regeneration failure has occurred during the last 150 years throughout the US Midwest (Abrams and Downs 1990, McCarthy et al. 2001), with a considerable shift from oak and hickory (Carya spp.) dominance to sugar maple dominance in forest habitats in Illinois (Miceli et al. 1977, Shotola et al. 1992, Cowell and Jackson 2002, Zaczek et al. 2002). If early springs that generate trophic mismatches similar to the patterns observed in this study become more frequent and more pronounced in the future, with no change in the relative abundance of caterpillars among tree species or resulting folivory levels, mismatches can be expected to drive changes in forest composition by promoting reduced forest growth and disrupting natural patterns of succession and regeneration.

The trophic mismatch also presents consequences for Neotropical–Nearctic songbirds that depend on the caterpillar food resources present at forest stopover habitats as birds move north to their breeding grounds in the spring (Graber and Graber 1983). In general, years with asynchrony can alter trophic cascades in terrestrial ecosystems (Hunter and Price 1992). In Europe, clutch sizes of some tit species vary with the density of winter moth larvae (*Operophtera brumata*), a common insect defoliator of oaks (Buse et al. 1999). The bottom-up effect can also be sexually selective. In migrant songbirds, adult males migrate earlier than females (Francis and Cooke 1986) and juveniles (Woodrey and Moore 1997, Stewart et al. 2002). Adult males are therefore at an advantage in early springs, whereas juveniles and females arrive at stopover habitats to rest and refuel when food resources are low—an age-class and sex disadvantage. Some species also migrate earlier than others and therefore are at an advantage in early springs.

In an earlier study, Strode (2003) found that the timing of spring, as indicated by a 300 DD accumulation, has not advanced in east central Illinois, where the current study was conducted. However, at higher latitudes in the upper US Midwest, climate change has promoted statistically significant trends toward earlier springs over the last 90 years, with no adjustment in the timing of migration for many species of long-distance migrants (Strode 2003). Climate models predict this warming of the upper latitudes to continue (Intergovernmental Panel on Climate Change 2007). Moreover, late winter drought conditions at the nonbreeding grounds of many Neotropical–Nearctic migrant songbirds have become more severe and variable with unpredictable food resources, constraining the photoperiodic response that initiates spring migration (Studds and Marra 2011). If later migrants do not advance the timing of their northward movements to match a changing climate, the relative abundance of Neotropical–Nearctic migrant songbird species may change, with some species becoming more common and others becoming more rare (Jones et al. 2003). There simply may not be enough variation among individuals in their photoperiodic response for populations of many migrating birds to evolve and adjust to earlier springs in the northern latitudes brought on by climate change. Indeed, declines in migrant birds due to phenological mistiming have already been documented among populations in Asia (Lee et al. 2011), Europe (Both et al. 2006, Møller et al. 2008), and North America (Anders and Post 2006).

It is unclear what effects changes in the relative abundance of species may have at population, community, and ecosystem levels of organization and interaction. Many species of Neotropical–Nearctic migrant songbirds have already been affected by habitat loss, degradation, and fragmentation (Robinson et al. 1995, Holmes and Sherry 2001, Parker et al. 2005). Songbird species impacted by habitat loss and, especially, late spring migrants may find it difficult to maintain viable populations in the future.

ACKNOWLEDGMENTS

The author wishes to thank S. Buck and the University of Illinois Committee on Natural Areas for providing unlimited access to the Trelease Woods research forest and A. Zawadzka and V. S. Zerwin for assistance in the field. R. Warner provided funding for the research equipment. C. Augspurger provided helpful comments on an earlier version of the manuscript. P. K. Strode was supported by a University of Illinois Graduate College Research Fellowship, an S. Charles Kendeigh research grant from the Champaign County Audubon Society, and a research grant from the College Research Board at the University of Illinois.

LITERATURE CITED

Abrams, M. D., and J. A. Downs. 1990. Successional replacement of old-growth white oak by mixed mesophytic hardwoods in southwestern Pennsylvania. Canadian Journal of Forest Resources 20:1864–1870.

Anders, A. D., and E. Post. 2006. Distribution-wide effects of climate on population densities of a declining migratory landbird. Journal of Animal Ecology 75:221–227.

Augspurger, C. K., and E. A. Bartlett. 2003. Differences in leaf phenology between juvenile and adult trees in a temperate deciduous forest. Tree Physiology 23:517–525.

Barber, N. A., and R. J. Marquis. 2009. Spatial variation in top-down direct and indirect effects on white oak (*Quercus alba* L.). American Midland Naturalist 162:169–179.

Bergès, L., V. Pellissier, C. Avon, K. Verheyen, and J. Dupouey. 2013. Unexpected long-range edge-to-forest interior environmental gradients. Landscape Ecology 28:439–453.

Boggess, W. R. 1964. Trelease Woods, Champaign County, Illinois: woody vegetation and stand composition. Transactions of the Illinois State Academy of Science 57:261–271.

Böhm, S. M., K. Wells, and E. K. V. Kalko. 2011. Top-down control of herbivory by birds and bats in the canopy of temperate broad-leaved oaks (*Quercus robur*). PLoS One 6:e17857.

Borer, E. T., B. S. Halpern, and E. W. Seabloom. 2006. Asymmetry in community regulation: effects of predators and productivity. Ecology 87:2813–2820.

Both, C., S. Bouwhuis, C. M. Lessells, and M. E. Visser. 2006. Climate change and population declines in a long-distance migratory bird. Nature 441:81–83.

Bray, J. R. 1964. Primary consumption in three forest canopies. Ecology 45:165–167.

Breuner, C. W., and J. C. Wingfield. 2000. Rapid behavioral response to corticosterone varies with photoperiod and dose. Hormones and Behavior 37:23–30.

Bryant, S. R., C. D. Thomas, and J. S. Bale. 2002. The influence of thermal ecology on the distribution of three nymphalid butterflies. Journal of Applied Ecology 39:43–55.

Buler, J. J., F. R. Moore, and S. Woltmann. 2007. A multi-scale examination of stopover habitat use by birds. Ecology 88:1789–1802.

Buse, A., S. J. Dury, R. J. W. Woodburn, C. M. Perrins, and J. E. G. Good. 1999. Effects of elevated temperature on multi-species interactions: the case of Pedunculate Oak, Winter Moth and Tits. Functional Ecology 13:74–82.

Buse, A., and J. E. G. Good. 1996. Synchronization of larval emergence in winter moth (*Operophtera brumata* L.) and budburst in pedunculate oak (*Quercus robur* L.) under simulated climate change. Ecological Entomology 21:335–343.

Cogliastro, A., D. Gagnon, D. Coderre, and P. Bhereur. 1990. Response of seven hardwood tree species to herbicide, rototilling, and legume cover at two southern Quebec (Canada) plantation sites. Canadian Journal of Forest Research 20:1172–1182.

Coley, P. D., and J. A. Barone. 1996. Herbivory and plant defenses in tropical forests. Annual Review of Ecology, Evolution, and Systematics 27:305–335.

Coppack, T., and C. Both. 2002. Predicting life-cycle adaptation of migratory birds to global climate change. Ardea 90:369–378.

Cowell, C. M., and M. T. Jackson. 2002. Vegetation change in a forest remnant at the eastern presettlement prairie margin, USA. Natural Areas Journal 22:53–60.

Dawson, A., V. M. King, G. E. Bentley, and G. F. Ball. 2001. Photoperiodic control of seasonality in birds. Journal of Biological Rhythms 16:365–380.

Diehl, R. H. 2003. Landscape associations of birds during migratory stopover. Ph.D. dissertation, University of Illinois at Urbana-Champaign, IL.

Dury, S. J., J. E. G. Good, C. M. Perrins, A. Buse, and T. Kaye. 1998. The effects of increasing CO_2 and temperature on oak leaf palatability and the implications for herbivorous insects. Global Change Biology 4:55–61.

Ewers, R. M., S. Thorpe, and R. K. Didham. 2007. Synergistic interactions between edge and area effects in a heavily fragmented landscape. Ecology 88:96–106.

Feeny, P. 1970. Seasonal changes in oak leaf tannins and nutrients as a cause of spring feeding by winter moth caterpillars. Ecology 51:565–581.

Finch, D. M., and W. Yong. 2000. Landbird migration in riparian habitats of the middle Rio Grande: a case study. Studies in Avian Biology 20:88–98.

Francis, C. M., and F. Cooke. 1986. Differential timing of spring migration in wood warblers (Parulinae). Auk 103:548–556.

Goodenough, A. E., A. G. Hart, and R. Stafford. 2010. Is adjustment of breeding phenology keeping pace with the need for change? Linking observed response in woodland birds to changes in temperature and selection pressure. Climatic Change 102:687–697.

Gosz, J. R., G. E. Likens, and F. H. Bormann. 1972. Nutrient content of litter fall on the Hubbard Brook experimental forest, New Hampshire. Ecology 53:770–784.

Graber, J. W., and R. R. Graber. 1983. Feeding rates of warblers in spring. Condor 85:139–150.

Graber, J. W., R. R. Graber, and E. L. Kirk. 1983. Illinois birds, Wood Warblers. State of Illinois, Department of Energy and Natural Resources, Natural History Survey Division, Champaign, IL.

Graber, J. W., R. R. Graber, and E. L. Kirk. 1985. Illinois birds, Vireos. State of Illinois, Department of Energy and Natural Resources, Natural History Survey Division, Champaign, IL.

Hagan, J. M., T. L. Lloyd-Evans, and J. L. Atwood. 1991. The relationship between latitude and the timing of spring migration of North American landbirds. Ornis Scandinavica 22:129–136.

Hairston, N. G., F. E. Smith, and L. B. Slobodkin. 1960. Community structure, population control, and competition. American Naturalist 94:421–425.

Halaj, J., and D. H. Wise. 2001. Terrestrial trophic cascades: how much do they trickle? American Naturalist 157:262–281.

Haney, J. C. 1999. Numerical response of birds to an irruption of elm spanworm (*Ennomos subsignarius* [Hbn.]; Geometridae: Lepidoptera) in old-growth forest of the Appalachian Plateau, USA. Forest Ecology and Management 120:203–217.

Harrington, R., I. Woiwod, and T. Sparks. 1999. Climate change and trophic interactions. Trends in Ecology and Evolution 14:146–150.

Hochwender, C. G., V. L. Sork, and R. J. Marquis. 2003. Fitness consequences of herbivory on *Quercus alba*. American Midland Naturalist 150:246–253.

Holmes, R. T., and S. K. Robinson. 1981. Tree species preferences of foraging insectivorous birds in a northern hardwoods forest. Oecologia 48:31–35.

Holmes, R. T., J. C. Schultz, and P. Nothnagle. 1979. Bird predation on forest insects: an exclosure experiment. Science 206:462–463.

Holmes, R. T., and T. W. Sherry. 2001. Thirty-year bird population trends in an unfragmented temperate deciduous forest: importance of habitat change. Auk 118:589–609.

Hunter, A. F. 1993. Gypsy moth population sizes and the window of opportunity in spring. Oikos 68:531–538.

Hunter, A. F., and M. J. Lechowicz. 1992a. Foliage quality changes during canopy development of some northern hardwood trees. Oecologia 89:316–323.

Hunter, A. F., and M. J. Lechowicz. 1992b. Predicting the timing of budburst in temperate trees. Journal of Applied Ecology 29:597–604.

Hunter, M. D. 2001. Multiple approaches to estimating the relative importance of top-down and bottom-up forces on insect populations: Experiments, life tables, and time-series analysis. Basic and Applied Ecology 2:295–309.

Hunter, M. D., and P. W. Price. 1992. Playing chutes and ladders: heterogeneity and the relative roles of bottom-up and top-down forces in natural communities. Ecology 73:723–732.

Hunter, M. D., G. C. Varley, and G. R. Gradwell. 1997. Estimating the relative roles of top-down and bottom-up forces on insect herbivore populations: a classic study revisited. Proceedings of the National Academy of Sciences of the USA 94:9176–9181.

Illinois Department of Agriculture. 2002. Illinois Agricultural Statistics 2002 Annual Summary. Illinois Agricultural Statistics Service, Springfield, IL.

Illinois State Water Survey. [online]. 2014. Illinois Climate Network Monthly Summary Data. <http://www.sws.uiuc.edu/warm/dataarchive.asp> (15 October 2013).

Intergovernmental Panel on Climate Change (IPCC). 2007. Climate change 2007: Impacts, adaptation and vulnerability: Working Group II Contribution to the Fourth Assessment Report of the IPCC Intergovernmental Panel on Climate Change. Parry, M. L. (editor), Cambridge University Press, Cambridge, UK.

Ivashov, A. V., G. E. Boyko, and A. P. Simchuk. 2002. The role of host plant phenology in the development of the oak leafroller moth, *Tortrix viridana* L. (Lepidoptera: Tortricidae). Forest Ecology and Management 157:7–14.

Jones, J., P. J. Doran, and R. T. Holmes. 2003. Climate and food synchronize regional forest bird abundances. Ecology 84:3024–3032.

Kendeigh, S. C. 1979. Invertebrate populations of the deciduous forest: fluctuations and relations to weather. Illinois Biological Monographs 50. University of Illinois Press, Champaign, IL.

Kendeigh, S. C. 1982. Bird populations in east central Illinois: fluctuations, variations, and development over a half-century. Illinois Biological Monographs 52. University of Illinois Press, Champaign, IL.

Lawrence, R. K., W. J. Mattson, and R. A. Haack. 1997. White spruce and the spruce budworm: defining the phenological window of susceptibility. Canadian Entomologist 129:291–318.

Lee, S., E. R. Ellwood, S. Park, and R. B. Primack. 2011. Late-arriving Barn Swallows linked to population declines. Biological Conservation 144:2182–2187.

Lichtenberg, J. S., and D. A. Lichtenberg. 2002. Weak trophic interactions among birds, insects and white oak saplings (*Quercus alba*). American Midland Naturalist 148:338–349.

Likert, R. 1932. A technique for the measurement of attitudes. Archives of Psychology 22:55.

Linit, M. J., P. S. Johnson, R. A. McKinney, and W. H. Kearby. 1986. Insects and leaf area losses of planted northern red oak seedlings in an Ozark forest. Forest Science 32:11–20.

Lovett, G. M., L. M. Christenson, P. M. Groffman, C. G. Jones, J. E. Hart, and M. J. Mitchell. 2002. Insect defoliation and nitrogen cycling in forests. BioScience 52:335–341.

Lowman, M. D. 1985. Temporal and spatial variability in insect grazing of the canopies of five Australian rainforest tree species. Australian Journal of Ecology 10:7–24.

Lysyk, T. J. 1989. Stochastic model of eastern spruce budworm (Lepidoptera: Tortricidae) phenology on white spruce and balsam fir. Journal of Economic Entomology 82:1161–1168.

MacArthur, R. H. 1958. Population ecology of some warblers of northeastern coniferous forests. Ecology 39:599–619.

Mäntylä, E., T. Klemola, and T. Laaksonen. 2011. Birds help plants: a meta-analysis of top-down trophic cascades caused by avian predators. Oecologia 165:143–151.

Marquis, R. J., and C. J. Whelan. 1994. Insectivorous birds increase growth of white oak through consumption of leaf-chewing insects. Ecology 75:2007–2014.

Martel, J., and Y. Mauffette. 1997. Lepidopteran communities in temperate deciduous forests affected by forest decline. Oikos 78:48–56.

Mattson, W. J., and N. D. Addy. 1975. Phytophagous insects as regulators of forest primary production. Science 190:515–522.

McCarthy, B. C., C. J. Small, and D. L. Rubino. 2001. Composition, structure and dynamics of Dysart Woods, an old-growth mixed mesophytic forest of southeastern Ohio. Forest Ecology and Management 140:193–213.

Miceli, J. C., G. L. Rolfe, D. R. Pelz, and J. M. Edgington. 1977. Brownfield Woods, Illinois: woody vegetation and changes since 1960. American Midland Naturalist 98:469–476.

Møller, A. P., D. Rubolini, and E. Lehikoinen. 2008. Populations of migratory bird species that did not show a phenological response to climate change are declining. Proceedings of the National Academy of Sciences of the USA 105:16195–16200.

Moore, F. R., and D. A. Aborn. 2000. Mechanisms of en route habitat selection: how do migrants make habitat decisions during stopover? Studies in Avian Biology 20:34–42.

Moore, F. R., S. A. Gauthreaux, Jr., P. Kerlinger, and T. R. Simons. 1995. Habitat requirements during migration: important link in conservation. Pp. 121–144 in T. E. Martin and D. M. Finch (editors), Ecology and management of Neotropical migratory birds, a synthesis and review of critical issues. Oxford University Press, New York, NY.

Moore, F. R., and W. Yong. 1991. Evidence of food-based competition among passerine migrants during stopover. Behavioral Ecology and Sociobiology 28:85–90.

Morse, D. H. 1978. Populations of Bay-Breasted and Cape May Warblers during an outbreak of the spruce budworm. Wilson Bulletin 90:404–413.

Moussus, J., R. Julliard, and F. Jiguet. 2010. Featuring 10 phenological estimators using simulated data. Methods in Ecology and Evolution 1:140–150.

National Climatic Data Center (NCDC). [online]. 2014. Data access. <http://ncdc.noaa.gov> (15 October 2013).

Pace, M. L., J. J. Cole, S. R. Carpenter, and J. F. Kitchell. 1999. Trophic cascades revealed in diverse ecosystems. Trends in Ecology and Evolution 14:483–488.

Parker, T. H., B. M. Stansberry, C. D. Becker, and P. S. Gipson. 2005. Edge and area effects on the occurrence of migrant forest songbirds. Conservation Biology 19:1157–1167.

Parmesan, C. 2006. Ecological and evolutionary responses to recent climate change. Annual Review of Ecology, Evolution, and Systematics 37:637–669.

Parry, D., J. R. Spence, and W. J. A. Volney. 1998. Budbreak phenology and natural enemies mediate survival of first-instar forest tent caterpillar (Lepidoptera: Lasiocampidae). Environmental Entomology 27:1368–1374.

Pelz, D. R., and G. L. Rolfe. 1977. Stand structure and composition of a natural mixed hardwood forest. Transactions of the Illinois State Academy of Science 69:446–454.

Piskorski, R., S. Ineichen, and S. Dorn. 2011. Ability of the Oriental Fruit Moth *Grapholita molesta* (Lepidoptera: Tortricidae) to detoxify juglone, the main secondary metabolite of the non-host plant walnut. Journal of Chemical Ecology 37:1110–1116.

Reichle, D. E., R. A. Goldstein, R. I. Van Hook, Jr., and G. J. Dodson. 1973. Analysis of insect consumption in a forest canopy. Ecology 54:1076–1084.

Robinson, S. K., F. R. Thompson III, T. M. Donovan, D. R. Whitehead, and J. Faaborg. 1995. Regional forest fragmentation and the nesting success of migratory birds. Science 267:1987–1990.

Rodewald, A. D., and M. D. Abrams. 2002. Floristics and avian community structure: implications for regional changes in eastern forest composition. Forest Science 48:267–272.

Rodríguez-Castañeda, G. 2013. The world and its shades of green: a meta-analysis on trophic cascades across temperature and precipitation gradients. Global Ecology and Biogeography 22:118–130.

Rotenberry, J. T., and C. R. Chandler. 1999. Dynamics of warbler assemblages during migration. Auk 116:769–780.

Sanz, J. J. 2001. Experimentally increased insectivorous bird density results in a reduction of caterpillar density and leaf damage to Pyrenean oak. Ecological Research 16:387–394.

Schowalter, T. D., W. W. Hargrove, and D. A. Crossley, Jr. 1986. Herbivory in forested ecosystems. Annual Review of Entomology 31:177–196.

Schwartz, M. D. 1998. Green-wave phenology. Nature 394:839–840.

Shotola, S. J., G. T. Weaver, P. A. Robertson, and W. C. Ashby. 1992. Sugar maple invasion of an old-growth oak-hickory forest in southwestern Illinois. American Midland Naturalist 127:125–138.

Sokal, R. R., and F. J. Rohlf. 1995. Biometry (3rd ed.), W. H. Freeman and Company, New York, NY.

Stenseth, N. C., and A. Mysterud. 2002. Climate, changing phenology, and other life history traits: nonlinearity and match–mismatch to the environment. Proceedings of the National Academy of Sciences of the USA 99:13379–13381.

Stewart, R. L. M., C. M. Francis, and C. Massey. 2002. Age-related differential timing of spring migration within sexes in passerines. Wilson Bulletin 114:264–271.

Strode, P. K. 2003. Implications of climate change for North American Wood Warblers (Parulidae). Global Change Biology 9:1137–1144.

Strode, P. K. 2009. Spring tree species use by migrating Yellow-rumped Warblers in relation to phenology and food availability. Wilson Journal of Ornithology 121:457–468.

Strong, A. M., T. W. Sherry, and R. T. Holmes. 2000. Bird predation on herbivorous insects: indirect effects on sugar maple saplings. Oecologia 125:370–379.

Studds, C. E., and P. P. Marra. 2011. Rainfall-induced changes in food availability modify the spring departure programme of a migratory bird. Proceedings of the Royal Society of London B 278:3437–3443.

van Asch, M., and M. E. Visser. 2007. Phenology of forest caterpillars and their host trees: the importance of synchrony. Annual Review of Entomology 52:37–55.

Varley, G. C., G. R. Gradwell, and M. P. Hassell. 1974. Insect population ecology: an analytical approach. University of California Press, Oakland, CA.

Wiedner, D. S., P. Kerlinger, D. A. Sibley, P. Holt, J. Hough, and R. Crossley. 1992. Visible morning flight of Neotropical landbird migrants at Cape May, New Jersey. Auk 109:500–510.

Wilson, W. H., Jr., D. Kipervaser, and S. A. Lilley. 2000. Spring arrival dates of Maine migratory breeding birds: 1994–1997 vs. 1899–1911. Northeastern Naturalist 7:1–6.

Woodrey, M. S., and F. R. Moore. 1997. Age-related differences in the stopover of fall landbird migrants on the coast of Alabama. Auk 114:695–707.

Yin, Z., S. Peng, H. Ren, Q. Guo, and Z. Chen. 2005. LogCauchy, log-sech and lognormal distributions of species abundances in forest communities. Ecological Modelling 184:329–340.

Zaczek, J. J., J. W. Groninger, and J. W. Van Sambeek. 2002. Stand dynamics in an old-growth hardwood forest in southern Illinois, USA. Natural Areas Journal 22:211–219.

Zar, J. H. 1999. Biostatistical analysis (4th ed.), Prentice Hall, Upper Saddle River, NJ.

Figure A.1 Black-throated Green Warbler (*Setophaga virens*), Pheasant Branch, Wisconsin, 3 May 2011 (spring migration). (Photo credit: Eric M. Wood).

Figure A.2 Chestnut-sided Warbler (*Setophaga pensylvanica*), Pheasant Branch, Wisconsin, 17 May 2014 (spring migration). (Photo credit: Mike McDowell).

Figure A.3   Yellow Warbler (*Setophaga petechia*), Magee Marsh Wildlife Area, Ohio, 2 May 2013 (spring migration). (Photo credit: Kimberly Hall).

Figure A.4   Jennifer McCabe releases a Black-capped Chickadee (*Poecile atricapillus*) at the Seawall banding station in Acadia National Park, Maine, 2010. (Photo courtesy University of Maine).

Figure A.5   Ruby-crowned Kinglet (*Regulus calendula*), Pittsfield, Massachusetts, 2014 (fall migration). (Photo credit: Ian Davies).

Figure A.6   Blackpoll Warbler (*Setophaga striata*), Amherst, Massachusetts, 2014 (fall migration). (Photo credit: Ian Davies).

Figure A.7   Dark-eyed Junco (*Junco hyemalis*), Amherst, Massachusetts, 2014 (fall migration). (Photo credit: Ian Davies).

Figure A.8   Le Conte's Sparrow (*Ammodramus leconteii*), Pheasant Branch, Wisconsin, 27 October 2012 (fall migration). (Photo credit: Mike McDowell).

CHAPTER SEVEN

Climatic Extremes Influence Spring Tree Phenology and Migratory Songbird Foraging Behavior[*]

Eric M. Wood and Anna M. Pidgeon

Abstract. In the Upper Midwest of the United States, fire suppression has resulted in succession of savanna and forests that differ in both plant community composition and vegetation structure from their condition prior to Euro-American settlement. Furthermore, variations in weather affect spring phenological events and potentially alter synchronous relationships of migratory songbirds with their seasonal resources. Our goal here was to understand how annual variation in phenology of four tree species—northern red oak (*Quercus rubra*), eastern white oak (*Q. alba*), sugar maple (*Acer saccharum*), and red maple (*A. rubrum*)—affect foraging behavior of migratory songbirds during spring migration. Oaks currently have poor regeneration, whereas maples have good regeneration in forests in the Upper Midwest. A typical temperature regime in 2009 coupled with a record warm winter and early spring in 2010 provided a natural experiment for addressing our goal. In the spring and early summer of 2009 and 2010, we monitored migratory songbird foraging behavior and collected data on tree flowering and leaf-out phenology for 160 replicate trees of the four study species at the Kickapoo Valley Reserve in southwest Wisconsin. In 2009, 15 species of migratory wood-warbler (F. Parulidae) arrived at the stopover study area in late April and were present until late May. Birds foraged heavily on flowering northern red oak and, to a lesser extent, on flowering eastern white oak and sugar maple. Red maple was not preferred by wood-warblers. In 2010, the arrival date and duration of stay among the 15 species of wood-warblers was similar to 2009, yet the frequency of use of the four tree species was reduced by 60%. Northern red oak, sugar maple, and red maple achieved summer condition 2 to 3 weeks earlier in 2010 than 2009, but these tree species were not preferred by the wood-warblers. Instead, eastern white oak, which flowered from early to late May, was the preferred foraging substrate in 2010. Our findings suggest that the flowering and early leaf-out phase of trees provides important resources to migrant wood-warblers that are apparently absent from trees that are more phenologically advanced. Our results also suggest that managing for heterogeneity in tree species, including early and late flowering species, as well as maintaining early successional tree species in the landscape, may be an important consideration in maintaining wood-warbler population levels under a variety of climate conditions.

Key Words: Acer, avian foraging, climate change, habitat selection, maple, oak, Quercus, tree composition, wood-warbler.

[*] Wood, E. M., and A. M. Pidgeon. 2015. Climatic extremes influence spring tree phenology and migratory songbird foraging behavior. Pp. 117–131 in E. M. Wood and J. L. Kellermann (editors), Phenological synchrony and bird migration: changing climate and seasonal resources in North America. Studies in Avian Biology (no. 47), CRC Press, Boca Raton, FL.

n the Upper Midwest of the United States, land use practices have resulted in succession of savanna and forests that differs in both plant community composition and vegetation structure from their condition prior to Euro-American settlement. In the mid- to late 1800s, large swaths of forests were cleared for timber production, prairie and savannas were converted to agriculture, and settlements were established (Schulte et al. 2007, Rhemtulla et al. 2009). These land use patterns created dramatic changes in land cover, with an ecological legacy that persists today. Over the past half-century, forest cover on nonurban lands or those not in production for agriculture has been regenerating (Iverson et al. 1997, Rhemtulla et al. 2007). Fire and other natural processes were a dominant factor shaping land cover but are now largely absent from the landscape, which has led to changes in vegetation (Abrams 1992). In southern Wisconsin, shade-intolerant trees such as oak (Quercus spp.) have declined in dominance, while shade-tolerant species such as maple (Acer spp.) have increased (Rogers et al. 2008). At a regional scale and since the mid-1800s, oak–hickory was the most common deciduous forest habitat, but it has been reduced by 50%, whereas northern hardwood forests dominated by maple have increased by 60% (Rhemtulla et al. 2009). The successional changes have resulted in the homogenization of forest composition and likely affect bird habitat use during migration (Wood et al. 2012).

In addition to long-term changes in forest structure, variation in climate potentially affects synchrony between migratory birds and their seasonal resources (Strode 2003, Visser and Both 2005). Migratory birds depend on seasonal resources at stopover habitats to replenish energy stores necessary for migration and breeding season demands (Moore et al. 2005). Yet, climate change induces earlier springs, which causes shifts in phenological events such as timing of flowering or leaf-out (Richardson et al. 2006, Ellwood et al. 2013). Phenological mismatches can be problematic for migratory birds, especially long-distance migrants for which departure from the wintering grounds is primarily governed by endogenous cues such as links between hormone activity and photoperiod (Holberton and Dufty 2005) or exogenous cues such as precipitation (Studds and Marra 2011). Since the early twentieth century in the North American Midwest, timing of spring migration among wood-warblers has varied little

from year to year (Strode 2003). However, onset of plant and resource phenology, such as leaf unfolding and caterpillar emergence, has become increasingly similar at different latitudes. These phenological changes may lead to a decoupling of synchronous events in which wood-warblers have less overlap with peak resource availability. In extreme cases, decoupling of phenological events could potentially affect both individual and population-level fitness (McKinney et al. 2012).

It is unclear whether future forest conditions in the Upper Midwest will provide necessary foraging substrates for migratory songbirds in the face of predicted trends in climate change and extreme annual weather variation. Our goal was to determine whether annual variation in tree phenology could affect migratory songbird foraging behavior during spring migration and thereby to gain insight into how greater variability in weather and phenology might impact a group of bird species in decline (Holmes 2007). We focused our study on four species of oaks and maple. The two species of oak—northern red oak (Quercus rubra) and eastern white oak (Q. alba)—are moderately shade-tolerant trees with poor rates of regeneration in southern Wisconsin forests. The two species of maples—sugar maple (A. saccharum) and red maple (A. rubrum)—are shade-tolerant trees that are regenerating well. We chose the four tree species because they are indicators of homogenization in forest composition throughout the region (Nowacki and Abrams 2008), but are currently abundant in the landscape (Wood et al. 2012). We had three study objectives:

1. Climatic conditions play an important role in plant phenology (Körner and Basler 2010), and warm temperatures induce earlier budburst (Richardson et al. 2006, Fu et al. 2012). Our first objective was to quantify the magnitude of difference in late winter and spring temperatures between our two study years of 2009 and 2010.

2. Synchrony between plant phenology, an indicator of food availability, and songbird use of plants as foraging substrates during spring migration are poorly known for migratory songbirds in forests in the Upper Midwest, but are needed to predict the potential effects of environmental change. For our second objective, we explored

patterns of phenological synchrony of the flowering phenophase of the four tree species used by migratory songbirds.

3. For our third objective, we quantified foraging success of migratory songbirds on the oak and maple tree species during two years of different environmental conditions, to understand whether seasonal variation in spring phenology can alter foraging conditions for migratory songbirds.

METHODS

Study Area and Sampling Design

Our study area was the 3,468-ha Kickapoo Valley Reserve in southwestern Wisconsin (Figure 7.1). The study site is located in the unglaciated Driftless Area, which is an important stopover region for migratory songbirds en route to breeding habitat in the boreal forest (Wilson 2008). The vegetation of the upland forest patches at the Kickapoo Valley Reserve, where the study was conducted, is dominated by dry- and southern-mesic forest (Curtis 1959). In a parallel study, we quantified the importance value of tree species in our study area, which is a measure of the availability of trees as foraging substrates for migratory birds, using the point-center quarter method (Wood et al. 2012). Common trees of the study area include northern red oak (17% relative importance value), eastern white oak (16%), sugar maple (16%), bitternut hickory (*Carya cordiformis*; 8%), American basswood (*Tilia americana*; 7%), and red maple (5%). Red maple was absent in the study area in the 1850s, but is now a common tree (Wood et al. 2012), similar to other forested areas throughout the region (Abrams 1998). We studied migratory songbird foraging behavior and tree phenology in four forest patches that ranged in size from 80 to 107 ha (Figure 7.1). We selected these patches because they were among the largest, unfragmented tracts

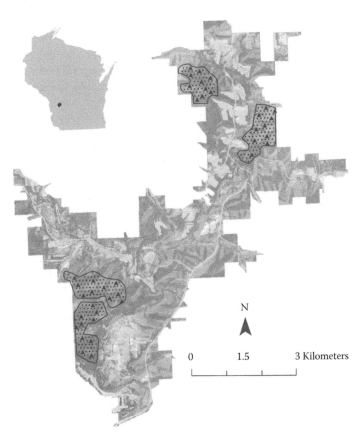

Figure 7.1. Location of the Kickapoo Valley Reserve, Wisconsin, and distribution of forested study areas (black polygons) within the reserve, with 258 lattice points superimposed. Black triangles represent randomly selected phenology-tree sampling stations.

of contiguous forest in southwest Wisconsin, representative of smaller forest patches of the region, and floristically diverse, including our four study species of deciduous trees. To select sampling locations, we digitized each forest patch in ArcGIS 9.3 (ESRI, Redlands, California, 2008) and plotted a lattice grid of points (hereafter, lattice points) separated by 100 m, with alternate rows offset by 50 m. We used each lattice point as a reference point during tree phenology and avian area-search surveys. We plotted 83, 52, 61, and 62 lattice points in the four patches for a total of 258 lattice points (Figure 7.1).

Climatic Data

To quantify differences in climatic conditions between 2009 and 2010, with a focus to inform patterns of tree phenology between years, we obtained maximum daily temperature data for the months of March, April, and May from a nearby National Oceanic and Atmospheric Administration weather station (Sparta, Wisconsin, NOAA Station ID: WI477997). The weather station was located ~15 miles north of our study area and was the closest station with climatic data. We obtained similar climatic data from 1950 to 2010 to compare long-term maximum daily temperature averages with the 2009 and 2010 data. We used maximum daily temperature during the months of March, April, and May as our weather variable of interest because tree budburst phenology is strongly influenced by late winter and early spring temperatures (Körner and Basler 2010).

Tree Phenology Measurements

To determine whether annual variation in tree phenology affects migratory songbird foraging behavior during spring migration, it was necessary to monitor tree phenology simultaneously. We randomly selected 10 lattice points within each forest patch, which were used as reference points for tree phenology measurements (Figure 7.1). At each reference point, we marked individuals from each of the four species of study trees: northern red oak, eastern white oak, red maple, and sugar maple ("phenology trees," hereafter). We selected and flagged the nearest individual of each of the four phenology trees, measured diameter at breast height (dbh) > 10 cm, and recorded GPS coordinates. We marked a total of 160 phenology trees,

with 40 in each forested patch (Figure 7.1). We later removed five phenology trees from analysis because we inadvertently included three eastern black oaks (Q. velutina), which are similar to northern red oak, and two bur oaks (Q. macrocarpa), which are similar to eastern white oak. We visited the four patches and all phenology trees on seven occasions from late April to late May in 2009 and 2010, with at least 4 days between consecutive visits. The Julian dates for our seven visits were 115, 119, 125, 132, 138, 142, and 147. We used this time interval between visits because we wanted to characterize tree phenological stages at a fine temporal scale and because the average stay for migratory birds at stopover habitats is less than 3 days (Moore and Kerlinger 1987). We revisited each phenology tree on the same Julian day in 2010 as in 2009, to control for timing of visits as a source of differences in tree phenology between years. Our sampling scheme allowed for a paired sampling design between years.

During a visit to a phenology tree, observers recorded tree phenology based on the proportion of the crown that was displaying any of the following nine tree phenophases: (1) winter condition, (2) budding, (3) bud swelling, (4) budburst, (5) young leaf, (6) mature leaf (summer condition), (7) flower emergence, (8) flowering, and (9) wilted flower. Observers stood 10 m from the base of the tree and used binoculars to carefully scan the portion of the tree crown in view before recording measurements. We collected four measurements at each tree, with one in each of the cardinal directions. The lead author trained and calibrated all observers throughout the month of April, practicing on tree species where average onset of budburst and leaf-out is similar or earlier than the phenology trees (e.g., quaking aspen, Populus tremuloides). We averaged the four measurements, resulting in a phenology-tree specific score for each phenophase during each visit. We further averaged data from each phenology-tree species by Julian day, throughout our study area, to understand broad phenological patterns of each tree species throughout the region.

Avian Foraging Observations

We quantified avian foraging behavior to determine whether variation in tree phenological stage affects migratory songbird tree use during spring migration. It was not possible to observe

TABLE 7.1

Scientific name, sample size of birds (n), and cumulative number of seconds of foraging observations for 15 migratory species of wood-warblers observed during spring migration in the Kickapoo Valley Reserve, Wisconsin, 2009 and 2010.

Common name	Scientific name	n (2009)	n (2010)
Golden-winged Warbler[a]	*Vermivora chrysoptera*	5^{595}	1^{49}
Blue-winged Warbler[a]	*Vermivora cyanoptera*	9^{424}	2^{170}
Tennessee Warbler	*Oreothlypis peregrina*	53^{4656}	18^{1900}
Nashville Warbler[a]	*Oreothlypis ruficapilla*	13^{1063}	8^{601}
American Redstart[a]	*Setophaga ruticilla*	3^{121}	2^{92}
Northern Parula	*Setophaga americana*	3^{304}	1^{140}
Chestnut-sided Warbler[a]	*Setophaga pensylvanica*	6^{390}	3^{60}
Magnolia Warbler	*Setophaga magnolia*	2^{177}	0
Cape May Warbler	*Setophaga tigrina*	3^{372}	0
Black-throated Blue Warbler	*Setophaga caerulescens*	1^{97}	0
Blackburnian Warbler	*Setophaga fusca*	19^{2460}	3^{167}
Yellow-rumped Warbler	*Setophaga coronata*	13^{1452}	8^{527}
Black-throated Green Warbler	*Setophaga virens*	10^{603}	10^{899}
Bay-breasted Warbler	*Setophaga castanea*	2^{103}	0
Blackpoll Warbler	*Setophaga striata*	3^{130}	0

[a] Species breeds in study area (Wisconsin Breeding Bird Atlas).

migratory songbird foraging behavior on individual phenology trees due to low frequency of use of these trees during the times we were observing them. Thus, we monitored migratory songbird foraging behavior on the phenology-tree species throughout the study area and related our observations to the study area-averaged phenology-tree measurements. We collected foraging data on 15 migratory species of Neotropical–Nearctic wood-warblers (hereafter, "wood-warblers"), which we selected because they are common at our study area and primarily use trees as foraging substrates during spring migration (Table 7.1). Wintering grounds for Yellow-rumped Warbler (*S. coronata*) are more northerly than those for the other 14 wood-warblers. We included this species in our analysis because they were commonly detected using phenology tree species along with other wood-warblers throughout our survey period. A few species were rare in our study area, such as Black-throated Blue Warblers (*S. caerulescens*). We nevertheless included rare species because they were similar to more common birds taxonomically and in their foraging habits (Whelan 2001). We grouped data from all wood-warblers to increase sample size necessary for analyses

and because our goal was to understand foraging behavior of this group of birds broadly relative to variations in annual spring tree phenology. We gathered foraging data on the same seven days on which we collected tree phenology data, which encompasses the migration period in Wisconsin for wood-warblers en route to northern breeding grounds (Temple et al. 1997). Between sunrise and 1 P.M., we performed extensive area searches of forest patches by walking on established routes designed to maximize coverage of a patch, actively searching for foraging flocks of the wood-warblers. Four trained observers collected data and rotated among forest patches with each visit in an effort to distribute observer variability as much as possible. We alternated starting points of the walking routes with each visit so that unique sections of the forest patches were visited at similar times among different visits.

After locating a flock, we followed and documented the foraging activities and movements of a wood-warbler for up to five minutes and within the boundaries of a forest patch using standardized methods (Holmes and Robinson 1981, Remsen and Robinson 1990). We often documented foraging behavior of an individual wood-warbler in multiple phenology trees over the course of a given observation bout. We restricted our arrival, use, and synchrony analyses to observations of foraging birds made only in the first tree to guard against potential autocorrelation (Gabbe et al. 2002, Wood et al. 2012). However, we included data from all phenology trees in which foraging was observed in analyses of foraging success to gain understanding of broad patterns of wood-warbler foraging behavior among phenology trees. We recorded behavioral data with a digital recorder with a built-in timer (Sony ICD-PX720 digital voice recorder). We recorded six variables for each individual: date, time of day, species of warbler, tree species used, search effort (e.g., number of hops, walks, flights), and type of prey attack (e.g., bud glean, flower glean, leaf glean, bark glean, hover, sally, and flush-chase; Remsen and Robinson 1990). In addition to monitoring wood-warbler foraging behavior, we recorded the dominant phenophase (defined as >50% of a tree in a particular phenophase) for all trees where birds were found foraging (hereafter, "foraging trees") to understand relationships of foraging-tree phenology and foraging behavior by wood-warblers. We collected foraging-tree phenophase measurements on a binary scale depending on the presence or absence of a dominant phenophase. We used a binary scale because determining phenophase of a tree is time intensive, and it was not possible to record conditions while simultaneously monitoring foraging behavior of wood-warblers. Once foraging data collection for a wood-warbler was concluded and a foraging-tree phenophase measurement was completed, we either moved to another wood-warbler in the immediate area—using a phenology tree, if possible—or moved back to the walking route in search of another foraging flock that was >300 m from the previous flock.

Statistical Analysis

For our first objective, we quantified differences in maximum temperatures during March, April, and May in 2009 and 2010 by calculating three paired two-sample t-tests, with one for each of March, April, and May, with daily maximum temperature as the response variable and year as the treatment. We also related the maximum daily temperature during March, April, and May in 2009 and 2010 with the average daily maximum temperature

from 1950 to 2010 in six additional two-sample t-tests (three for March, April, and May in 2009 and 2010, respectively). We checked for normality of the residuals of the dependent variables using QQ-norm plots, which revealed that no transformations were necessary. We used the critical value of $\alpha \leq 0.05$ to determine significance.

To address our second objective, we explored patterns of phenological synchrony of the flowering phenophase of the four phenology-tree species with wood-warbler foraging behavior on those trees. First, to understand wood-warbler use of the phenology trees of the study area within each sampling season, we calculated the proportion of tree use—defined as active feeding or searching by wood-warblers among the seven sampling periods—and then compared this proportional distribution in 2009 and 2010. Second, we calculated the percentage of tree crown in the flowering phenophase on phenology trees among the seven sampling periods and then compared estimates between years. We focused on the flowering phenophase because flowering is a proximate cue to the availability of food (McGrath et al. 2008) and because we commonly noticed birds using flowering trees as foraging substrate during pilot field work for our project.

Furthermore, we quantified foraging success of wood-warblers relative to phenophase of foraging trees. We constructed an attack index response variable, which is a measure of foraging success and is calculated as the ratio of the total attacks by wood-warblers, divided by total search maneuvers, scaled per minute (Wood et al. 2012). We explored differences in the attack index related to the dominant phenophase of the foraging trees (which we categorized as budburst, flowering, or mature leaf) for 2009 and 2010 combined, using a Kruskal-Wallis test, with phenophase of a foraging tree as the treatment. In a handful of cases, a particular foraging tree was categorized by a different phenology event from the three treatments we used, such as young leaves with wilted flowers. We removed these cases to ensure that phenophase categories in the analysis were distinct. Furthermore, although the timing of flowering and leaf-out is variable among the phenology trees, the phenological physiognomy is similar within oaks and maples (e.g., catkin flowers in oaks). We pooled data from the two oak species versus the two maple species to understand foraging success of wood-warblers among the two tree groups. When Kruskal-Wallis tests were significant, we employed a nonparametric multiple comparison procedure, based on relative contrast effects, using nparcomp (Konietschke 2011). We used a Bonferroni adjustment to the critical α value of $0.05/3 = 0.02$ to assess significance for multiple tests.

For our third objective, to determine whether variations in annual spring phenology alter foraging conditions for members of this family of birds en route to their breeding grounds, we quantified foraging success of wood-warblers, among the four study tree species, between years. To do so, we explored differences in the attack index on the phenology-tree species for 2009, 2010, and both years combined, by using a Kruskal-Wallis test, with phenology tree as the treatment. This test was similar to the previous analysis, except that we used a Bonferroni adjustment to the critical α value of $0.05/6 = 0.01$ to assess significance. We used a Wilcoxon rank sum test for 2010 data because only northern red oak and eastern white oak were used by the wood-warblers in that year. For the Wilcoxon rank sum test analysis, we used a significance threshold of $\alpha \leq 0.05$.

RESULTS

Patterns of Annual Late-Winter and Spring Temperatures

The maximum daily high temperature in 2009 was similar to the average monthly high temperature from 1950 to 2010 in March ($t_{30} = 0.46$, $P = 0.65$) and April ($t_{29} = 1.42$, $P = 0.17$), but was significantly lower in May ($t_{30} = 2.35$, $P = 0.02$; Figure 7.2). In contrast, the maximum daily high temperature in 2010 was significantly warmer than the average monthly high temperature from 1950 to 2010 in March ($t_{30} = -5.83$, $P < 0.01$) and April ($t_{29} = -3.41$, $P < 0.01$), but similar in May ($t_{30} = -0.75$, $P = 0.46$; Figure 7.2). Maximum daily temperatures were significantly warmer in 2010 than 2009 in March ($t_{30} = -5.91$, $P < 0.01$), April ($t_{29} = -3.28$, $P < 0.01$), and May ($t_{30} = -2.03$, $P = 0.05$; Figure 7.2).

Tree Phenology and Migratory Bird Synchrony

Northern red oak was the largest species in our sample of phenology trees with an average dbh of 63.8 ± 2.7 cm. Eastern white oak and sugar maple were slightly smaller: 56.3 ± 3.2, and 53.2 ± 2.1 cm, respectively, followed by red maple at 43.2 ± 3.1 cm. We monitored foraging behavior of 15 wood-warbler species in 2009, for a total of 216 min and 27 s (Table 7.1). In 2010, we monitored foraging behavior of 10 wood-warbler species, for a total of 79 min and 58 s. The average foraging observation was 1 min and 29 s ± 0.10 in 2009, and 1 min and 21 s ± 0.09 in 2010. In both years, wood-warblers used phenology trees at the stopover study area from late April (Julian day 119) until late May (Julian day 147 in 2009 and 142 in 2010; Figure 7.3). However, the frequency of bird-use of the phenology trees in the study area was 60% less in 2010 than 2009 (Figure 7.3), despite similar survey effort.

In 2009, there were apparent patterns of synchrony between wood-warbler foraging behavior and the flowering phenophase of northern red oak, eastern white oak, and, to a lesser extent, sugar maple (Figure 7.4). Red maple flowered prior to wood-warbler arrival and was not preferred by wood-warblers (Figures 7.4 and 7.5).

In 2010, northern red oak, sugar maple, and red maple flowered 2 to 3 weeks earlier than in 2009, and these tree species were not preferred by the wood-warblers (Figures 7.4 and 7.5). On the other hand, eastern white oak, which flowered from late April to mid-May in 2010, was the preferred foraging substrate in that year (Figures 7.4 and 7.5). When flowering of eastern white oaks was complete, wood-warbler use of the phenology trees subsided dramatically, which was in contrast to 2009, where birds were observed foraging among the phenology trees until late May (Figure 7.4).

In 2009 and 2010, wood-warbler foraging success was significantly greater on oak trees that were in flower, compared to trees at the budburst or mature leaf stage ($H_2 = 23.23$, $P < 0.01$; Figure 7.6). In contrast, foraging success of wood-warblers on maples did not differ among the phenological stages of budburst, flowering, or mature leaf ($H_2 = 1.49$, $P = 0.48$; Figure 7.6).

Migratory Bird Foraging Success

Foraging success differed among phenology trees within (2009: $H_3 = 16.52$, $P < 0.01$; and 2010: $W_{58} = 351.5$, $P = 0.02$) and between ($H_3 = 19.29$, $P < 0.01$; Figure 7.7) years. In 2009, average foraging success of wood-warblers was highest on northern red oak (0.34 ± 0.04), followed by eastern

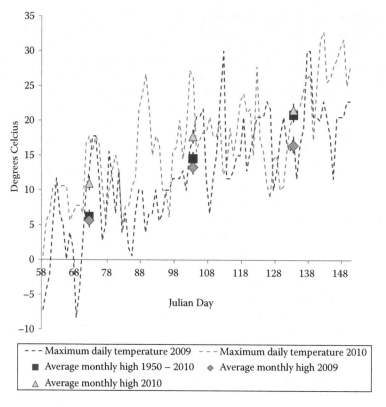

Figure 7.2. Average monthly high temperature for March, April, and May during 1950–2010 (black squares), 2009 (gray diamonds), and 2010 (gray triangles), ±1 SE, and daily maximum high temperatures for 2009 and 2010 represented by dotted lines. Weather data obtained from NOAA Station ID: WI477997, Sparta, Wisconsin.

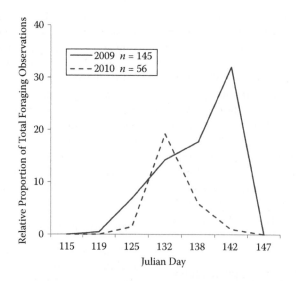

Figure 7.3. Relative number of foraging observations (number of observed foraging bouts per Julian day divided by the total for both years combined) on four tree species—*Quercus rubra*, *Q. alba*, *Acer rubrum*, and *A. saccharum*—by 15 species of wood-warblers, from 24 April to 26 May (Julian day: 115–147) at the Kickapoo Valley Reserve, Wisconsin.

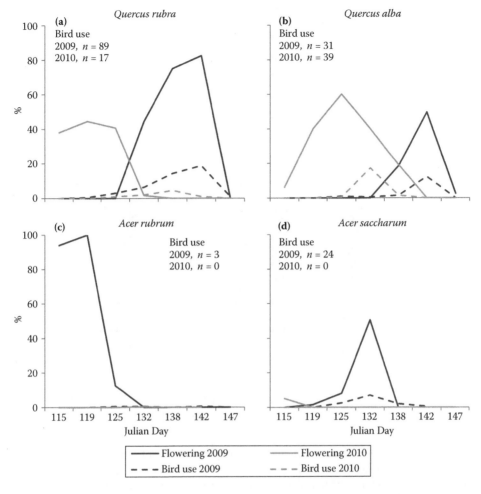

Figure 7.4. The relationship between the relative use (number of observed foraging bouts per Julian day, per tree, divided by the total foraging bouts in 2009 and 2010 combined) by 15 species of wood-warblers, and percentage of flowering of four species of trees—*Quercus rubra*, *Q. alba*, *Acer rubrum*, and *A. saccharum*—from 24 April to 26 May (Julian day: 115–147), 2009 and 2010.

white oak (0.27 ± 0.05), sugar maple (0.25 ± 0.06), and red maple (0.13 ± 0.08). Wood-warbler foraging success on northern red oak was similar to that on eastern white oak in 2009, but was significantly higher than on sugar maple (similar to eastern white oak) or red maple (Figure 7.7). In 2010, wood-warblers did not use red maple or sugar maple. Wood-warbler foraging success was highest on eastern white oak (0.38 ± 0.05), which was significantly higher than northern red oak (0.21 ± 0.05; Figure 7.7). When the two years were considered together, wood-warbler foraging success was highest on eastern white oak (0.34 ± 0.04), followed by northern red oak (0.32 ± 0.03), sugar maple (0.21 ± 0.06), and red maple (0.13 ± 0.08; Figure 7.7). Wood-warbler foraging success

on the two oak species was similar between years, yet was significantly higher than on the maple species (Figure 7.7).

DISCUSSION

Our results suggest that extreme differences in annual late-winter and spring temperatures impacted spring tree phenology, which in turn affected foraging quality of dominant trees at a stopover location in a forest in the Upper Midwest and influenced foraging behavior of wood-warblers. We found that flowering northern red oak and eastern white oak, which are both trees that are regenerating poorly in the region, are preferred as foraging substrates by

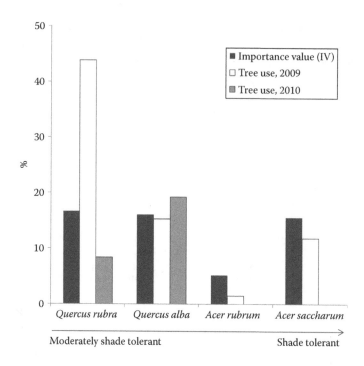

Figure 7.5. Relative importance value, which is a measure of the proportional availability of each of four tree species ordered by shade tolerance—*Quercus rubra*, *Q. alba*, *Acer rubrum*, and *A. saccharum*—as foraging substrates versus the proportion of all foraging observations by 15 species of wood-warblers on the four species of trees from 24 April to 26 May (Julian day: 115–147) at the Kickapoo Valley Reserve in 2009 and 2010. Relative use was calculated as the total number of wood-warbler observations for a tree species, in a given year, divided by the sum of all observations among the four tree species, in both study years.

wood-warblers. Eastern white oak generally flowers later in May than northern red oak throughout our study region. Therefore, if both tree species persist in this landscape, the foraging needs of spring migrant wood-warblers are likely to be met. Ensuring long-term viability of oak species is one step that can be taken to buffer the negative impacts to foraging quality of variable climate conditions. Wood-warblers also used flowering sugar maple, but red maple was rarely used. During warm springs, both species of maples are poor foraging substrates for wood-warblers. Our observations raise a conservation concern because maples are regenerating throughout forests of the Upper Midwest, with a substantial increase in red maple in particular (Abrams 1998, Wood et al. 2012).

Tree phenology is influenced by a variety of environmental cues interacting throughout the winter and spring period. Winter chilling, photoperiod, and late-winter and spring temperatures affect budburst of both early and late successional tree species (Körner and Basler 2010). For some plant species, once the winter chilling

requirement has been met, late-winter and early spring temperatures largely influence budburst (Körner and Basler 2010). For example, common lilac (*Syringa vulgarisi*) is an ornamental shrub that advances budburst in warm springs (Körner and Basler 2010) and has been the subject of ongoing research on effects of climate change on spring green-up in North America (Schwartz and Reiter 2000). In New England hardwood forests, onset of sugar maple budburst is ~9 days earlier than in 1957 due to increasing spring temperatures (Richardson et al. 2006). In a controlled experiment, saplings of three temperate trees species advanced budburst with increased warming temperatures (*Betula*, *Fagus*, and *Quercus* spp.; Fu et al. 2012).

Our 2-year sample size was too low to explore relationships between spring climatic conditions and tree phenology, but our results suggest that annual variation in late-winter and spring temperatures likely affect budburst of oak and maple species. Late-winter and spring temperatures were significantly warmer in 2010 than in 2009, and the long-term average for these months appears

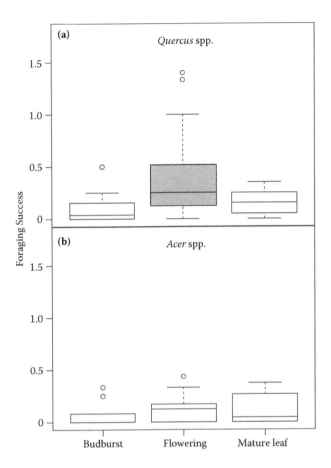

Figure 7.6. Box plots of the attack index (total attacks/total searches, per minute), a measure of foraging success, for 15 species of wood-warblers using two genera of trees (a-b), *Quercus* and *Acer*, in 2009 and 2010. Attack index is shown for three distinct phenophases: budburst, flowering, and mature leaf. Box plots with different colors differ significantly based on a Kruskal-Wallis test with nonparametric multiple comparisons procedure with a Bonferroni adjusted $P \leq 0.05/3 = 0.016$.

to be associated with a 2- to 3-week phenology advancement of the four tree species. In contrast to Körner and Basler (2010), our results suggest that, for late-successional tree species such as red maple and sugar maple, temperature appears to play a larger role in determining budbust than photoperiod. We matched Julian days in the two years of our study to control for differences in day length between years, thus allowing for direct comparisons of phenology. Photoperiod appeared to be an important factor determining spring budburst for some tree species in our study area, such as black walnut (*Juglans nigra*; pers. obs.), but was likely not as important as temperature in determining budburst of oaks and maples. Taken together, our findings suggest that warming trends and increased variability in spring temperatures (Schär et al. 2004) will likely alter

spring tree phenology and possibly resource availability, such as caterpillar emergence, for migratory bird species.

Changes in phenology are a concern because we found patterns of phenological synchrony between wood-warblers and the flowering period of northern red oak and white oak and, to a lesser extent, sugar maple. Our observations of synchrony between migratory birds and seasonal resources are similar to patterns reported elsewhere: Spring arrival of Orange-crowned Warblers (*Oreothlypis celata*) overlaps with flowering of honey mesquite found in Arizona riparian habitats (McGrath et al. 2008), Broad-tailed Hummingbird (*Selasphorus platycerus*) migration is coincident with abundance of nectar resources in Arizona pine–oak woodlands (McKinney et al. 2012), migratory landbirds concentrate in habitats

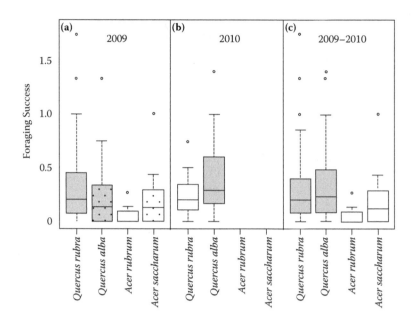

Figure 7.7. Box plots of foraging success, measured by the attack index (total attacks/total searches, per minute), for 15 species of wood-warblers using four tree species as foraging substrates during spring migration: *Quercus rubra, Q. alba, Acer rubrum,* and *A. saccharum* in (a) 2009, (b) 2010, and (c) both years combined. Box plots with different colors or patterns differ significantly based on a Kruskal-Wallis test with nonparametric multiple comparison procedures with a Bonferroni adjusted $\alpha \leq 0.05/4 = 0.013$). Data from 2010 were evaluated based on a two-way Wilcoxon rank test, $P = 0.05$.

where plants are in early leaf development in Pennsylvania suburbs and forests (Rodewald and Brittingham 2007), and wood-warbler migration corresponds with peaks in Lepidopteran availability in Illinois woodlots (Graber and Graber 1983, Strode 2009). Lepidopteran abundance on oak trees is greatest during the spring flowering and leaf-out period, prior to an increase in tree chemical defenses during the summer (Forkner et al. 2004). Maples and oaks have different levels of chemical defenses; this, in turn, likely affects the abundance of insects available as food resources to birds (Adams et al. 2009). Chemical defenses may be a possible mechanism explaining the foraging patterns we observed. We did not measure food availability among the four phenology trees throughout our study period, but our findings of higher foraging success on northern red oak and eastern white oak during flowering, rather than during budburst or mature leaf-out stages, suggests the importance of synchrony between timing of flowering and use by wood-warblers. The flowering duration of northern red oak and eastern white oak is ~three weeks throughout our study area. Thus, if tree phenology is advanced by two to three weeks, as in 2010, and wood-warblers are not able to adjust their arrival to

coincide with flowering (Strode 2003), the birds will miss the flowering phenophase and peak food availability altogether, and foraging success will likely be negatively impacted.

A plausible explanation for lower wood-warbler foraging success on northern red oak and eastern white oak trees that were in summer condition with mature leaves rather than flowering is the differences in food accessibility (Wood et al. 2012). During spring, trees undergo rapid physiognomic changes, which likely affects accessibility of food items found on flower or leaf surfaces. In 2009, we observed some use by wood-warblers of sugar maple during the flowering and emerging leaf-out period. In 2010, the majority of sugar maple achieved summer condition by April 28 (Julian day 119), which is two to three weeks earlier than peak wood-warbler migration throughout southern Wisconsin (Temple et al. 1997). We did not observe wood-warblers foraging on sugar maple in 2010. In general, leaves of shade-tolerant trees, such as sugar maple, are supported by long petioles, which support the large leaves necessary for improved light-gathering in shady conditions (Takenaka et al. 2001). To acquire food items on leaf surfaces of tree species with long leaf petioles, bird species must use energetically costly foraging maneuvers (Whelan 2001). Under optimal foraging theory, migrating wood-warblers should forage to maximize caloric intake while minimizing energetic output and risk, and they might be predicted to avoid unsuitable foraging substrates, such as red maple or sugar maple, in summer conditions. Foraging success of migratory songbirds is negatively correlated with leaf-petiole length a measure of food accessibility that is characteristic of shade-tolerant tree species that are regenerating throughout forests in the Upper Midwest (Wood et al. 2012). In addition to red maple and sugar maple, other shade-tolerant tree species with long leaf petioles, such as American basswood, are also regenerating regionally (Wood et al. 2012). Without management for early or mid-successional tree species, forest homogenization will likely continue, resulting in lower quality stopover habitat for wood-warbler species. Our results suggest that these effects will be exacerbated in warm springs when shade-tolerant tree species will be in summer condition throughout the migratory period, creating poor-quality foraging conditions for songbird species in Upper Midwestern forests.

Our findings provide evidence that large variation in late-winter and spring temperatures likely affects tree phenology, resulting in asynchrony of migratory wood-warbler species with ephemeral tree phenological stages necessary for optimal foraging at stopover habitats. We show that this asynchrony affects foraging success and, in extreme cases, may result in deleterious effects to the condition and breeding success of individuals (McKinney et al. 2012). Migratory bird species are constantly encountering environmental variability, which affects stopover habitat quality and food resource availability (Hutto 1985). Yet, we show that the combination of forest succession and climate variability will likely continue to impact spring stopover habitat quality negatively for wood-warblers throughout the Driftless Area.

As a strategy for maintaining stopover habitat for migratory songbirds under a variety of climate conditions in forests in the Upper Midwest, we advocate for management that encourages forest tree heterogeneity. The Kickapoo Valley was historically a mesic forest island within a savanna matrix, shaped by relatively cool summer temperatures and high annual rainfall, in contrast to the surrounding oak savanna landscape (Kline and Cottam 1979). Sugar maple and white oak historically dominated the Kickapoo Valley, and these species remain common today (Kline and Cottam 1979). Yet, the presence of red maple, which was historically an uncommon tree species in southern Wisconsin forests, as well as poor recruitment rates of oak and other early successional species, suggests a change in forest composition toward further dominance by shade-tolerant species throughout the region (Wood et al. 2012). We suggest that forest management plans for the Driftless Area and other similar forested habitats in the Upper Midwest include promoting early and late flowering species such as red and white oak, as well as trees adapted to low or moderate shade such as quaking aspen. Furthermore, we recommend carefully managing recruitment of shade-tolerant tree species, especially on south-facing slopes, and ridge tops, where these species were most likely never common (Kline and Cottam 1979).

ACKNOWLEDGMENTS

We would like to thank our field assistants, H. Llanas, S. Grover, P. Kearns, and A. Derose-Wilson, for their help in collecting foraging observations as well as tree phenology measurements. M. West, E. Epstein, and B. Johnston provided valuable logistical support while in the field. We are grateful to the McIntire-Stennis Formula Fund and the Kickapoo Valley Reforestation Fund for funding this research.

LITERATURE CITED

Abrams, M. D. 1992. Fire and the development of oak forests. BioScience 42:346–353.

Abrams, M. D. 1998. The red maple paradox. BioScience 48:355–364.

Adams, J. M., B. Rehill, Y. Zhang, and J. Gower. 2009. A test of the latitudinal defense hypothesis: herbivory, tannins and total phenolics in four North American tree species. Ecological Research 24:697–704.

Curtis, J. T. 1959. The vegetation of Wisconsin: an ordination of plant communities. P. 704. University of Wisconsin Press, Madison, WI.

Ellwood, E. R., S. A. Temple, R. B. Primack, N. L. Bradley, and C. C. Davis. 2013. Record-breaking early flowering in the eastern United States. PLoS One 8:e53788.

ESRI California, USA, R. 2008. ArcGIS 9.3. 2008. ESRI, Redlands, CA.

Fu, Y. H., M. Campioli, G. Deckmyn, and I. A. Janssens. 2012. The impact of winter and spring temperatures on temperate tree budburst dates: results from an experimental climate manipulation. PloS One 7:e47324.

Gabbe, A. P., S. K. Robinson, and J. D. Brawn. 2002. Tree-species preferences of foraging insectivorous birds: implications for floodplain forest restoration. Conservation Biology 16:462–470.

Graber, J. W., and R. R. Graber. 1983. Feeding rates of warblers in spring. Condor 85:139–150.

Holberton, R. L., and A. M. Dufty, Jr. 2005. Hormones and variation in life history strategies of migratory and nonmigratory birds. Pp. 290–302 in R. Greenberg and P. P. Marra (editors), Birds of two worlds: the ecology and evolution of migration. Johns Hopkins University Press, Baltimore, MD.

Holmes, R. T. 2007. Understanding population change in migratory songbirds: long-term and experimental studies of Neotropical migrants in breeding and wintering areas. Ibis 149:2–13.

Holmes, R. T., and S. K. Robinson. 1981. Tree species preferences of foraging insectivorous birds in a northern hardwoods forest. Oecologia 48:31–35.

Hutto, R. L. 1985. Habitat selection by nonbreeding, migratory land birds. Pp. 455–476 in M. L. Cody (editor), Habitat selection in birds. Academic Press, Orlando, FL.

Iverson, L. R., M. E. Dale, C. T. Scott, and A. Prasad. 1997. A GIS-derived integrated moisture index to predict forest composition and productivity of Ohio forests (USA). Landscape Ecology 12:331–348.

Kline, V. M., and G. Cottam. 1979. Vegetation response to climate and fire in the Driftless Area of Wisconsin. Ecology 60:861–868.

Konietschke, F. [online]. 2011. nparcomp: nparcomp-package. R package. version 1.0-1 <http://CRAN.R-project.org/package=nparcomp>.

Körner, C., and D. Basler. 2010. Phenology under global warming. Science 327:1461–1462.

McKinney, A. M., P. J. CaraDonna, D. W. Inouye, B. Barr, C. D. Bertelsen, and N. M. Waser. 2012. Asynchronous changes in phenology of migrating Broad-tailed Hummingbirds and their early-season nectar resources. Ecology 93:1987–1993.

McShea, W. J., W. M. Healy, P. Devers, T. Fearer, F. H. Koch, D. Stauffer, and J. Waldon. 2007. Forestry matters: decline of oaks will impact wildlife in hardwood forests. Journal of Wildlife Management 71:1717–1728.

Moore, F. R., and P. Kerlinger. 1987. Stopover and fat deposition by North American wood-warblers (Parulinae) following spring migration over the Gulf of Mexico. Oecologia 74:47–54.

Moore, F. R., R. J. Smith, and R. Sandberg. 2005. Stopover ecology of intercontinental migrants: en route problems and consequences for reproductive performance. Pp. 251–262 in R. Greenberg and P. P. Marra (editors), Birds of two worlds: the ecology and evolution of migration. Johns Hopkins University Press, Baltimore, MD.

Nowacki, G. J., and M. D. Abrams. 2008. The demise of fire and "mesophication" of forests in the eastern United States. BioScience 58:123–138.

Remsen, J. V., and S. K. Robinson. 1990. A classification scheme for foraging behavior of birds in terrestrial habitats. Studies in Avian Biology 13:144–160.

Rhemtulla, J. M., D. J. Mladenoff, and M. K. Clayton. 2007. Regional land-cover conversion in the US Upper Midwest: magnitude of change and limited recovery (1850–1935–1993). Landscape Ecology 22:57–75.

Rhemtulla, J. M., D. J. Mladenoff, and M. K. Clayton. 2009. Legacies of historical land use on regional forest composition and structure in Wisconsin, USA (mid-1800s–1930s–2000s). Ecological Applications 19:1061–1078.

Richardson, A. D., A. S. Bailey, E. G. Denny, C. W. Martin, and J. O'Keefe. 2006. Phenology of a northern hardwood forest canopy. Global Change Biology 12:1174–1188.

Rodewald, P. G., and M. C. Brittingham. 2007. Stopover habitat use by spring migrant landbirds: the roles of habitat structure, leaf development, and food availability. Auk 124:1063–1074.

Rogers, D. A., T. P. Rooney, D. Olson, and D. M. Waller. 2008. Shifts in southern Wisconsin forests canopy and understory richness, composition, and heterogeneity. Ecology 89:2482–2492.

Schär, C., P. L. Vidale, D. Lüthi, C. Frei, C. Häberli, M. A. Liniger, and C. Appenzeller. 2004. The role of increasing temperature variability in European summer heatwaves. Nature 427:332–336.

Schulte, L. A., D. J. Mladenoff, T. R. Crow, L. C. Merrick, and D. T. Cleland. 2007. Homogenization of northern US Great Lakes forests due to land use. Landscape Ecology 22:1089–1103.

Schwartz, M. D., and B. E. Reiter. 2000. Changes in North American spring. International Journal of Climatology 932:929–932.

Strode, P. K. 2003. Implications of climate change for North American wood warblers (Parulidae). Global Change Biology 9:1137–1144.

Strode, P. K. 2009. Spring tree species use by migrating Yellow-rumped Warblers in relation to phenology and food availability. Wilson Journal of Ornithology 121:457–468.

Studds, C. E., and P. P. Marra. 2011. Rainfall-induced changes in food availability modify the spring departure programme of a migratory bird. Proceedings of the Royal Society of London B 278:3437–3443.

Takenaka, A., K. Takahashi, and T. Kohyama. 2001. Optimal leaf display and biomass partitioning for efficient light capture in an understorey palm, *Licuala arbuscula*. Functional Ecology 15:660–668.

Temple, S. A., J. R. Cary, and R. E. Rolley. 1997. Wisconsin birds: a seasonal and geographical guide. P. 320. University of Wisconsin Press, Madison, WI.

Visser, M. E., and C. Both. 2005. Shifts in phenology due to global climate change: the need for a yardstick. Proceedings of the Royal Society of London B 272:2561–2569.

Way, D. A. 2011. Tree phenology responses to warming: spring forward, fall back? Tree Physiology 31:469–471.

Whelan, C. J. 2001. Foliage structure influences foraging of insectivorous forest birds: an experimental study. Ecology 82:219–231.

Wilson, D. C. 2008. Managing from a landscape perspective: a guide for integrating forest interior bird habitat considerations and forest management planning in the Driftless Area of the upper Mississippi River basin. Version 1.1, pp. 1–29. Wisconsin Department of Natural Resources, Madison, WI.

Wood, E. M., A. M. Pidgeon, F. Liu, and D. J. Mladenoff. 2012. Birds see the trees inside the forest: the potential impacts of changes in forest composition on songbirds during spring migration. Forest Ecology and Management 280:176–186.

CHAPTER EIGHT

Phenological Synchrony of Bird Migration with Tree Flowering at Desert Riparian Stopover Sites[*]

Jherime L. Kellermann and Charles van Riper III

Abstract. Small-bodied songbirds replenish fat reserves during migration at stopover sites where they continually encounter novel and often unpredictable environmental conditions. The ability to select and utilize high-quality habitats is critical to survival and fitness. Vegetation phenology is closely linked with emergence of insect prey and may provide valid cues of food availability for stopover habitat selection. Climate change is disrupting phenological synchrony across trophic levels with negative impacts on bird populations. However, whether synchrony or mismatch indicates historic or disrupted systems remains unclear. Many Neotropical migratory songbirds of western North America must cross arid regions where drought conditions related to climate change and human water use are expected to increase. We studied migrant abundance and the diversity (niche breadth) and proportional use of vegetation species as foraging substrates and their synchrony with vegetation flowering during spring migration along the lower Colorado River in the Sonoran Desert of the United States and Mexico. Peak migrant abundance in late March–early April coincided with the period of narrowest niche breadth. Abundance of migrants increased with willow and mesquite flowering, but significantly only in 2003, a severe drought year. Annual niche breadth was negatively correlated with flowering and with total monsoon precipitation. Birds proportionally foraged most in willow and mesquite, shifting from one species to the next, temporally synchronous with their flowering. The period of greatest overlap in willow and mesquite flowering and use also coincided with peak migrant abundance and narrowest niche breadth. Our results show that migration timing and foraging habitat selection is highly synchronized with flowering of large woody perennials, particularly in dry years. However, if food availability declines in dry years, increased selection of fewer species for foraging could result in greater competition and reduced fitness. Therefore, increased synchrony rather than mismatch could indicate negative climate-driven impacts on some systems.

Key Words: climate change, drought, mesquite, niche breadth, phenological mismatch, *Prosopis*, *Salix*, Sonoran Desert, willow, wood-warbler.

[*] Kellermann, J. L., and C. van Riper III. 2015. Phenological synchrony of bird migration with tree flowering at desert riparian stopover sites. Pp. 133–144 in E. M. Wood and J. L. Kellermann (editors), Phenological synchrony and bird migration: changing climate and seasonal resources in North America. Studies in Avian Biology (no. 47), CRC Press, Boca Raton, FL.

Neotropical songbirds must make frequent stops to rest and feed during migration, a phenological period with potentially high mortality (Moore and Kerlinger 1987, Sillett and Holmes 2002, Seewagen and Guglielmo 2010). During stopover, birds repeatedly encounter novel habitats and unpredictable environmental conditions (Moore et al. 1990, Parrish 2000, Petit 2000). An ability to identify, select, and utilize high-quality stopover habitat elements across a wide range of habitat types and conditions would benefit reproduction and survival (Balbontin et al. 2009, Bauchinger et al. 2009, LaManna et al. 2012). Understanding factors that influence stopover habitat selection and use is important for improving migratory bird conservation (Faaborg et al. 2010a, b).

Food availability is an intrinsic component of stopover habitat quality (Rodewald and Brittingham 2007, Bauchinger et al. 2009, Carlisle et al. 2012), but can only be directly assessed once birds have entered a stopover site (Aborn and Moore 2004; Paxton et al. 2007, 2008). Songbirds rely on insect prey during spring migration (Aborn and Moore 1997, MacDade et al. 2011), and emergence and abundance of invertebrates often track the phenology of plant resources, resulting in phenological synchrony between both trophic levels (Visser and Holleman 2001; Visser et al. 2006; van Asch et al. 2007, 2010). Vegetation phenology may provide valuable cues of food availability and habitat quality that allow en route migrants to maximize stopover habitat selection among and within stopover sites (Buler et al. 2007, McGrath et al. 2009, Strode 2009). Bird migration can thus become spatially and temporally synchronized with plant phenology due to tropic relationships with prey. Climate-driven phenological mismatches have been linked to trophic cascades that disrupt migration patterns and induce population declines (Both et al. 2009, Thackeray et al. 2010, Saino et al. 2011).

The ecological consequences of phenological synchrony or mismatch are often unclear. Few studies directly assess synchrony of bird migration with vegetation phenology at stopover sites and the demographic impacts can be species and system dependent (Jones and Cresswell 2010, Miller-Rushing et al. 2010). It is easy to assume that climate-driven mismatch has necessarily negative consequences for populations (Both et al. 2006), but mismatch may be the historical condition or even be evolutionarily adaptive (Singer and Parmesan 2010, Visser et al. 2012). Many studies of bird migration and phenological mismatch have focused on how temporal patterns of bird abundance at a site coincide with resource phenology, such as correlations between the temporal peaks in migrant detections or capture rates with flowering of plants at seasonally used sites (Bertin 1982, Hart et al. 2011) or the correlation between onset dates of plant flowering and bird arrival (McKinney et al. 2012). Although these seasonal patterns provide an index of synchrony, they do not directly link use of specific foraging substrates with their phenology of flowering or other key resources. In addition to examining patterns of relative migrant abundance and plant phenology, we directly assessed how the use of vegetation species as foraging substrates was temporally correlated with flowering phenology to gain greater insight into the implications of observed levels of synchrony.

Many Neotropical migratory songbirds in the western United States must cross large regions of the arid Southwest (Kelly and Hutto 2005). Riparian forests and woodlands that border large river systems can provide vital stopover habitat to a range of species (Skagen et al. 2005). Under climate change models, arid lands of North America are expected to experience more extreme conditions, becoming increasingly hot and dry with extended severe drought periods (Seager et al. 2007, Weiss et al. 2009, Balling and Goodrich 2010). Bird migration can be synchronized by climate as well as food availability. The interactions of climate, plant phenology, and food availability could have important implications for birds that rely on riparian stopover habitat in desert regions (Jones et al. 2003, Kelly and Hutto 2005, Skagen et al. 2005).

Our goals were to examine temporal patterns and phenological synchrony of migrant abundance, the diversity and proportional selection of different plant species as foraging substrates, and the flowering phenology of these species at spring stopover sites along the lower Colorado River in southwestern Arizona and northwestern Sonora, Mexico. We also assessed the influence of annual variation in monsoon precipitation and drought on phenological synchrony.

METHODS

Study Sites

We assessed bird migration and vegetation phenology at four sites along the lower Colorado River (LCR) in the southern United States and northern Mexico. The LCR watershed includes a vast region of western North America and bridges the US and Mexican border en route to the Gulf of California. Like many large rivers, the waters of the Colorado River are intensively managed (Poff et al. 1997, Rajagopalan et al. 2009). Our study sites were located at Cibola and Bill Williams National Wildlife Refuges, Arizona, and El Doctor, Sonora, and Alto Golfo de California Biosphere Reserve, Baja California, Mexico (Figure 8.1).

Climate

We obtained climate data on precipitation and drought from the National Oceanic and Atmospheric Association National Climatic Data Center for Arizona Climate Division 5 (NOAA 2013). The majority of annual precipitation in this region falls during the North American monsoons (Adams and Comrie 1997), and we used the total amount of precipitation (millimeters) recorded over the monsoon season during July through September in the year prior to each spring migration season. We assessed annual drought conditions leading up to each migration season using the Palmer drought severity index (PDSI) for March through February (NOAA 2013).

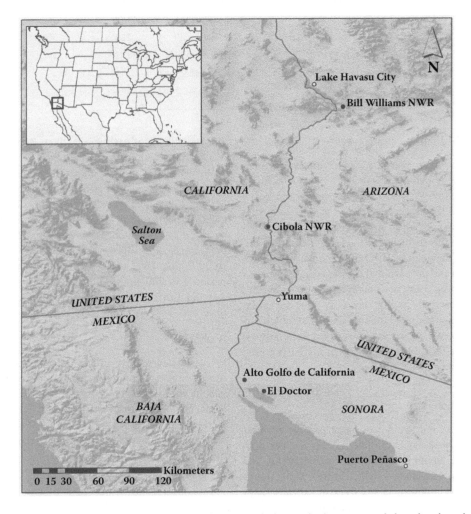

Figure 8.1. Locations of four study sites (black dots) where we studied spring bird migration and plant phenology along the lower Colorado River in the southwestern United States and northwestern Mexico, 2000–2003.

TABLE 8.1

Focal species of neotropical migratory songbirds and the number of foraging observations made along the lower Colorado River in southwestern Arizona and northwestern Sonora, Mexico, during Spring migration, 2000–2003.

Common name	Scientific name	No. Foraging observations
Wilson's Warbler	*Cardellina pusilla*	3969
Orange-crowned Warbler	*Oreothlypis celata*	1935
Black-throated Gray Warbler	*Setophaga nigrescens*	1207
Nashville Warbler	*Oreothlypis ruficapilla*	1165
Warbling Vireo	*Vireo gilvus*	942
Yellow-rumped Warbler	*Setophaga coronata*	693
Townsend's Warbler	*Setophaga townsendi*	592
Lucy's Warbler	*Oreothlypis luciae*	562
Yellow Warbler	*Setophaga petechia*	338
MacGillivray's Warbler	*Oreothlypis tolmiei*	236
Hermit Warbler	*Setophaga occidentalis*	163
Ruby-crowned Kinglet	*Regulus calendula*	43
Bell's Vireo	*Vireo bellii*	21

Focal Bird Species and Surveys

We selected an assemblage of 12 species of insectivorous songbirds (Table 8.1). Of these birds, only Yellow Warbler (*Setophaga petechia*) and Lucy's Warbler (*Oreothlypis luciae*) breed at our study sites (Corman and Wise-Gervais 2005). We used point count transects to estimate relative migrant abundance (Reynolds et al. 1980). To determine niche breadth and proportional use of vegetation species as foraging substrates during each survey period, we walked the same transects used for point counts and recorded the tree species in which foraging activity was first observed for each individual of a focal bird species encountered (Remsen and Robinson 1990).

Vegetation Phenology

We measured the flowering phenology of mesquite (*Prosopis* sp.), Gooding's willow (*Salix goodingii*), Fremont cottonwood (*Populus fremontii*), and tamarisk (*Tamarix* sp.), which represent the dominant and largest species of trees and shrubs at our study sites. Along each survey transect where we performed point counts and bird foraging surveys, we also estimated the percentage of flowering of

100 individuals of each species (van Riper 1980, McGrath et al. 2009).

We performed point counts, foraging, and vegetation phenology surveys between 1 March and 31 May 2000–2003. We divided survey dates into eight 10-day periods within which no site was surveyed more than once. We were unable to conduct point-count surveys during periods 1 and 8 in 2002 and 2003 and foraging surveys during period 4 in 2001, period 2 in 2002, and periods 1–3 and 8 in 2003.

Analyses

We calculated relative migrant abundance as the mean number of migrants per point at each site per survey period. We used the Shannon and Weaver (1964) entropy statistic as a measure of niche breadth ($\hat{H}$) for individual focal bird species during each survey period using the following function:

$$\hat{H} = -\Sigma p_i \ln(p_i)$$

where p = the proportional use of plant species i as a foraging substrate (Strode 2009). We use the term "niche breadth" to refer only to the diversity of vegetation species used as foraging substrates at the stopover sites we surveyed, recognizing that diversity represents only one dimension of a bird's overall niche. We calculated mean values across study dates to assess annual patterns of relative abundance and niche breadth.

We used generalized linear models (GLMs) to examine variation in relative migrant abundance, niche breadth, and proportional use of focal tree species among years and across survey dates and to explore correlations of niche breadth with migrant abundance, monsoon precipitation, and tree flowering. To assess phenological synchrony, we also used GLMs to examine the relationships of migrant abundance and proportional tree use with percentage of flowering, including tree species and survey period as factors. We treated survey date as an integer to assess trends in abundance, niche breadth, and phenological synchrony through time and included second-order terms and two-way interactions in analyses. We examined differences among factor levels with Tukey's multiple comparison tests with Bonferroni adjustments. We performed all analyses in program R 2.1.5 (R Development Core Team 2012).

RESULTS

Migrant Abundance

Relative migrant abundance per site did not differ among years (F = 2.3, df = 3, P = 0.12). Within years, relative abundance had a negative quadratic relationship with survey date (F = 5.9, df = 1, P < 0.05), whereby abundance generally peaked in early April (Figure 8.2).

Niche Breadth

Niche breadth was different among years (F = 4.0, df = 3, P < 0.01); it was broadest in 2000 (z = 2.9, P < 0.01) and narrowest in 2002 (z = 2.3, P < 0.05; Figure 8.3). Niche breadth had a positive quadratic relationship with survey date (F = 4.9, df = 1, P < 0.05; Figure 8.2), with birds foraging in the narrowest diversity of species in late March. Niche breadth was also inversely correlated with migrant abundance (F = 5.7, df = 1, P < 0.05; Figure 8.2).

Proportional Tree Use

Proportional use of focal tree species as foraging substrates was not different among years (F = 2.5, df = 9, P = 0.12). Birds disproportionately foraged

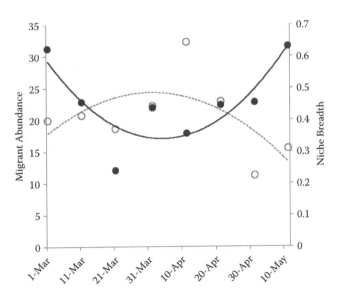

Figure 8.2. Abundance (circles) and niche breadth (dots) of 13 species of migratory birds combined during spring migration along the lower Colorado River in the southwestern United States and northwestern Mexico, plotted by 10-day time periods from 2000 to 2003. Niche breadth represents a Shannon-Weaver diversity index (0–1) of tree species birds used as foraging substrates during stopover; values closer to one indicate greater niche breadth and the use of more tree species. Best-fit polynomial lines are shown for abundance (dashed) and niche breadth (solid).

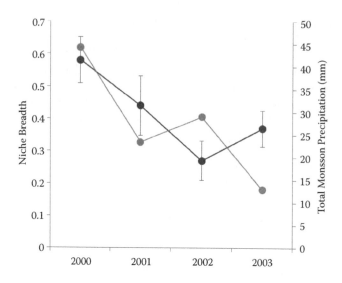

Figure 8.3. Annual mean niche breadth (black) ± SE of migratory birds during spring migration (March–May) and total monsoon precipitation (July–September, gray) of the preceding year along the lower Colorado River in the southwestern United States and northwestern Mexico. Precipitation is for Arizona climate division 5 (NOAA 2013).

in tree species ($F = 33.0$, df = 3, $P < 0.0001$). Migrants used mesquite more than all other species ($z = 9.9$, $P < 0.0001$), and willow more than cottonwood and tamarisk ($z = 2.4$, $P < 0.05$; Table 8.2).

TABLE 8.2

Plant species used as foraging substrates for focal migratory bird species and the mean proportion of foraging observations (p) per 10-day survey period along the lower Colorado River in the United States and Mexico, March–May, 2000–2003.

Plant species	Common name	p
Prosopis sp.	Mesquite	0.72
Salix gooddingii	Gooding's willow	0.40
Populus fremontii	Fremont cottonwood	0.28
Tamarix sp.	Tamarisk	0.24
Pluchea sericea	Arrow weed	0.18
Parkinsonia sp.	Palo verde	0.12
Baccarus sp.	Baccarus	0.10
Larea tridentata	Creosote	0.07
Lycium berlandieri	Wolfberry	0.02
Typha latifolia	Cattails	0.02
Atriplex canescens	Salt bush	0.01

NOTE: The four most used species represent our focal species for which flowering phenology was recorded.

Tree Flowering

Mean percent flowering was not different among years for willow and mesquite combined ($F = 1.5$, df = 2, $P = 0.24$) or for mesquite alone ($F = 0.5$, df = 2, $P = 0.63$). However, flowering of willow was greater in 2001 ($23.2 \pm 9.3\%$, $t = 2.7$, $P < 0.01$) than in 2002 ($3.9 \pm 1.3\%$) and 2003 ($8.1 \pm 2.8\%$). Flowering of willow and mesquite combined was significantly correlated with survey date ($F = 10.7$, df = 1, $P < 0.01$).

Phenological Synchrony

Migrant abundance was not correlated with the percentage of flowering of focal tree species combined ($F = 1.2$, df = 1, $P = 0.31$). However, there was an interaction of flowering and year ($F = 8.4$, df = 2, $P < 0.05$); only in 2003 was the positive linear increase in migrant abundance with flowering significant ($\beta = 2.2$, $F = 19.2$, df = 1, $P < 0.05$; Figure 8.4a). The relationship between abundance and flowering was also positive but not significant in 2001 ($\beta = 0.4$, $F = 1.0$, df = 1, $P = 0.36$) and 2002 ($\beta = 0.6$, $F = 1.3$, df = 1, $P = 0.33$; Figure 8.4a). Niche breadth had a negative linear relationship with percentage of flowering of willow and mesquite ($F = 12.2$, df = 1, $P < 0.05$; Figure 8.5).

Proportional use of mesquite and willow for foraging was highly correlated with the percentage

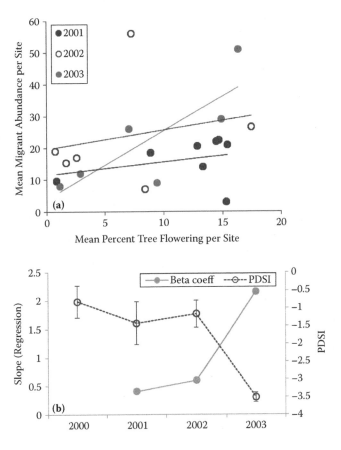

Figure 8.4. (a) The relationship between mean relative abundance of migratory birds and percentage of flowering of willow (*Salix gooddingii*) and mesquite (*Prosopis* sp.) at four sites along the lower Colorado River from 1 March–31 May, 2001–2003. Best-fit regression lines for each year are 2001 (solid black), 2002 (dashed), 2003 (solid gray). (b) The regression coefficients (β) from the annual fitted linear models in panel a (solid gray line and circles) and the mean Palmer drought severity index (PDSI) over a 12-month period (dashed black line with open circles) March through February.

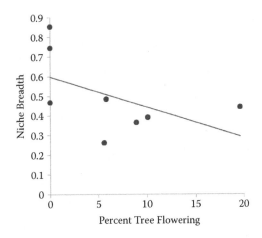

Figure 8.5. Relationship between niche breadth of migratory birds and percentage of flowering of willow and mesquite at stopover sites along the lower Colorado River in the southwestern United States and northwestern Mexico during spring migration, 2001–2003.

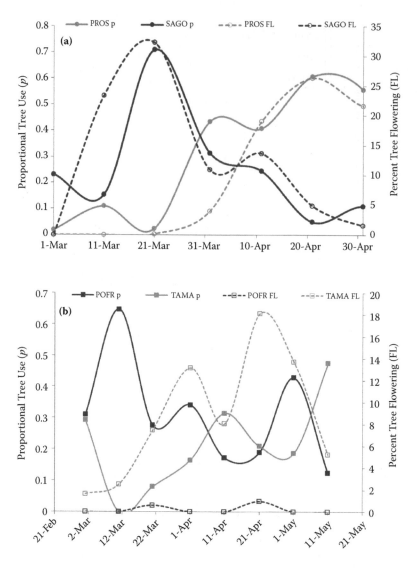

Figure 8.6. Phenological synchrony between mean percentage of flowering (FL) and the mean proportional use (p) as foraging substrates by migratory birds of (a) *Prosopis* mesquite species (PROS) and Gooding's willow (*Salix gooddingii*, SAGO) and (b) Fremont cottonwood (*Populus fremontii*, POFR) and tamarisk (*Tamarix* sp., TAMA) during spring migration at four sites along the lower Colorado River in the southwestern United States and northwestern Mexico, 2001–2003.

of flowering of these species ($F = 9.8$, $df = 1$, $P < 0.01$), and proportional use of each species was highly synchronized with flowering over the migration period (Figure 8.6a). There was an interaction of proportional use with survey date ($F = 5.0$, $df = 1$, $P < 0.05$), and mesquite use was negatively correlated with willow use ($F = 7.1$, $df = 1$, $P < 0.05$), representing a shift in foraging from willow to mesquite (Figure 8.6a). There was no significant relationship between proportional use of cottonwood and tamarisk and their

percentage of flowering ($F = 0.4$, $df = 1$, $P = 0.52$; Figure 8.6b).

Precipitation and Drought

Total monsoon precipitation was lowest the summer prior to the 2003 migration season (Figure 8.3) and had the lowest mean monthly PDSI value (Figure 8.4b), indicating a severe drought (NOAA 2013). Niche breadth had a positive linear relationship with total monsoon precipitation the preceding summer (F = 129.2, df = 3, P < 0.01), with annual niche breadth and precipitation showing a decreasing trend from 2000 to 2003 (Figure 8.3). The broadest mean niche breadth occurred in 2000, following the highest total monsoon precipitation during our study (Figure 8.3), and the largest β coefficient for the relationship between migrant abundance and tree flowering was in 2003, following a severe drought (PDSI < −3, NOAA 2013, Figure 8.4b).

DISCUSSION

Temporal patterns of migrant abundance and the diversity (niche breadth) and proportional selection of plant species as foraging substrates were highly synchronized with flowering of these plant species during spring migratory stopover along the lower Colorado River. Peak migrant abundance during late March and early April coincided with the period of narrowest niche breadth when migrants were primarily foraging in willow and mesquite. Timing of migration further coincided with the greatest temporal overlap in willow and mesquite flowering, suggesting that this may be the period of peak flowering phenology and hence food availability and habitat quality for this riparian habitat (McGrath et al. 2009).

The strong patterns of phenological synchrony we observed—particularly between niche breadth, proportional substrate use, and flowering in relation to precipitation and climate—question whether synchrony in this case is always a benefit that represents the historic or unmodified condition of the ecosystem. High concentrations of migrants at stopover sites can rapidly decrease food abundance and limit birds' ability to replenishment energy reserves (Moore and Yong 1991). In drier years, migrant abundance had a stronger positive correlation with willow and mesquite flowering, while niche breadth decreased as migrants concentrated their foraging within these few species, possibly abandoning more peripheral habitat areas that could contain fewer preferred prey under dry conditions. Increased selection of willow and mesquite for foraging could lead to greater densities of birds in these high-quality substrates and have negative density-dependent effects on migrants—resulting in greater competition for prey, decreased replenishment of energy reserves, and decreased fitness as birds'

ability to complete migration and breed successfully is compromised (Moore and Yong 1991, Smith and Moore 2003, Norris and Marra 2007). Furthermore, if the emergence of preferred prey were to become decoupled from flowering phenology, flowering would no longer represent a valid indicator of habitat quality (Johnson 2007). Unsuitable cues could result in a greater concentration of birds selecting foraging habitat that has less available food.

The concentration of proportional foraging on willow and mesquite trees and shrubs, especially during drier years, may indicate greater predictability of preferred prey availability in these species in addition to their synchronous timing of flowering. Variation in plant growth and productivity in arid lands is strongly influenced by precipitation, the majority of which comes during summer monsoons (Adams and Comrie 1997, Ogle and Reynolds 2004). Large, woody perennials such as willow and mesquite can access deeper ground water and may be less susceptible to annual variation in precipitation and thus could sustain greater densities of preferred prey species (Ehleringer et al. 1991, Stromberg et al. 1996, McGrath et al. 2009). Altered flood frequency, reduced water tables, and increased salinity have already drastically reduced the abundance of large woody perennials and riparian forest (Busch and Smith 1995). Increasing human demands on the limited water available in the Colorado River Basin due to climate change-induced drought and continued exurban development, agriculture, and recreation further threaten riparian habitat (Christensen et al. 2004). Maintaining sufficient willow and mesquite stopover habitat along the lower Colorado will likely require active management strategies that address flow regimes, forest restoration, and water conservation in the region (Stromberg 2001, Garfin et al. 2007, van Riper et al. 2008).

Latitudinal differences in phenology and synchrony among our sites was beyond the scope of this chapter. Spring phenology seems to be advancing most rapidly at more northern latitudes (Parmesan 2007, Ellwood et al. 2013); however, our most distant sites were only about 250 km apart, were within the same bioregion and climate zone, and had the same vegetation communities. Perhaps more important than latitude, our sites spanned the international border between the United States and Mexico. Water policies in the United States strongly impact the amount of water in the Colorado River flowing into Mexico and the Gulf of California and thus the condition of riparian habitats there (Glenn et al. 2001, Zamora-Arroyo et al. 2001). Future research in the US–Mexican borderlands would benefit from international consideration of how political control and legislation of water use and access may affect the presence, distribution, and condition of riparian stopover habitats.

ACKNOWLEDGMENTS

Thanks to Laura McGrath. Support came from the Desert Southwest Cooperative Ecosystem Studies Unit, USGS-SBSC-Sonoran Desert Research Station, and the University of Arizona. Any use of trade, product, or firm names is for descriptive purposes only and does not imply endorsement by the U.S. Government.

LITERATURE CITED

Aborn, D. A., and F. R. Moore. 1997. Pattern of movement by Summer Tanagers (*Piranga rubra*) during migratory stopover: A telemetry study. Behaviour 134:1077–1100.

Aborn, D. A., and F. R. Moore. 2004. Activity budgets of Summer Tanagers during spring migratory stopover. Wilson Bulletin 116:64–68.

Adams, D. K. and A. C. Comrie. 1997. The North American monsoon. Bulletin of the American Meteorological Society 78:2197–2213.

Balbontin, J., A. P. Moller, I. G. Hermosell, A. Marzal, M. Reviriego, and F. de Lope. 2009. Individual responses in spring arrival date to ecological conditions during winter and migration in a migratory bird. Journal of Animal Ecology 78:981–989.

Balling, R. C., Jr., and G. B. Goodrich. 2010. Increasing drought in the American Southwest? A continental perspective using a spatial analytical evaluation of recent trends. Physical Geography 31:293–306.

Bauchinger, U., T. Van't Hof, and H. Biebach. 2009. Food availability during migratory stopover affects testis growth and reproductive behaviour in a migratory passerine. Hormones and Behavior 55:425–433.

Bertin, R. I. 1982. The Ruby-throated Hummingbird and its major food plants: ranges, flowering phenology, and migration. Canadian Journal of Zoology 60:210–219.

Both, C., S. Bouwhuis, C. M. Lessells, and M. E. Visser. 2006. Climate change and population declines in a long-distance migratory bird. Nature 441:81–83.

Both, C., M. van Asch, R. G. Bijlsma, A. B. van den Burg, and M. E. Visser. 2009. Climate change and unequal phenological changes across four trophic levels: constraints or adaptations? Journal of Animal Ecology 78:73–83.

Buler, J. J., F. R. Moore, and S. Woltmann. 2007. A multi-scale examination of stopover habitat use by birds. Ecology 88:1789–1802.

Busch, D. E., and S. D. Smith. 1995. Mechanisms associated with decline of woody species in riparian ecosystems of the southwestern U.S. Ecological Monographs 65:347–370.

Carlisle, J. D., K. L. Olmstead, C. H. Richart, and D. L. Swanson. 2012. Food availability, foraging behavior, and diet of autumn landbirds in the Boise foothills of southwestern Idaho. Condor 114:449–461.

Christensen, N. S., A. W. Wood, N. Voisin, D. P. Lettenmaier, and R. N. Palmer. 2004. The effects of climate change on the hydrology and water resources of the Colorado River basin. Climatic Change 62:337–363.

Corman, T. E., and C. Wise-Gervais (editors). 2005. The Arizona breeding bird atlas. UNM Press, Albuquerque, NM.

Ehleringer, J. R., S. L. Phillips, W. S. F. Schuster, and D. R. Sandquist. 1991. Differential utilization of summer rains by desert plants. Oecologia 88:430–434.

Ellwood, E. R., S. A. Temple, R. B. Primack, N. L. Bradley, and C. C. Davis. 2013. Record-breaking early flowering in the eastern United States. PloS One 8:e53788.

Faaborg, J., R. T. Holmes, A. D. Anders, K. L. Bildstein, K. M. Dugger, S. A. Gauthreaux, P. Heglund, K. A. Hobson, A. E. Jahn, D. H. Johnson, S. C. Latta, D. J. Levey, P. P. Marra, C. L. Merkord, E. Nol, S. I. Rothstein, T. W. Sherry, T. S. Sillett, F. R. Thompson, and N. Warnock. 2010a. Conserving migratory land birds in the New World: do we know enough? Ecological Applications 20:398–418.

Faaborg, J., R. T. Holmes, A. D. Anders, K. L. Bildstein, K. M. Dugger, S. A. Gauthreaux, P. Heglund, K. A. Hobson, A. E. Jahn, D. H. Johnson, S. C. Latta, D. J. Levey, P. P. Marra, C. L. Merkord, E. Nol, S. I. Rothstein, T. W. Sherry, T. S. Sillett, F. R. Thompson, and N. Warnock. 2010b. Recent advances in understanding migration systems of New World land birds. Ecological Monographs 80:3–48.

Garfin, G., M. A. Crimmins, and K. L. Jacobs. 2007. Drought, climate variability, and implications for water supply and management. Pp. 61–78 in B. G. Colby and K. L. Jacobs (editors), Arizona water policy: management innovations in an urbanizing, arid region. Resources for the Future, Washington, DC.

Glenn, E. P., F. Zamora-Arroyo, P. L. Nagler, M. Briggs, W. Shaw, and Karl Flessa. 2001. Ecology and conservation biology of the Colorado River Delta, Mexico. Journal of Arid Environments 49:5–15.

Hart, P. J., B. L. Woodworth, R. J. Camp, K. Turner, K. McClure, K. Goodall, C. Henneman, C. Spiegel, J. Lebrun, E. Tweed, and M. Samuel. 2011. Temporal variation in bird and resource abundance across an elevational gradient in Hawaii. Auk 128:113–126.

Johnson, M. D. 2007. Measuring habitat quality: a review. Condor 109:489–504.

Jones, J., P. J. Doran, and R. T. Holmes. 2003. Climate and food synchronize regional forest bird abundances. Ecology 84:3024–3032.

Jones, T., and W. Cresswell. 2010. The phenology mismatch hypothesis: are declines of migrant birds linked to uneven global climate change? Journal of Animal Ecology 79:98–108.

Kelly, J. F., and R. L. Hutto. 2005. An east–west comparison of migration in North American wood warblers. Condor 107:197–211.

LaManna, J. A., T. L. George, J. F. Saracco, M. P. Nott, and D. F. DeSante. 2012. El Niño southern oscillation influences annual survival of a migratory songbird. Auk 129:734–743.

MacDade, L. S., P. G. Rodewald, and K. A. Hatch. 2011. Contribution of emergent aquatic insects to refueling in spring migrant songbirds. Auk 128:127–137.

McGrath, L. J., C. van Riper, and J. J. Fontaine. 2009. Flower power: tree flowering phenology as a settlement cue for migrating birds. Journal of Animal Ecology 78:22–30.

McKinney, A. M., P. J. CaraDonna, D. W. Inouye, B. Barr, C. D. Bertelsen, and N. M. Waser. 2012. Asynchronous changes in phenology of migrating Broad-tailed Hummingbirds and their early-season nectar resources. Ecology 93:1987–1993.

Miller-Rushing, A. J., T. T. Høye, D. W. Inouye, and E. Post. 2010. The effects of phenological mismatches on demography. Philosophical Transactions of the Royal Society B 365:3177–3186.

Moore, F., and P. Kerlinger. 1987. Stopover and fat deposition by North American wood-warblers (Parulinae) following spring migration over the Gulf of Mexico. Oecologia 74:47–54.

Moore, F. R., P. Kerlinger, and T. R. Simons. 1990. Stopover on a Gulf Coast barrier island by spring trans-Gulf migrants. Wilson Bulletin 102:487–500.

Moore, F. R., and W. Yong. 1991. Evidence of food-based competition among passerine migrants during stopover. Behavioral Ecology and Sociobiology 28:85–90.

NOAA. [online]. 2013. National Oceanic and Atmospheric Administration: National Climatic Data Center. <http://www.ncdc.noaa.gov/cag/> (accessed 7 August 2013).

Norris, D. R., and P. P. Marra. 2007. Seasonal interactions, habitat quality, and population dynamics in migratory birds. Condor 109:535–547.

Ogle, K., and J. F. Reynolds. 2004. Plant responses to precipitation in desert ecosystems: integrating functional types, pulses, thresholds, and delays. Oecologia 141:282–294.

Parmesan, C. 2007. Influences of species, latitudes and methodologies on estimates of phenological response to global warming. Global Change Biology 13:1860–1872.

Parrish, J. D. 2000. Behavioral, energetic, and conservation implications of foraging plasticity during migration. Studies in Avian Biology 20:53–70.

Paxton, K. L., C. Van Riper, and C. O'Brien. 2008. Movement patterns and stopover ecology of Wilson's Warblers during spring migration on the lower Colorado River in southwestern Arizona. Condor 110:672–681.

Paxton, K. L., C. Van Riper, T. C. Theimer, and E. H. Paxton. 2007. Spatial and temporal migration patterns of Wilson's Warbler (Wilsonia pusilla) in the southwest as revealed by stable isotopes. Auk 124:162–175.

Petit, D. R. 2000. Habitat use by landbirds along Nearctic–Neotropical migration routes: implications for conservation of stopover habitats. Studies in Avian Biology 20:15–33.

Poff, N. L., J. D. Allan, M. B. Bain, J. R. Karr, K. L. Prestegaard, B. D. Richter, R. E. Sparks, and J. C. Stromberg. 1997. The natural flow regime. BioScience 47:769–784.

R Development Core Team. 2012. R: A language and environment for statistical computing. R Foundation for Statistical Computing, Vienna, Austria.

Rajagopalan, B., K. Nowak, J. Prairie, M. Hoerling, B. Harding, J. Barsugli, A. Ray, and B. Udall. 2009. Water supply risk on the Colorado River: can management mitigate? Water Resources 45:W08201.

Remsen, J. V., Jr., and S. K. Robinson. 1990. A classification scheme for foraging behavior of birds in terrestrial habitats. Studies in Avian Biology 13:144–160.

Reynolds, R. T., J. M. Scott, and R. A. Nussbaum. 1980. A variable radius circular plot method for estimating bird numbers. Condor 82:309–313.

Rodewald, P. G., and M. C. Brittingham. 2007. Stopover habitat use by spring migrant landbirds: The roles of habitat structure, leaf development, and food availability. Auk 124:1063–1074.

Saino, N., R. Ambrosini, D. Rubolini, J. von Hardenberg, A. Provenzale, K. Huppop, O. Huppop, A. Lehikoinen, E. Lehikoinen, K. Rainio, M. Romano, and L. Sokolov. 2011. Climate warming, ecological mismatch at arrival and population decline in migratory birds. Proceedings of the Royal Society of London B 278:835–842.

Seager, R., M. F. Ting, I. Held, Y. Kushnir, J. Lu, G. Vecchi, H. P. Huang, N. Harnik, A. Leetmaa, N. C. Lau, C. H. Li, J. Velez, and N. Naik. 2007. Model projections of an imminent transition to a more arid climate in southwestern North America. Science 316:1181–1184.

Seewagen, C. L., and C. G. Guglielmo. 2010. Effects of fat and lean body mass on migratory landbird stopover duration. Wilson Journal of Ornithology 122:82–87.

Shannon, C. E., and W. Weaver. 1964. The mathematical theory of communication. University of Illinois Press, Urbana, IL.

Sillett, T. S., and R. T. Holmes. 2002. Variation in survivorship of a migratory songbird throughout its annual cycle. Journal of Animal Ecology 71:296–308.

Singer, M. C., and C. Parmesan. 2010. Phenological asynchrony between herbivorous insects and their hosts: signal of climate change or pre-existing adaptive strategy? Philosophical Transactions of the Royal Society B 365:3161–3176.

Skagen, S. K., J. F. Kelly, C. Van Riper, R. L. Hutto, D. M. Finch, D. J. Krueper, and C. P. Melcher. 2005. Geography of spring landbird migration through riparian habitats in southwestern North America. Condor 107:212–227.

Smith, R. J., and F. R. Moore. 2003. Arrival fat and reproductive performance in a long-distance passerine migrant. Oecologia 134:325–331.

Strode, P. K. 2009. Spring tree species use by migrating Yellow-rumped Warblers in relation to phenology and food availability. Wilson Journal of Ornithology 121:457–468.

Stromberg, J. C. 2001. Restoration of riparian vegetation in the south-western United States: importance of flow regimes and fluvial dynamism. Journal of Arid Environments 49:17–34.

Stromberg, J. C., R. Tiller, and B. Richter. 1996. Effects of groundwater decline on riparian vegetation of semiarid regions: the San Pedro, Arizona. Ecological Applications 6:113–131.

Thackeray, S. J., T. H. Sparks, M. Frediksen, S. Burthe, P. J. Bacon, J. R. Bell, M. S. Botham, T. M. Brereton, P. W. Bright, L. Carvalho, T. Clutton-Brock, A. Dawson, M. Edwards, J. M. Elliott, R. Harrington, D. Johns, I. D. Jones, J. T. Jones, D. I. Leech, D. B. Roy, W. A. Scott, M. Smith, R. J. Smithers, I. J. Winfield, and S. Wanless. 2010. Trophic level asynchrony in rates of phenological change for marine, freshwater and terrestrial environments. Global Change Biology 16:3304–3313.

van Asch, M., R. Julkunen-Tiito, and M. E. Visser. 2010. Maternal effects in an insect herbivore as a mechanism to adapt to host plant phenology. Functional Ecology 24:1103–1109.

van Asch, M., P. H. Tienderen, L. J. M. Holleman, and M. E. Visser. 2007. Predicting adaptation of phenology in response to climate change, an insect herbivore example. Global Change Biology 13:1596–1604.

van Riper, C., III. 1980. The phenology of the dryland forest of Mauna Kea Hawaii USA and the impact of recent environmental perturbations. Biotropica 12:282–291.

van Riper, C., K. L. Paxton, C. O'Brien, P. B. Shafroth, and L. J. McGrath. 2008. Rethinking avian response to *Tamarix* on the lower Colorado River: a threshold hypothesis. Restoration Ecology 16:155–167.

Visser, M. E., and L. J. M. Holleman. 2001. Warmer springs disrupt the synchrony of oak and winter moth phenology. Proceedings of the Royal Society of London B 268:289–294.

Visser, M. E., L. J. M. Holleman, and P. Gienapp. 2006. Shifts in caterpillar biomass phenology due to climate change and its impact on the breeding biology of an insectivorous bird. Oecologia 147:164–172.

Visser, M. E., L. te Marvelde, and M. E. Lof. 2012. Adaptive phenological mismatches of birds and their food in a warming world. Journal of Ornithology 153:75–84.

Weiss, J. L., C. L. Castro, and J. T. Overpeck. 2009. Distinguishing pronounced droughts in the southwestern United States: seasonality and effects of warmer temperatures. Journal of Climate 22:5918–5932.

Zamora-Arroyo, F., P. L. Naglerb, M. Briggs, D. Radtke, H. Rodriquez, J. Garcia, C. Valdes, A. Huete, and E. P. Glenn. 2001. Regeneration of native trees in response to flood releases from the United States into the delta of the Colorado River, Mexico. Journal of Arid Environments 49:49–64.

CHAPTER NINE

Shorebird Migration in the Face of Climate Change*

POTENTIAL SHIFTS
IN MIGRATION PHENOLOGY
AND RESOURCE AVAILABILITY

Ryan J. Stutzman and Joseph J. Fontaine

Abstract. Changes in temperature and seasonality resulting from climate change are heterogeneous, potentially altering important sources of natural selection acting on species phenology. Some species have apparently adapted to climate change but the ability of most species to adapt remains unknown. The life history strategies of migratory animals are dictated by seasonal factors, which makes these species particularly vulnerable to heterogeneous changes in climate and phenology. Here, we examine the phenology of migratory shorebirds, their habitats, and primary food resources, and we hypothesize how climate change may affect migrants through predicted changes in phenology. Daily abundance of shorebirds at stopover sites was correlated with local phenology and peaked immediately prior to peaks in invertebrate food resources. A close relationship between migrant and invertebrate phenology indicates that shorebirds may be vulnerable to changes in seasonality driven by climate change. It is possible that shifts in migrant and invertebrate phenology will be congruent in magnitude and direction, but because migration phenology is dependent on a suite of ecological factors, any response is likely to occur at a larger temporal scale and may lag behind the response of invertebrate food resources. The resulting lack of sufficient access to food at stopover habitats may cause migrants to extend migration and have cascading effects throughout their life cycle. If the heterogeneous nature of climate change results in uneven changes in phenology between migrants and their prey, it may threaten the long-term viability of migratory populations.

Key Words: Calidris, climate change, food availability, habitat selection, phenology, stopover habitat.

Global climate change is proceeding at an unprecedented rate, creating known and unknown challenges for conservation and research professionals (IPCC 2007). Climate change is spatially and temporally heterogeneous, which makes predicting ecological consequences difficult and designing effective mitigation strategies challenging. Spatial and temporal disparity in changes to seasonality, resource availability, and phenology are predicted to have far-reaching

---

* Stutzman, R. J., and J. J. Fontaine. 2015. Shorebird migration in the face of climate change: potential shifts in migration phenology and resource availability. Pp. 145–159 in E. M. Wood and J. L. Kellermann (editors), Phenological synchrony and bird migration: changing climate and seasonal resources in North America. Studies in Avian Biology (no. 47), CRC Press, Boca Raton, FL.

implications for biodiversity (Sala et al. 2000, Thomas et al. 2004, Botkin et al. 2007), particularly for species that occupy large geographic areas and have complex life history strategies such as long-distance migrants (Both and Visser 2001, Robinson et al. 2009, Both et al. 2010). Understanding the degree to which life history events like migration are dependent on intertwined phenological cues such as trees beginning to flower or seasonal insect blooms is essential for wildlife professionals to mitigate the effects of climate change. Seasonality has been important in shaping life history evolution, such as Neotropical songbirds that preferentially forage on trees with more flowers (McGrath et al. 2009) or the apparent ability of some species to adapt to changes in phenology (Walther et al. 2002, Root et al. 2003, Jonzén et al. 2006). Nevertheless, general information concerning the phenological sensitivity and progression for most species is lacking. Addressing the implications of climate change for species phenology is of growing interest, but few studies have considered these relationships in the context of additional sources of anthropogenic change (Opdam and Wascher 2004).

Avian migration is a well-studied life-history event, but our understanding of the phenological cues driving migratory phenology, and the potential for climate change and other sources of anthropogenic change to influence migration behaviors remains limited (Ahola et al. 2004, Gordo 2007, Petersen 2009). Avian species often show preference for stopover habitats with greater food availability (Hutto 1985, Russell et al. 1992, Kelly et al. 2002, van Gils et al. 2005), and variation in food availability at stopover sites affects body condition and, ultimately, individual fitness (Moore et al. 1995, Pfister et al. 1998, Drent et al. 2003, Baker et al. 2004). However, anthropogenic change, be it from climate change or other forces such as land-use changes, can change the cues that predict food resources, the food resources themselves, or both, potentially leading to an ecological trap (Battin 2004, Robertson and Hutto 2006). Moreover, heterogeneity in the rate of climate change across the range of many migratory bird species has the potential to affect habitats and resources differently at various locations throughout the migration cycle (Visser et al. 2004, Fontaine et al. 2009, Jones and Cresswell 2010). Strong selection pressure and a reliance on predictable spatial and temporal relationships

have resulted in stopover events that often occur during optimal resource availability at a single location en route (McGrath et al. 2009), despite the fact that migratory timing is dictated in part by conditions at earlier stages of the migratory cycle and that migrants make local habitat decisions without prior knowledge of habitat conditions (Hutto 1985, Loria and Moore 1990, Moore et al. 1990, Moore and Aborn 2000, Petit 2000).

Some migratory species are flexible in their response to changes in seasonality, with variability in arrival dates among years (Crick et al. 1997, Hüppop and Hüppop 2003, Jenni and Kéry 2003, Lehikoinen et al. 2004, Stervander et al. 2005, Jonzén et al. 2006, Tøttrup et al. 2006, Swanson and Palmer 2009). In other cases, phenological responses are variable and inconsistent among species (Inouye et al. 2000, Both and Visser 2001, Gordo et al. 2005, Weidinger and Král 2007, Wilson 2007, Møller et al. 2008, Both 2010). Given inconsistency among species, it is unknown how most species will respond to changes in food availability or phenology driven by climate change, land-use change, or the interactions among them. However, species that are not able to adapt migratory patterns effectively to changing conditions at stopover sites may experience population declines. Food availability prior to and during migration clearly has the potential to impact the timing and duration of migration (Piersma 1987, Russell et al. 1992, Yong and Moore 1997, Newton 2006). Furthermore, populations that have responded to changes in resource phenology through advanced migration phenology may be less prone to declines than populations unable to advance the timing of their migration (Strode 2003, Møller et al. 2008). Differential responses may result in higher rates of population decline among long-distance migrants than among resident species (Sherry and Holmes 1996, Sanderson et al. 2006, Both et al. 2010). Migratory populations are likely to be affected negatively when migration events and periods of peak resources that were once synchronized become decoupled due to independent changes in phenology (Both 2010, Jones and Cresswell 2010).

Two factors—degree of phenological mismatch and migratory distance—influence the effect that changes in phenology will have on migratory populations (Jones and Cresswell 2010). Decoupling between migrant arrival and availability of resources can occur one of four ways:

changes in migration phenology, changes in resource phenology, changes in cue phenology, or a combination of factors (Jones and Cresswell 2010). For example, it is possible that changes in phenology of resources or cues are occurring in the Prairie Pothole Region of North America as the region is experiencing warmer winters (Swanson and Palmer 2009), which may cause earlier peaks in green-up of vegetation or invertebrate abundance. While resources and cues are dependent on local climatic conditions, migrant arrival at stopover sites is dependent on endogenous and external factors at overwintering locations, previous stopover sites, and predicted phenological conditions at breeding grounds (Gwinner 1996, Yong and Moore 1997, Marra et al. 1998, Ottick and Dierschke 2003, Studds and Marra 2011). Given the heterogeneous nature of climate and climate change, it is possible that migrants will not respond in the same manner to local phenological conditions at one or more stopover locations (Rosenzweig et al. 2008, Fontaine et al. 2009, Both 2010). If there is not a corresponding shift in avian migration, it will likely lead to a mismatch in timing of migration and resource availability that ultimately leads to a decrease in stopover success through reduced fat deposition, prolonged stopover duration, or direct mortality.

Here, we make predictions for how shorebird populations may respond to climate change by examining a number of possible climate change-induced phenological shifts. We then test our predictions with empirical data to examine the influence of local phenological factors on shorebird migration and invertebrate abundance to compare the potential sensitivity of shorebirds and their prey to climate change and other phenological factors.

SCENARIO DEVELOPMENT

We hypothesized patterns between shorebird migration and invertebrate food resources based on changes to the predicted historical relationship given hypothetical changes in phenology (Miller-Rushing et al. 2010). The Prairie Pothole Region's spring temperatures are expected to increase and result in advancing phenology, and all scenarios involve either no change or advances in phenology. Furthermore, our scenarios contain an invertebrate phenology comparison between agricultural lands with reduced food availability versus grassland wetlands as the assumed historical condition.

Migratory shorebirds are known to select agricultural wetlands during stopover (Elphick and Oring 1998, Niemuth et al. 2006, Taft and Haig 2006) and may even prefer these habitats. However, agricultural wetlands often have lower food availability than grassland wetlands (Euliss and Mushet 1999, but see Taft and Haig 2005). Migrants may be able to buffer against the effects of using novel habitats through behavioral modification, but it is worth exploring how climate change might affect resource and migration phenology at preferred habitats because the degree of behavioral modification and, subsequently, the ability of migrants to adapt to change may be limited.

It is possible that shorebirds and other migrants may adapt to changing conditions brought about by climate change through behavioral modification or dietary flexibility or by making adjustments to migration routes. However, climate change and the corresponding changes in phenology may compound the impacts of land-use changes on shorebird stopover success in the midcontinent region, eventually resulting in population-level effects. If resource phenology shifts to earlier in the migration season and migrants do not adapt, shifts would likely prolong migration through increases in stopover duration and number, and they could delay arrival to the breeding grounds, which can reduce recruitment and lead to population declines (Piersma 1987, Kuenzi et al. 1991, van Eerden et al. 1991, Russell et al. 1992, Moore et al. 1995, Yong and Moore 1997).

Scenario 1: No Change

Here, we show the expected historical relationship between migration and invertebrate phenology with the added effect of migrants using habitats with reduced food availability (Figure 9.1a). Midcontinental migratory shorebirds prefer using agricultural wetlands for stopover, despite the likelihood that they have a lower abundance of benthic invertebrates. We predict this pattern if climate change does not affect the phenology of migrants or invertebrates in our study area.

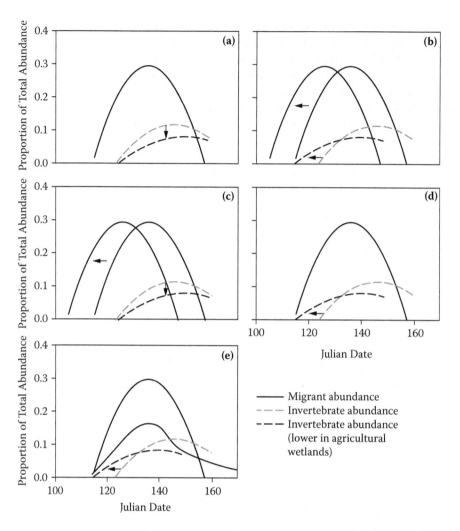

Figure 9.1. (a) Scenario 1: current conditions; available food resources in agricultural fields represented by lower dotted line. (b) Scenario 2: peaks in migration and food resources both occur earlier but the relationship remains unchanged. (c) Scenario 3: peak in migration occurs earlier but invertebrate phenology is unchanged. (d) Scenario 4: peak in migration remains unchanged but invertebrate food resources peak earlier. (e) Scenario 5: peak in migration remains the same but duration is extended. Earlier peak in food resources.

Scenario 2: Matched Advances in Migration and Invertebrate Phenology

Scenario 2 assumes that migration and invertebrate food resources both respond to changing climatic conditions by peaking earlier than under current conditions (Figure 9.1b). Here, invertebrates respond quickly to local changes in wetland conditions and migrants are able to respond at an equal rate. In this scenario, the relationship between migration and invertebrate phenology is unchanged, although migrants still face reduced food availability through a continued preference for agricultural fields. An additional potentially negative effect of advancing migration phenology is increased exposure to extreme weather events (Moore et al. 1995, 2005; Decker and Conway 2009).

Scenario 3: Advancing Migration Phenology, but No Change in Invertebrate Phenology

Scenario 3 represents the relationship between migrant and invertebrate phenology if only migration advances (Figure 9.1c). This set of circumstances is likely to occur if southern stopover or overwintering locations warm at a faster rate than stopover sites in the Prairie Pothole Region. The timing of migration is dependent on many complex factors, including endogenous factors, photoperiod, and conditions at overwintering grounds (Gwinner 1996, Marra et al. 1998, Studds and Marra 2011). Still, extreme late-winter warming in the southern latitudes may drive migrants to depart earlier, causing migrants to arrive prior to the peak in food resources.

Scenario 4: No Change in Migration Phenology, but Invertebrate Phenology Advances

Scenario 4 represents the phenological relationship if only invertebrate phenology changes (Figure 9.1d). Given that our study area is expected to experience warmer temperatures, such a response would likely manifest as an earlier peak in food resources (Walther et al. 2002, Parmesan and Yohe 2003, Root et al. 2003). Conversely, because previous stopover sites occur nearer the equator, they may not experience congruent changes in climate (Fontaine et al. 2009). Migration arrival dates are dependent on a suite of ecological conditions (Gwinner 1996, Yong and Moore 1997, Marra et al. 1998, Ottick and Dierschke 2003, Studds and Marra 2011), and any response is likely to occur at a larger temporal scale and may lag behind the response of invertebrate food resources at any one location in the migratory cycle (Jones and Cresswell 2010). Here, migration abundance peaks after the predicted peak in food resources, which may preclude migrants from achieving optimal migratory condition. In this case, migrants face depressed food availability in concert with the potentially negative effects of foraging in agricultural habitats.

Scenario 5: Migration Phenology Is Extended as Migratory Success Is Constrained by Advancing Invertebrate Phenology

Here, we show the same change to invertebrate phenology as in scenario 4, with the peak in food resources occurring earlier (Figure 9.1e). However, because migrants are likely to experience the effects of climate change at each stop during migration, the response to this suite of changing conditions would be cumulative. As a consequence of continuously missing periods of

peak resources at stopover sites, migrants would likely have to extend their stopover duration at each site, leading to a prolonged period of migration for individuals and the population.

METHODS

Study Area

We collected data in the Prairie Pothole Region of north-central North America, specifically McPherson, Edmunds, and Brown Counties in northeast South Dakota (Figure 9.2). This region of north-central North America is characterized by millions of small depressional wetlands left by receding ice sheets in the late Pleistocene and by a seasonal, relatively dry climate punctuated by severe droughts and deluges (Johnson et al. 2005). The region experiences daily average high temperatures between 14.1°C and 21.2°C and receives an average of 11.5 cm of precipitation during the 3-month study period (April–June). The region has high wetland density and diverse land-use practices, including row crops, rangelands, hay fields, and conservation reserve grassland as well as many native prairie remnants. Shorebirds use shallow water for foraging (Skagen and Knopf 1994, Davis and Smith 1998), and sampling was restricted to wetlands with seasonal and temporary hydrologic regimes to avoid sampling of unsuitable habitat (Stewart and Kantrud 1971). All sampling was done from early April through mid-June of 2010 and 2011 to encompass the entire migration period of all northbound migratory shorebirds in the region (Skagen et al. 2008).

Study Species

We limited our surveys of migratory shorebirds to seven species of arctic-nesting sandpipers (*Calidris* spp., Table 9.1). We established sample wetlands along nine road transects within the study area and surveyed shorebirds at 155 and 163 wetlands in 2010 and 2011, respectively, and visited 85% of the wetlands in both years. We selected transects following a systematic random sampling protocol, constrained by logistics like road passability and safety, and all transects were between 15 and 30 km long. We surveyed transects every 7–10 days, as this time exceeds average stopover duration for shorebirds in the region (Skagen and Knopf 1994) and reduces the likelihood of

Figure 9.2. Map of South Dakota, with inset of study area.

TABLE 9.1

Species of migratory sandpipers and sample size (n) of birds
observed during Spring migration in South Dakota, 2010 and 2011.

Species		2010	2011
Baird's Sandpiper	*Calidris bairdii*	46	170
Dunlin	*Calidris alpina*	1	6
Least Sandpiper	*Calidris minutilla*	54	217
Pectoral Sandpiper	*Calidris melanotos*	21	231
Semipalmated Sandpiper	*Calidris pusilla*	49	250
Stilt Sandpiper	*Calidris himantopus*	2	25
Unknown small *Calidris*	*Calidris* spp.		254
Unknown large *Calidris*	*Calidris* spp.		26
White-rumped Sandpiper	*Calidris fuscicollis*	258	364
Total		431	1543

resampling individuals. To maximize detection of shorebirds, we only sampled wetlands that were located within 150 m of the transect. Wetlands along transects were separated by a minimum of 0.8 km, creating a sample of wetlands randomly distributed in different land-use types. Observers began surveys within an hour of sunrise and did not continue counts after 2 P.M. Using binoculars and a spotting scope, a single observer identified and enumerated all *Calidris* that were visually detected either on the ground or in the air before alighting at the wetland during a 10-minute sampling window. We used a standardized sampling window in an effort to control for sampling effort and detection probability.

Due to small sample sizes for individual species, we analyzed pooled counts for all *Calidris* spp. Different species had subtle differences in microhabitat use and foraging technique; however, differences were negligible in the scope of this study because migratory species of *Calidris* shorebirds occupy the same ecological niche whereby they forage in shallow water and mud-flat habitats for benthic invertebrates (Skagen and Oman 1996, Davis and Smith 2001, Skagen 2006). Stopover periods overlap, but do not coincide between species (Skagen et al. 2008), and by including all species in subsequent analysis, we improved the scope of inference of the study.

Migration Phenology

We compared migration phenology with wetland phenology as indicated by local characteristics. Total daily bird abundance across the study area was used as an index of migration phenology. We used generalized linear models (hereafter GLMs) with a Poisson distribution and included date as a covariate to examine the relationship between migration phenology (total daily abundance), mean daily values for green vegetation, invertebrate abundance, and daily minimum water temperature. We did an independent analysis for each year due to high variability in local conditions between years. However, we tested for the influence of green vegetation and water temperature on daily migrant abundance across 2010–2011, using year as a factor. We estimated the proportion of green vegetation of all nonsubmerged vegetation and shore within 10 m of the water's edge for each wetland. Timing of spring green-up is an indicator of wetland phenology and migrating species use vegetative characteristics as a cue to select sites with favorable foraging conditions (McGrath et al. 2009). We measured green vegetation as a potential cue because it changes predictably through the season and we hypothesized that it may indicate food availability.

Food availability is a primary concern for migratory species and is often cited as the limiting resource during stopover (Hutto 1985, Moore et al. 1995, Newton 2006). The relationship between migration phenology and benthic invertebrate abundance is important because the two are influenced by climatological variables at different scales that are not expected to change uniformly with climate (Cresswell and McCleery 2003, Visser et al. 2004, Fontaine et al. 2009, Jones and Cresswell 2010). We included water temperature in the migration phenology model

as a predictor and potential driver of local phenology. Temperature is known to influence the overall phenological progression of invertebrates (Corbet 1964, Wiggins et al. 1980, Hogg and Williams 1996) and may act as a reliable indicator of food potential. Furthermore, invertebrates are expected to be sensitive to changes in temperature associated with climate change (Bale et al. 2002). Both vegetation and temperature have the potential to change in response to climate and create a mismatch in the cue–resource relationship. A key distinction, however, is that migrants are likely responding to vegetation as a cue, whereas the invertebrate community is likely responding to water temperature to assess optimal emergence conditions.

We deployed temperature loggers (HOBO pendant loggers, Onset Instruments) in the water column using a weight and buoy system that ensured that they remain at a consistent depth (2010: n = 21; 2011: n = 51). Data loggers recorded water temperature every hour and were deployed before migration began (late April) and retrieved after northward migration through the region was completed (late June). We performed all analyses using the minimum mean daily temperature as an indicator of biophenological progression.

Invertebrate Phenology

In 2011, we measured benthic invertebrate availability at 26 wetlands. We resampled each wetland up to three times every 10–14 days or until dry, resulting in 70 wetland sampling visits. Three soil cores were taken within a 3 × 3 m sample plot to a depth of 5 cm using a 5-cm-diameter corer (Sherfy et al. 2000). We selected three to five plot locations at randomly selected compass bearings from the wetland's center point for a total of 9–15 soil cores per wetland per sampling session. In all cases, we ensured that plots were separated by >10 m. We then washed core samples through a 0.5-mm soil sieve and enumerated invertebrates at the wetland to establish relative abundance. We did not classify benthic invertebrates because *Calidris* sandpipers exhibit high dietary plasticity across invertebrates (Skagen and Oman 1996), allowing individuals to feed opportunistically as they move across latitudes and encounter different communities. Invertebrate biomass may be a better index of total caloric availability, but invertebrate abundance is generally correlated with biomass and responds similarly to changing conditions (Whiles and Goldowitz 2005, Hamer et al. 2006). We predicted that shorebirds are more likely to forage on larger prey items, which would create a scenario where a lower abundance of individual prey items would result in disproportionately lower biomass available for foraging.

Using data collected in 2011, we compared local conditions with invertebrate abundance using a GLM that included date as a covariate. The importance of food availability to migrants is well documented and invertebrate populations are sensitive to changes in temperature (Wiggins et al. 1980, Bale et al. 2002), creating a potential for the primary resource of migrant shorebirds to undergo relatively rapid changes in phenology that may result in a disparity of the cue–resource relationship. We compared estimates of invertebrate abundance to water temperature values from the wetland data loggers and with estimates of dissolved chlorophyll a from wetlands where invertebrates were sampled. We hypothesized that these parameters would influence invertebrate abundance and might be even more sensitive as indicators of changing phenology. Before sampling invertebrates at each wetland, we measured the dissolved chlorophyll a fluorescence using an in vivo probe (Aquaflor handheld fluorometer, Turner Designs). Water samples were placed in the probe whereby a relative chlorophyll a reading is returned. Chlorophyll a is an indicator of phytoplankton growth and is a sensitive index of overall wetland productivity (Desortova 1981, Canfield et al. 1984).

RESULTS

Daily abundance of shorebirds was correlated with local conditions in both years. In 2010, migration phenology was significantly correlated with both green vegetation and water temperature (green vegetation: $F_{1,36} = 378.4$, $P < 0.001$; water temperature: $F_{1,36} = 13.3$, $P < 0.001$; date: $F_{1,36} = 104.4$, $P < 0.001$). Data from 2011 produced a similar pattern as all three phenological variables were significant (green vegetation: $F_{1,26} = 523.4$, $P < 0.001$; water temperature: $F_{1,26} = 62.7$, $P < 0.001$; invertebrate abundance: $F_{1,26} = 99.4$, $P < 0.001$; date: $F_{1,26} = 4.5$, $P = 0.035$). Across years, green vegetation and date were significantly correlated with daily migrant abundance (green vegetation: $F_{1,69} = 1743.4$, $P < 0.001$; water temperature:

$F_{1,69} = 14.1$, $P = 0.294$; date: $F_{1,69} = 51.7$, $P < 0.001$; year: $F_{1,69} = 0.7$, $P = 0.41$). Invertebrate abundance was not significantly correlated with either water temperature or dissolved chlorophyll a but was significantly correlated with date (water temperature: $F_{1,30} = 0.1$, $P = 0.708$; chlorophyll a: $F_{1,30} = 0.2$, $P = 0.644$; date: $F_{1,30} = 5.2$, $P = 0.028$).

In both years, migrant daily abundance was positively correlated with water temperature early in the season before peaking and eventually became negatively correlated (Figure 9.3a). The relationship between green vegetation and daily migrant abundance showed an initial positive correlation in both years, before the peak of migration fell off (Figure 9.3b). Last, the relationship between bird migration and food availability showed that peak shorebird migration occurred immediately prior to peak resource availability (Figure 9.3c).

DISCUSSION

We provide a preliminary examination of the relationships between migratory shorebird phenology and local phenological factors, and we examine a number of scenarios and how they may affect shorebird populations. The potential consequences of climate change and the resulting changes in phenology to migratory shorebirds remain unclear. Migratory shorebirds use widely distributed habitats and the nature of migration requires individuals to make habitat decisions repeatedly in novel environments under temporal constraints (Moore et al. 1990, Moore and Aborn 2000, Petit 2000). Given the nature of the shorebird migratory strategies and their reliance on specialized habitats in midcontinental flyways, shorebirds may be particularly vulnerable to the effects of climate change. Alternatively, because migrants encounter a wide range of habitats and climatic conditions, they may be well suited to adapt to changing conditions. For example, it is well known that shorebirds use agricultural fields (Elphick and Oring 1998, Niemuth et al. 2006, Taft and Haig 2005) and may even prefer these habitats despite lower resource availability (but see Taft and Haig 2005). Thus, even under current conditions (Figures 9.1a and 9.2c), migrants still face the potentially negative effects of using a habitat type with lower food availability. However, migrants have seemingly adapted to a new suite of conditions by compensating for the limited

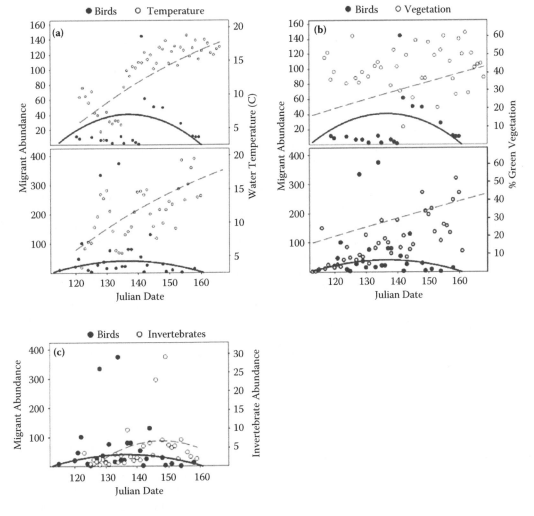

Figure 9.3. (a) In 2010 (top panel) and 2011 (bottom panel), total daily bird abundance increased with water temperature early in the season before declining. (b) Total daily abundance was positively correlated with green vegetation in both 2010 and 2011 until daily abundances peaked. (c) Total daily migrant abundance is correlated with invertebrate abundance (mean number per sample at each wetland) and peaked just prior to invertebrate peak, consistent with our predictions and suggesting a linkage between migration and invertebrate phenology.

food resources available in these habitats through behavioral modifications that optimize trade-offs with predation risk.

We considered the potential risks of changing phenology by examining the relationship between migration phenology and local phenological factors. Predictably, bird abundance increased throughout the early migration period before declining at the end of May. All three local phenological variables we examined were significantly correlated with shorebird abundance, but the relationship between abundance and green vegetation was inconsistent (Figure 9.3b). Bird migration apparently coincides with increasing water temperature through the migration period and was significantly correlated with invertebrate phenology. In both cases, the pattern follows predictions if migrants are to optimize foraging opportunities, with the peak in migration occurring immediately prior to the peak in food resources (Figure 9.3c). By arriving slightly before the peak in resources (macroinvertebrates) or the abiotic factors driving resource phenology (temperature), migrating shorebirds ensure access to adequate food resources if stopover duration is extended. Early arrival is the most important phenological pattern because it indicates that bird migration is closely linked to timing of limiting resources such as food (McGrath et al. 2009). We did not find any significant correlation between invertebrate abundance and the local phenological conditions that we measured. The relationship was slightly positive for chlorophyll *a* and slightly negative for temperature, which is inconsistent with what theory would predict, but may be a consequence of our small sample sizes. Given that landscape-level changes driven by agriculture have already caused shorebirds to prefer habitat with lower resource availability (J. J. Fontaine, unpubl. data), any changes to either invertebrate or migration phenology that is not congruent in both magnitude and direction could have severe impacts on migrant populations. Further negative impacts are possible because climate change is heterogeneous and migrants respond at different phenological scales than the resources on which they depend (Both and Visser 2001, Both et al. 2006, but see Marra et al. 2005).

Climate change is expected to be spatially and temporally heterogeneous and has been shown as such in the context of North American migratory bird species (IPCC 2007, Fontaine et al. 2009). Some species have shifted their phenology to match changing climatic conditions, but patterns are inconsistent across taxa (Root et al. 2003). We developed a number of predictions that represent possible scenarios of how invertebrates and migrants might respond to changing climatic conditions. While these are certainly simplified scenarios in the scope of global climate change, they explore a wide range of circumstances under which shorebird migration is likely to occur in the future. Our data show that migrant abundance peaks immediately prior to the peak in food availability (Figure 9.3c), a result that is consistent with the predicted relationship of scenario 1 (Figure 9.1a). We cannot reject the possibility that both migrant and invertebrate phenology has advanced with climate change as predicted under scenario 2 (Figure 9.1b). However, as the effects of climate change increase, it is possible that the phenological relationship between migrants and their food resources will be subject to further changes that could result in patterns shown by our scenarios and ultimately affect populations. Warming is known to lead to earlier migrant arrival in some species, suggesting that migratory species are flexible in their phenology (Crick et al. 1997, Hüppop and Hüppop 2003, Jenni and Kéry 2003, Lehikoinen et al. 2004, Stervander et al.

2005, Jonzén et al. 2006). It is unclear if migrants will be able to adapt to changes in resource phenology at various locations along their migratory route, especially when that phenology does not change at the same amplitude in all locations.

Successful stopover depends on many environmental and behavioral factors and is driven by a multitude of selection pressures (Petit 2000, Newton 2006). Given that the primary reason for stopover is the acquisition of energy, adequate access to food resources is critical (Hutto 1985, Moore et al. 1995). Due to a collection of human impacts, migratory birds are often required to migrate through highly altered landscapes that may have reduced resource availability (Niemuth et al. 2006), and loss of stopover habitat is predicted to result in declines of migratory species (Skagen 1997, Weber et al. 1999, Harrington et al. 2002). However, some migrants, including arctic-nesting shorebirds, have shown the ability to adapt to alterations to stopover habitats (Krapu et al. 1984, Taft and Haig 2005). One possible reason for the persistence of migratory species despite habitat alteration is a strong phenological link between resources and migration. However, migrants may be less able to buffer against the consequences of using novel habitats if migration and resource phenology are no longer congruent due to climate change (Visser et al. 2004, Both et al. 2006, Both 2010, Jones and Cresswell 2010). Given that spring temperature changes may be more extreme at stopover locations than at breeding sites at the times when birds are using them (Fontaine et al. 2009), migrants may encounter novel trade-offs in resource availability en route. For example, migrants may advance migration to track similarly advancing invertebrate food resources at stopover sites (Figure 9.3b). However, birds may then be more likely to encounter adverse weather events en route or to reach the breeding grounds before adequate food resources are available (Alerstam 1991, Decker and Conway 2009). This scenario would seem unlikely if invertebrates are more sensitive to local conditions and if impacts of climate change are heterogeneous. However, it is the best-case scenario for conservation planners and, in that sense, is worth documenting.

Earlier peaks in migration may also allow individuals to extend stopover beyond the historical norm to take advantage of invertebrate peaks, given no change to food resource phenology (Figure 9.1c). Although the peaks in migration and food availability become decoupled under this scenario, the fact that migrants still arrive prior to the peak in food resources may allow individuals to counteract the negative effects by changing stopover dynamics. Migratory species are highly adaptable in regard to stopover duration (reviewed by Newton 2006) and are known to stay longer when food resources are lower (Piersma 1987, Ydenberg et al. 2002) or when individuals have inadequate body reserves (Moore and Kerlinger 1987, Kuenzi et al. 1991). However, lean birds are unlikely to stay at stopovers with inadequate food reserves, prompting potentially risky flights that may result in mortality (Newton 2006). Extending stopover duration is not without costs, including increased risk of predation (Ydenberg et al. 2004). Given that stopover initiation is earlier in this scenario, individuals may not incur the potential costs of late arrival to the breeding grounds (Potti 1998, Currie et al. 2000, Weggler 2006). However, if resource phenology shifts to earlier in the year independently of migration phenology (Both et al. 2006; Figure 9.1d of this study), then migrants may not be able to obtain adequate energy reserves and population viability will be threatened. Such a conclusion is supported by the fact that migratory species that advance their arrival date are less likely to decline than those that do not (Møller et al. 2008).

While it is possible that climate change will negatively affect the integration of migration phenology and resource availability, resulting in population declines, the extent of such effects is unknown relative to more long-standing stressors such as habitat loss (Opdam and Wascher 2004). The likely scenario is that the effects of multiple stressors will interact and compound one another (Robinson et al. 2009). For example, climate change is predicted to alter precipitation and evapotranspiration rates and that is expected to alter wetland habitat in the midcontinental region (Johnson et al. 2005, IPCC 2007). However, such a change will also have implications on which crops are planted in the region and how they are cultivated, potentially leading to further land-use changes.

Furthermore, the push for alternative energy sources, such as corn-based ethanol, may motivate land owners to alter farming practices. The additive influence of continued land-use change and changing climatic conditions will obviously alter the wetland habitat upon which migrating shorebirds rely (Euliss and Mushet 1999, Gleason et al. 2003, Johnson et al. 2005) and, in doing so, further to affect shorebird migration in the region. Although migrants in general (Krapu et al. 1984, Stervander et al. 2005, Jonzén et al. 2006) and shorebirds in particular (Taft and Haig 2005) may be particularly adaptable to changing ecological conditions, it is unknown if they will be afforded the evolutionary time needed to adapt to such a suite of negative impacts. Further research is needed to examine the potential for land-use changes and changing climate conditions to act in concert to drive migrant species declines.

ACKNOWLEDGMENTS

We are grateful to field and lab technicians G. Rozhon, J. Gehant, C. Welch, J. Walker, P. Nepp, and V. Simonsen. We thank S. Skagen, D. Riveros-Iregui, J. Brandle, N. Alhadeff, and K. Decker for comments on previous drafts and Wren, Barley, and Caddis for their considerable support. Funding for this project was received from the US Geological Survey National Climate and Wildlife Science Center. Any use of trade, firm, or product names is for descriptive purposes only and does not imply endorsement by the US government. The Nebraska Cooperative Fish and Wildlife Research Unit is supported by a cooperative agreement among the US Geological Survey, the Nebraska Game and Parks Commission, the University of Nebraska, the US Fish and Wildlife Service, and the Wildlife Management Institute.

LITERATURE CITED

Ahola, M., T. Laaksonen, K. Sippola, T. Eeva, and K. Rainio. 2004. Variation in climate warming along the migration route uncouples arrival and breeding dates. Global Change Biology 10:1610–1617.

Alerstam, T. 1991. Bird migration, Cambridge University Press, Cambridge, UK.

Baker, A. J., P. M. Gonzalez, T. Piersma, L. J. Niles, I. de Lima Serrano do Nascimento, P. W. Atkinson, N. A. Clark, C. D. T. Minton, M. K. Peck, and G. Aarts. 2004. Rapid population decline in Red Knots: fitness consequences of decreased refueling rates and late arrival in Delaware Bay. Proceedings of the Royal Society of London B 271:875–882.

Bale, J. S., G. J. Masters, I. D. Hodkinson, C. Awmack, T. M. Bezemer, V. K. Brown, J. Butterfield, A. Buse, J. C. Coulson, J. Farrar, J. E. G. Good, R. Harrington,

S. Hartley, T. H. Jones, R. L. Lindroth, M. C. Press, I. Symrnioudis, A. D. Watt, and J. B. Whittaker. 2002. Herbivory in global climate change research: direct effects of rising temperature on insect herbivores. Global Change Biology 8:1–16.

Battin, J. 2004. When good animals love bad habitats: ecological traps and the conservation of animal populations. Conservation Biology 18:1482–1491.

Both, C. 2010. Flexibility of timing of avian migration to climate change masked by environmental factors. Current Biology 20:243–248.

Both, C., S. Bouwhuis, C. M. Lessells, and M. E. Visser. 2006. Climate change and population declines in a long-distance migratory bird. Nature 441:81–83.

Both, C., C. A. M. Van Turnhout, R. G. Bijlsma, H. Siepel, A. J. Van Strien, and R. P. B. Foppen. 2010. Avian population consequences of climate change are most severe for long-distance migrants in seasonal habitats. Proceedings of the Royal Society of London B 277:1259–1266.

Both, C., and M. E. Visser. 2001. Adjustment to climate change is constrained by arrival date in a long-distance migrant bird. Nature 411:296–298.

Botkin, D. B., H. Saxe, M. B. Araûji, R. Betts, R. H. W. Bradshaw, T. Cedhagen, P. Chesson, T. P. Dawson, J. R. Etterson, D. P. Faith, S. Ferrier, A. Guisan, A. S. Hansen, D. W. Hilbert, C. Loehle, C. Margules, M. New, M. J. Sobel, and D. R. B. Stockwell. 2007. Forecasting the effects of global warming on biodiversity. BioScience 57:227–236.

Canfield, D. E., J. V. Shireman, D. E. Colle, W. T. Haller, C. E. Watkins, and M. J. Maceina. 1984. Prediction of chlorophyll-a concentrations in Florida lakes: Importance of aquatic macrophytes. Canadian Journal of Fisheries and Aquatic Science 44:497–501.

Corbet, P. S. 1964. Temporal patterns of emergence in aquatic insects. Canadian Entomologist 96:264–279.

Cresswell, W., and R. McCleery. 2003. How Great Tits maintain synchronization of their hatch date with food supply in response to long-term variability in temperature. Journal of Animal Ecology 72:356–366.

Crick, H. Q. P., C. Dudley, D. E. Glue, and D. L. Thomson. 1997. UK birds are laying eggs earlier. Nature 388:526–527.

Currie, D., D. B. A. Thompson, and T. Burke. 2000. Patterns of territory settlement and consequences for breeding success in the Northern Wheatear Oenanthe oenanthe. Ibis 142:389–398.

Davis, C. A., and L. M. Smith. 1998. Behavior of migrant shorebirds in playas of the southern High Plains, Texas. Condor 100:266–276.

Davis, C. A., and L. M. Smith. 2001. Foraging strategies and niche dynamics of coexisting shorebirds at stopover sites in the southern Great Plains. Auk 118:484–491.

Decker, K. L., and C. J. Conway. 2009. Effects of an unseasonal snowstorm on Red-faced Warbler nesting success. Condor 111:392–395.

Desortova, B. 1981. Relationship between chlorophyll-a concentration and phytoplankton biomass in several reservoirs in Czechoslovakia. Internationale Revue der Gesamten Hydrobiologie 66:153–122.

Drent, R., C. Both, M. Green, J. Madsen, and T. Piersma. 2003. Pay-offs and penalties of competing migratory schedules. Oikos 103:274–292.

Elphick, C. S., and L. W. Oring 1998. Winter management of Californian rice fields for waterbirds. Journal of Applied Ecology 35:95–108.

Euliss, N. H., Jr., and D. M. Mushet. 1999. Influence of agriculture on aquatic invertebrate communities of temporary wetlands in the Prairie Pothole Region of North Dakota, USA. Wetlands 19:578–583.

Fontaine, J. J., K. L. Decker, S. K. Skagen, and C. van Riper. 2009. Spatial and temporal variation in climate change: A bird's eye view. Climatic Change 97:305–311.

Gleason, R. A., N. H. Euliss, Jr., D. E. Hubbard, and W. G. Duffy. 2003. Effects of sediment load on emergence of aquatic invertebrates and plants from wetland soil egg and seed banks. Wetlands 23:26–34.

Gordo, O. 2007. Why are bird-migration dates shifting? A review of weather and climate effects on avian migratory phenology. Climate Research 35:37–58.

Gordo, O., L. Brotons, X. Ferrer, and P. Comas. 2005. Do changes in climate patterns in wintering areas affect the timing of spring arrival of trans-Saharan migrant birds? Global Change Biology 11:12–21.

Gwinner, E. 1996. Circadian and circannual programmes in avian migration. Journal of Experimental Biology 199:39–48.

Hamer, G. L., E. J. Heske, J. D. Brawn, and P. W. Brown. 2006. Migrant shorebird predation on benthic invertebrates along the Illinois River, Illinois. Wilson Journal of Ornithology 118:152–163.

Harrington, B. A., S. C. Brown, J. Corven, and J. Bart. 2002. Collaborative approaches to the evolution of migration and the development of science-based conservation in shorebirds. Auk 119:914–921.

Hogg, W. D., and D. D. Williams. 1996. Response of stream invertebrates to a global-warming thermal regime: an ecosystem-level manipulation. Ecology 77:395–407.

Hüppop, O., and K. Hüppop. 2003. North Atlantic oscillation and timing of spring migration in birds. Proceedings of the Royal Society of London B 270:233–240.

Hutto, R. L. 1985. Habitat selection by nonbreeding, migratory land birds. Pp. 455–476 in M. L. Cody (editor), Habitat selection in birds. Academic Press, New York, NY.

Inouye, D. W., B. Barr, K. B. Armitage, and B. D. Inouye. 2000. Climate change is affecting altitudinal migrants and hibernating species. Proceeding of the National Academy of Sciences of the USA 97:1630–1633.

IPCC. 2007. Climate change 2007: the physical science basis. Cambridge University Press, Cambridge, UK.

Jenni, L., and M. Kéry. 2003. Timing of autumn bird migration under climate change: advances in long-distance migrants, delays in short-distance migrants. Proceedings of the Royal Society of London B 270:1467–1471.

Johnson, W. C., B. V. Millett, T. Gilmanov, R. A. Voldseth, G. R. Guntenspergen, and D. E. Naugle. 2005. Vulnerability of northern prairie wetlands to climate change. BioScience 55:863–872.

Jones, T., and W. Cresswell. 2010. The phenology mismatch hypothesis: are declines of migrant birds linked to uneven global climate change? Journal of Animal Ecology 79:98–108.

Jonzén, N., A. Linden, T. Ergon, E. Knudsen, J. O. Vik, D. Rubolini, D. Piacentini, C. Brinch, F. Spina, L. Karlsson, M. Stervander, A. Andersson, J. Waldenstrom, A. Lehikoinen, E. Edvardsen, R. Solvang, and N. C. Stenseth. 2006. Rapid advance of spring arrival dates in long distance migratory birds. Science 312:1959–1961.

Kelly, J. F., V. Atudorei, Z. D. Sharp, and D. M. Finch. 2002. Insights into Wilson's Warbler migration from analyses of hydrogen stable-isotope ratios. Oecologia 130:216–221.

Krapu, G. L., D. E. Facey, E. K. Fritzell, and D. H. Johnson. 1984. Habitat use by migrant Sandhill Cranes in Nebraska. Journal of Wildlife Management 48:407–417.

Kuenzi, A. J., F. R. Moore, and T. R. Simons. 1991. Stopover of Neotropical landbird migrants on East Ship Island following trans-Gulf migration. Condor 93:869–883.

Lehikoinen, E., T. H. Sparks, and M. Zalakevicius. 2004. Arrival and departure dates. Pp. 1–31 in A. P. Møller, W. Fiedler, and P. Berthold (editors), Birds and climate change (vol 35). Elsevier, Amsterdam, Netherlands.

Loria, D. E., and F. R. Moore. 1990. Energy demands of migration on Red-eyed Vireos (Vireo olivaceus). Behavioral Ecology 1:24–35.

Marra, P. P., C. M. Francis, R. S. Mulvihill, and F. R. Moore. 2005. The influence of climate on the timing and rate of spring bird migration. Oecologia 142:307–315.

Marra, P. P., K. A. Hobson, and R. T. Holmes. 1998. Linking winter and summer events in a migratory bird by using stable-carbon isotopes. Science 282:1884–1886.

McGrath, L. J., C. van Riper, and J. J. Fontaine. 2009. Flower power: tree flowering phenology as a settlement cue for migrating birds. Journal of Animal Ecology 78:22–30.

Miller-Rushing, A. J., T. T. Høye, D. W. Inouye, and E. Post. 2010. The effects of phenological mismatches on demography. Philosophical Transactions of the Royal Society B 365: 3177–3186.

Møller, A. P., D. L. Rubolini, and E. Lehikoinen. 2008. Populations of migratory bird species that did not show a phenological response to climate change are declining. Proceedings of the National Academy of Science of the USA 105:16195–16200.

Moore, F. R., Smith, R. J., and R. Sandberg. 2005. Stopover ecology of intercontinental migrants. Pp. 251–261 in R. Greenberg and P. P. Marra (editors). Birds of two worlds: the ecology and evolution of migration. Johns Hopkins University Press, Baltimore, MD.

Moore, F. R., and D. A. Aborn. 2000. Mechanisms of en route habitat selection: how do migrants make habitat decisions during stopover? Studies in Avian Biology 20:34–42.

Moore, F. R., S. A. Gauthreaux, Jr., P. Kerlinger, and T. R. Simons, 1995. Habitat requirements during migration: important link in conservation. Pp. 121–144 in T. E. Martin and D. M. Finch (editors), Ecology and management of Neotropical migratory birds. Oxford University Press, New York, NY.

Moore, F. R., and P. Kerlinger. 1987. Stopover and fat deposition by North American wood-warblers (Parulinae) following spring migration over the Gulf of Mexico. Oecologia 74:47–54.

Moore, F. R., P. Kerlinger, and T. R. Simons. 1990. Stopover on Gulf coast barrier island by spring trans-Gulf migrants. Wilson Bulletin 102:487–500.

Newton, W. E. 2006. Can conditions experienced during migration limit the population levels of birds? Journal of Ornithology 147:146–166.

Niemuth, N. D., M. E. Estey, R. E. Reynolds, C. R. Loesch, and W. A. Meeks. 2006. Use of wetlands by spring-migrant shorebirds in agricultural landscapes of North Dakota's drift prairie. Wetlands 26:30–39.

Opdam, P., and D. Wascher. 2004. Climate change meets habitat fragmentation: linking landscape and biogeographical scale levels in research and conservation. Biological Conservation 117:285–297.

Ottick, I., and V. Dierschke. 2003. Exploitation of resources modulates stopover behaviour of passerine migrants. Journal für Ornithologie 144:307–316.

Parmesan, C., and G. Yohe. 2003. A globally coherent fingerprint of climate change impacts across natural systems. Nature 421:37–42.

Petersen, M. R. 2009. Multiple spring migration strategies in a population of Pacific Common Eiders. Condor 111:59–70.

Petit, D. R. 2000. Habitat use by landbirds along Nearctic–Neotropical migration routes: implications for conservation of stopover habitats. Studies in Avian Biology 20:15–33.

Pfister, C., M. J. Kasprzyk, and B. A. Harrington. 1998. Body-fat levels and annual return in migrating Semipalmated Sandpipers. Auk 115: 904–915.

Piersma, T. 1987. Hop, skip or jump? Constraints in migration of arctic waders by feeding, fattening and flight speed. Limosa 60:185–194.

Potti, J. 1998. Arrival time from spring migration in male Pied Flycatchers: individual consistency and familial resemblance. Condor 100:702–708.

Robertson, B. A., and R. L. Hutto. 2006. A framework for understanding ecological traps and an evaluation of existing evidence. Ecology 87:1075–1085.

Robinson, A., H. Q. P. Crick, J. A. Learmonth, I. M. D. Maclean, C. D. Thomas, F. Bairlein, M. C. Forchhammer, C. M. Francis, J. A. Gill, B. J. Godley, J. Harwood, G. C. Hays, B. Huntley, A. M. Hutson, G. J. Pierce, M. M. Rehfisch, D. W. Sims, M. C. Vieira dos Santos, T. H. Sparks, D. Stroud, and M. E. Visser. 2009. Travelling through a warming world: climate change and migratory species. Endangered Species Research 7:87–99.

Root, T. L., J. T. Price, K. R. Hall, S. H. Schneider, C. Rosenzweig, and J. A. Pounds. 2003. Fingerprints of global warming on wild animals and plants. Nature 421:57–60.

Rosenzweig, C., D. Karoly, M. Vicarelli, P. Neofotis, Q. Wu, G. Casassa, A. Menzel, T. L. Root, N. Estrella, B. Seguin, P. Tryjanowski, C. Liu, S. Rawlins, and A. Imeson. 2008. Attributing physical and biological impacts to anthropogenic climate change. Nature 453:353–357.

Russell, R. W., F. L. Carpenter, M. A. Hixon, and D. C. Paton. 1992. The impact of variation in stopover habitat quality on migrant Rufous Hummingbirds. Conservation Biology 8:483–490.

Sala, O. E., F. S. Chapin III, J. J. Armesto, E. Berlow, J. Bloomfield, R. Dirzo, E. Huber-Sawald, L. F. Huenneke, R. B. Jackson, A. Kinzig, R. Leemans, D. M. Lodge, H. A. Mooney, M. Oesterheld, N. L. Poff, M. T. Sykes, B. H. Walker, M. Walker, and D. H. Wall. 2000. Global biodiversity scenarios for the year 2100. Science 287:1770–1774.

Sanderson, F. J., P. F. Donald, D. J. Pain, I. J. Burfield, and F. P. J. van Bommel. 2006. Long-term declines in Afro-Palearctic migrant birds. Biological Conservation 131:93–105.

Sherfy, M. H., R. L. Kirkpatrick, and K. D. Richkus. 2000. Benthos core sampling and chironomid vertical distribution: implications for assessing shorebird food availability. Wildlife Society Bulletin 28:124–130.

Sherry, T. W., and R. T. Holmes. 1996. Winter habitat quality, population limitation, and conservation of Neotropical–Nearctic migrant birds. Ecology 77:36–48.

Skagen, S. K. 2006. Migration stopovers and the conservation of arctic-breeding calidridine sandpipers. Auk 123:313–322.

Skagen, S. K., D. A. Granfors, and C. P. Melcher. 2008. Determining the significance of ephemeral continental wetlands to North American migratory shorebirds. Auk 125:20–29.

Skagen, S. K. 1997. Stopover ecology of transitory populations: the case of migrant shorebirds. Ecological Studies 125:244–269.

Skagen, S. K., and F. L. Knopf. 1994. Migrating shorebirds and habitat dynamics at a prairie wetland complex. Wilson Bulletin 106:91–105.

Skagen, S. K., and H. D. Oman. 1996. Dietary flexibility of shorebirds in the Western Hemisphere. Canadian Field-Naturalist 110:419–444.

Stervander, M., K. Lindström, N. Jonzén, and A. Andersson. 2005. Timing of spring migration in birds: long-term trends, North Atlantic oscillation and the significance of different migration routes. Journal of Avian Biology 36:210–221.

Stewart, R. E., and H. A. Kantrud. 1971. Classification of natural ponds and lakes in the glaciated prairie region. Resource publication 92, Bureau of Sport Fisheries and Wildlife, US Fish and Wildlife Service, Washington, DC.

Strode, P. K. 2003. Implications of climate change for North American wood warblers (*Parulidae*). Global Change Biology 9:1137–1144.

Studds, C. E., and P. P. Marra. 2011. Nonbreeding habitat occupancy and population processes: an upgrade experiment with a migratory bird. Ecology 86:2380–2385.

Swanson, D. L., and J. S. Palmer. 2009. Spring migration phenology of birds in the Northern Prairie region is correlated with local climate change. Journal of Field Ornithology 80:351–363.

Taft, T. W., and S. M. Haig. 2005. The value of agricultural wetlands as invertebrate resources for wintering shorebirds. Agriculture, Ecosystem and Environment 110:249–256.

Thomas, C. D., A. Cameron, R. E. Green, M. Bakkenes, L. J. Beaumont, Y. C. Collingham, B. F. N. Erasmus, M. Ferreira de Siqueira, A. Grainger, L. Hannah, L. Hughes, B. Huntley, A. S. van Jaarsveld, G. F. Midgley, L. Miles, M. A. Ortega-Huerta, A. T. Peterson, O. L. Phillips, and S. E. Williams. 2004. Extinction risk from climate change. Nature 427:145–148.

Tøttrup, A. P., K. Thorup, and C. Rahbek. 2006. Patterns of change in timing of spring migration in North European songbird populations. Journal of Avian Biology 37:84–92.

van Eerden, M. R., M. Zylstra, and M. J. J. E. Loonen. 1991. Individual patterns of staging during autumn migration in relation to body composition in Greylag Geese *Anser anser* in the Netherlands. Ardea 79:261–264.

van Gils, J. A., A. Dekinga, B. Spaans, W. K. Vahl, and T. Piersma. 2005. Digestive bottleneck affects foraging decisions in Red Knots *Calidris canutus*. II. Patch choice and length of working day. Journal of Animal Ecology 74:120–130.

Visser, M. E., C. Both, and M. M. Lambrechts. 2004. Global climate change leads to mistimed avian reproduction. Pp. 89–110 in A. P. Møller, W. Fiedler, and P. Berthold (editors), Birds and climate change (vol 35), Elsevier, Amsterdam, Netherlands.

Walther, G. R., E. Post, P. Convey, A. Menzel, C. Parmesan, T. J. C. Beebee, J. Fromentin, O. Hoegh-Guldberg, and F. Bairlein. 2002. Ecological responses to recent climate change. Nature 416:389–395.

Weber, T. P., A. I. Houston, and B. J. Ens. 1999. Consequences of habitat loss at migratory stopover sites: a theoretical investigation. Journal of Avian Biology 30:416–426.

Weggler, M. 2006. Constraints on, and determinants of, the annual number of breeding attempts in the multi-brooded Black Redstart *Phoenicurus ochruros*. Ibis 148:273–284.

Weidinger, K., and M. Král. 2007. Climatic effects on arrival and laying dates in a long-distance migrant, the Collared Flycatcher *Ficedula albicollis*. Ibis 149:836–847.

Whiles, M. R., and B. S. Goldowitz. 2005. Macroinvertebrate communities in central Platte River wetlands: patterns across a hydrologic gradient. Wetlands 25:462–472.

Wiggins, G. B., R. J. Mackay, and I. M. Smith. 1980. Evolutionary and ecological strategies of animals in annual temporary pools. Archiv für Hydrobiologie Suppl. 58:97–206.

Wilson, W. H. 2007. Spring arrival dates of migratory breeding birds in Maine: sensitivity to climate change. Wilson Journal of Ornithology 119:665–677.

Ydenberg, R. C., R. W. Butler, D. B. Lank, C. G. Guglielmo, M. Lemon, and N. Wolf. 2002. Trade-offs, condition dependence and stopover site selection by migrating sandpipers. Journal of Avian Biology 33:47–55.

Ydenberg, R. C., R. W. Butler, D. B. Lank, B. D. Smith, R. C. Ydenberg, and J. Ireland. 2004. Western Sandpipers have altered migration tactics as Peregrine Falcon populations have recovered. Proceedings of the Royal Society of London B 271:1263–1269.

Yong, W., and F. R. Moore. 1997. Spring stopover of intercontinental migratory thrushes along the northern coast of the Gulf of Mexico. Auk 114:263–278.

Fall Migration

Matching Ephemeral Resources on Autumnal Stopover and the Potential for Mismatch[*]

Brian J. Olsen, Jennifer D. McCabe, Evan M. Adams, David P. Grunzel, and Adrienne J. Leppold

Abstract. Climate change can alter the timing of biological events differently across members of an ecosystem, changing the nature of ecological interactions and the efficacy of evolutionary strategies. Mismatches in phenology have been reported between breeding songbirds and their food resources in north temperate ecosystems, especially for long-distance migrants and dietary specialists. Far less attention, however, has been paid to periods outside the breeding season. The success of songbird migration, like breeding, should also be determined by the degree of matching between bird and resource phenologies. Here, we tested if habitat patch use for songbirds along the migratory route was predicted by (a) habitat characteristics that varied throughout the migration season, (b) static characteristics, or (c) a combination of the two variable types. We tested for these patterns in four families of songbirds that ranged from specialists to generalists and across four stopover sites along the Gulf of Maine. Dynamic habitat characteristics alone predicted patch use in all four families better than static variables or a combination of dynamic and static characteristics. Regardless of mean dietary niche breadth, mist net captures indicated that all four families shifted their apparent patch use throughout the season as a function of ripe fruit abundance, invertebrate biomass, vegetation defoliation, and soil moisture. The best predictors of patch use were different for short- versus long-distance migrants across the two most commonly captured families, but all species and migration strategies were sensitive to phenological changes. Our results suggest that migratory songbirds assess patch quality within a stopover site differently through time. Variation in the timing of stopover arrival thus results in different habitat selection behaviors. The potential therefore exists for timing mismatches that reduce individual fitness. We recommend that conservationists prioritize stopover habitats where landscape-scale phenological heterogeneity maintains benefits to migrants throughout the season.

Key Words: flycatchers, habitat use, phenological mismatch, sparrows, stopover behavior, vireos, warblers.

[*] Olsen, B. J., J. D. McCabe, E. M. Adams, D. P. Grunzel, and A. J. Leppold. 2015. Matching ephemeral resources on autumnal stopover and the potential for mismatch. Pp. 163–176 in E. M. Wood and J. L. Kellermann (editors), Phenological synchrony and bird migration: changing climate and seasonal resources in North America. Studies in Avian Biology (no. 47), CRC Press, Boca Raton, FL.

Temperate ecosystems are dominated by seasonal ephemera. The evolutionary success of boreal migratory birds stems in part from their ability to benefit from high-latitude summer abundance while avoiding winter scarcity. Success of migratory birds is dependent upon a match between the timing of bird behavior and the local phenology of ecological resources (Lack 1950). The fitness costs of mismatches are readily apparent on the breeding grounds of many north temperate species (Møller 1994, Sanz et al. 2003). Less attention has been paid to the potential and importance of phenological matching during migratory movements and stopover dynamics. Given the substantial energetic requirements (Berthold 1975, Blem 1980) and mortality risk experienced by songbirds during the migratory period (Lack 1968, Alerstam and Hedenström 1998, Sillett and Holmes 2002), it is reasonable to hypothesize that phenological mismatch during migratory stopover is as important as ecological conditions encountered on the breeding grounds. Here, we test for the ability of ephemeral versus static ecological resources to predict bird habitat use within four autumnal migratory stopover sites along the US coast of the Gulf of Maine.

THE CASE FOR PHENOLOGICAL MISMATCH DURING STOPOVER

On the breeding grounds, ecosystem phenology is a strong determinant of fitness for migratory birds. The timing of migratory arrival is predicted by the average timing of resources required for territory establishment and reproduction (Both 2010a). Migratory behaviors worldwide show great flexibility in timing (Sokolov et al. 1998, Marra et al. 2005, Rubolini et al. 2007), likely due to the gain in fitness achieved by matching interannual variability in local phenology. When the temporal advance or delay of breeding resources outpaces the behavioral flexibility of bird species ("mismatch": Both and Visser 2001; Both 2010a, b), birds can suffer both in terms of offspring productivity (Møller 1994, Verboven et al. 2001, Sanz et al. 2003, Both and Visser 2005, Drent 2006) or long-term species viability (McLaughlin et al. 2002, Both et al. 2006, Møller et al. 2008). On the European breeding grounds, dietary specialists—and especially insectivore specialists—appear more sensitive to phenological mismatch than dietary generalists

and long-distance migrants appear more sensitive than short-distance migrants (Both and Visser 2001, Moussus et al. 2011).

While mismatch with resources at migratory stopover sites has received less attention, the same ecological scenario may apply. As with the breeding grounds, arrival of individuals to stopover locations can be predicted by local environmental conditions (Hüppop and Hüppop 2003, Marra et al. 2005) and arrival is broadly timed to the availability of local food resources (Castro and Myers 1993, Tsipoura and Burger 1999, Lank et al. 2003, McKinney et al. 2012). Temporal mismatch can increase fitness costs by negatively impacting refueling rate or increasing stopover duration (Baker et al. 2004, Balbontín et al. 2009). For songbirds specifically, migration likely represents a majority of their annual mortality risk (Sillett and Holmes 2002, Newton 2007). Thus, mismatched timing on migratory stopover could have large consequences for individual fitness and population viability through its effect on survival, time spent migrating, and individual condition upon arrival (Lehikoinen and Sparks 2010).

Detecting Sensitivity for Phenological Mismatch on Stopover

Intra-annual habitat variation within a stopover site can be attributed broadly to three sources: static habitat structure, dynamic habitat structure, and dynamic food availability. The first two sources of variation are related to nonfood resources used by birds during migratory stopover. Static and dynamic habitat structures provide protection from predators and inclement climatic conditions, but may also provide coarse habitat cues for migrating birds unfamiliar with a specific locale. The first source of spatial variation, "static habitat structure," is unchanging throughout the migration season. This source of variation differentiates habitat patches with different plant communities such as tree- versus shrub-dominated patches or patches dominated by one shrub species over another. The second source of variation, "dynamic habitat structure," changes through time by plant defoliation, varying soil moisture throughout the migration season, or other processes. Food resources can be both static and dynamic, but for migratory species in north temperate ecosystems, food availability at the prey–species scale is almost exclusively dynamic.

Few plant species produce fruit or seeds continuously and few invertebrate species are consistently available for bird predation throughout the migration season. Further, the distributions of dynamic resources have predicted small-scale habitat selection for a few songbird species during the post-fledging period or autumnal migration (Tietz and Johnson 2007, Streby et al. 2011).

In this study, we investigated the potential for phenological mismatches to occur during migration by evaluating habitat selection within a stopover site. Specifically, we tested whether birds selected habitat patches based on static or dynamic characteristics. Temporally variable habitat use with selection for different patches at different points within a migratory season determines a taxon's risk for phenological mismatch. Taxa with temporally variable habitat use within a stopover site clearly show behavioral plasticity in the face of resource change. Behavioral plasticity itself should prevent birds from choosing a patch without adequate resources and should buffer a species from climate-change-induced resource fluctuations at a site. For taxa to be at risk of phenological mismatch, they must first match habitat use to local resource variability. Matching can become mismatched if there is a threshold beyond which behavioral plasticity can no longer adjust to temporal variation at breeding or nonbreeding sites. If matching during temporally dynamic patch use is symptomatic of behavioral plasticity at one level of temporal variability, it can also be symptomatic of a threshold in habitat variability beyond which reaction norms cannot continue to maintain benefits to individuals and lead to a mismatch. In this study, we measured matching among patches within a stopover site as small-scale patch selection (Hutto 1985). Temporal dynamics could also be at play at larger regional or subcontinental scales, especially if the relative quality of different migratory routes varies across the migration season.

Here, we examine variation in the within-stopover site, among-patch habitat use for four families of migratory songbird. We chose families that vary in their mean dietary niche breadth and possess species with varying migration distances, because these characteristics predict sensitivity to mismatch on the breeding grounds (Both and Visser 2001, Moussus et al. 2011). Our study taxa included sparrows that were generalists with a granivore to invertivore diet (F. Emberizidae,

n = 7 species); vireos (F. Vireonidae, n = 3) and warblers (F. Parulidae, n = 19), which were generalists with a frugivore to invertivore diet; and flycatchers (F. Tyrannidae, n = 5) as an invertivore specialist family (but see Parrish 1997 and Suomala et al. 2010). For each of these four families of birds, we tested (a) whether among-patch variability in habitat use was best predicted by static habitat structure, dynamic habitat structure, dynamic food availability, or any linear combination of those three sources of patch variation; and (b) whether these patterns varied for long- versus short-distance migrants within families.

METHODS

Study Area and Sampling Design

We gathered habitat-characteristic and bird-habitat-use data from 30 August to 18 October 2011 at four migratory stopover sites along the Gulf of Maine, where the dominant plant community type transitions from the northern hardwood to boreal forests. The period of observation covered the peak abundance of all groups of migratory songbirds in our region, although it did miss the beginning of migration for warblers and the end of migration for sparrows. By not sampling outside this temporal window, we lowered the probability that preferred habitats would be unoccupied merely due to low numbers of individuals.

Study sites included the two small islands of Great Duck Island (105 ha) and Metinic Island (130 ha), the southern headland of Mount Desert Island (280 km²), and the narrow peninsular headland of Petit Manan Point. The four sites are located in a series along a southwest to northeast transect. Metinic Island is the southwest site, Mount Desert Island and Great Duck Island were 10 km apart and 76 km to the northeast of Metinic, and Petit Manan Point was located 40 km farther to the northeast. Within each site, we measured bird use using mist net captures in habitats dominated by shrubs (Alder, Aronia, Sorbus, and Ilex spp.) or trees (mainly Picea spp., with Abies, Larix, Betula, Acer, and Populus spp.).

At each site, we defined a series of habitat patches that were centered on paired pitfall and hanging invertebrate traps placed in shrubs. Number of habitat patches per site ranged from one (Great Duck) to seven (Metinic) to eight (Mount Desert and Petit Manan) patches. We defined patches as the 25-m radius circle around each invertebrate collection location and sampled these patches with one to six mist nets placed randomly with regard to the traps. Each patch was centered on one or more shrubs, but varied in the plant community across the 1.7-ha patch. Our study design standardized comparisons of invertebrate communities by sampling species only within shrub habitats and by allowing patches to vary in both static and dynamic habitat structure.

Quantifying Habitat Structure and Bird Use

We quantified habitat structure along two 4-m-wide belt transects along each side of the mist nets and averaged values across all nets within each habitat patch. Static habitat structure was then defined at the end of the migration season (October) with seven metrics: the mean, per-meter woody stem density along the center of the belt transects; the relative cover of each plant species for the shrub, tree, and herbaceous structural layers; and both the layer-specific maximum vegetation height and the modal vegetation height across all vegetative layers at each third meter along the 12-m-long transects. Relative cover was assessed based on either woody tissues only (shrub and tree layers) or final size at the end of the summer growing season (herbaceous layer). While the herbaceous layer is more dynamic than the woody structural layers in our temperate ecosystem, our single end-season measure of cover is an index for community type and does not capture the small dynamic changes in the cover of this structural layer from August to October.

We calculated the first principal components for both the relative cover measures (hereafter "cover-PC") and vegetative heights (hereafter "height-PC"), which accounted for 41% and 63% of the variance in each of the two sets of measurements, respectively. Cover-PC represents an ecological gradient from a plant community dominated by dense shrubs over a forb- and litter-dominated ground layer (low values are shrub habitat) to a more forested community with a grass-dominated ground layer and little shrub understory (high values are woodland habitat). The height-PC captures the different heights possible within each of these community types. Low values indicated a tall shrub and herbaceous layers with short trees, whereas high values indicated a short shrub and herbaceous layer with a canopy of tall trees.

We assessed dynamic habitat structure and food availability throughout the observed migratory period. Soil moisture, leaf greenness, and fruit availability were measured weekly along the two belt transects near each mist net. Soil moisture was identified as one of four categories (dry, moist but not saturated, saturated, and standing water); leaf greenness was measured as the percentage of leaves in four categories (>90% green, <90% green and <50% brown, >50% brown, and bare stems) separately for the shrub and tree structural layers, and fruit availability was measured as the relative abundance of all ripe fruit in eight broad categories (Smith and McWilliams 2009). We excluded plants in the genus *Myrica* due to unpalatability of fruit for our study species of birds (Place and Stiles 1992). We calculated a single weighted average for each of these three dynamic habitat variables per transect. We multiplied the number of samples in each category by the median value of each category's range and then summed these products. Number of samples included the percentage of leaves or fruiting plants or the raw number of soil samples. High values of the indices indicated more saturated soils, green leaves, or ripe fruits. We then calculated single values for each habitat patch by averaging the transect values within the patch.

We captured invertebrates using pitfall traps placed under shrubs and yellow cup traps hanging in shrubs at the center of each patch. All traps were filled with soapy water and deployed for 4 days at a time, and the resulting dried biomass of captured invertebrates was used as an index of availability for the habitat patch for the period from 2 days prior to 2 days subsequent to each sampling period of 8 days. We only operated invertebrate traps during periods without significant precipitation (<1 cm) to avoid collection bias due to water-filled traps. The periods of the migration cycle for which we have data, therefore, possessed more similar weather conditions than occur over the entire season, and we also avoided migrant "fall-out" conditions associated with rain events.

To assess the other dynamic habitat variables on a similar time frame, we interpolated the mean soil moisture, leaf greening, and fruit abundance variables for the median day of invertebrate sampling by assuming that the other dynamic variables changed linearly between weekly assessments. If soil moisture had been assessed during the middle of the invertebrate sampling period, then the patch was given that value for the entire period, but if soil moisture had been assessed on the first and eighth days of the invertebrate sampling period, then the patch was assigned the soil moisture value halfway between the two samples.

We operated mist nets in suitable capture conditions and closed nets if temperatures were outside the range of 0°C–27°C, if winds were >20 km per hour, or if precipitation or fog was heavy enough for water to collect on the nets. Nets were monitored within each patch from 30 minutes before sunrise for 6 hours or until conditions deteriorated. Most species of birds captured in mist nets were from one of four families (Parulidae, Emberizidae, Vireonidae, and Tyrannidae), and we summed number of captures daily within each habitat patch (Table 10.1). We did not include Yellow-rumped Warblers (*Setophaga coronata*) to avoid overrepresentation within warblers (>66% of all warbler captures) and their preference for *Myrica* fruits, which were not included in our surveys of fruit abundance (Place and Stiles 1992, Borgmann et al. 2004).

Statistical Analysis

To test for the ability of dynamic and static habitat characteristics to predict bird habitat use, we developed a series of repeated-measures, negative-binomial, generalized linear models (Proc Genmod, SAS 9.2). We repeatedly measured daily captures in 24 habitat patches through time (366 patch-days) and used the negative binomial distribution to account for overdispersion and zero inflation in our capture data. All models included the effects of capture effort (total net hours per meters of net), stopover site (to account for site-specific capture probabilities), taxonomic family (to account for family-specific capture probabilities), and both linear and quadratic terms for the calendar day of capture nested within taxonomic family. Exploratory analysis showed that the 8-day moving averages of capture rate across all sites exhibited a quadratic relationship with Julian date nested with taxonomic family (linear term: $F_{4,8} = 63.9$, $P < 0.0001$; quadratic term: $F_{4,8} = 64.0$, $P < 0.0001$). By estimating a quadratic relationship for each family, we controlled for the broad seasonal phenology of migration within each family and thus availability to occupy patches within each site as a function of regional abundance.

We considered eight candidate variables nested within taxonomic family of birds in three categories: static habitat structure (stem density, cover-PC, and height-PC), dynamic habitat structure (soil moisture, shrub-layer leaf greenness, and tree-layer leaf greenness), and dynamic food availability (ripe fruit abundance and invertebrate biomass). We also included a quadratic term for stem density to test the possibility that intermediate values of density might be preferred by a given bird taxon. We considered quadratic terms for the remaining variables, but quadratic patterns were not supported by initial data exploration, and we dropped these terms to simplify our models. We scaled all of our candidate variables by dividing them by the mean value of each measurement across all sites and habitat patches. Likewise, we scaled our capture rates by dividing each daily capture total by the mean daily capture total for that taxonomic family across all sites. The adjustments allowed for direct comparison of parameter estimates among our candidate variables and predicted capture rates among our taxonomic families.

We then created eight separate models, including:

1. Three models that each used only the static habitat structure variables, the dynamic habitat structure variables, or the dynamic food availability variables

2. Three models that included the pairwise combinations of the three variable categories

3. The full model with all of the static and dynamic variables

4. The null model with just the five covariates

Final model selection used the quasi-likelihood under the independence model criterion corrected for parsimony (QICu; Pan 2001, Hardin and Hilbe 2007). We considered models with ΔQICu < 2.0 as equivalent.

To test for the sensitivity of the final model to the migration distance of a species, we performed a phylogenetically controlled contrast by estimating the top model parameters for both a long-distance and short-distance migrant group within each family. We assessed relative migratory distance by first measuring the latitudinal distance between the centers of the breeding and nonbreeding ranges for each species (Erickson et al. 2011). We then classified each species within the

TABLE 10.1

Total individuals captured and temporal range of capture (in calendar days) for four migratory stopover sites along the Gulf of Maine during the Fall of 2011.

Taxon	Individuals captured	First detection	Median detection	Last detection	Migration distance
Tyrannidae					
Eastern Phoebe	9	270	270	270	
Least Flycatcher	8	242	250	250	
Traill's (Alder or Willow) Flycatcher	45	242	244	268	
Yellow-breasted Flycatcher	8	244	244.5	268	
Vireonidae					
Blue-headed Vireo	25	268	269.5	290	
Philadelphia Vireo	4	268	269	269	
Red-eyed Vireo	187	242	268	284	
Parulidae					
American Redstart	87	242	244	291	Long
Black-and-white Warbler	16	245	258	270	Long
Black-throated Blue Warbler	14	243	270	271	Short
Black-throated Green Warbler	12	243	257.5	271	Long
Blackpoll Warbler	172	244	269	276	Long
Canada Warbler	5	244	244	244	Long
Cape May Warbler	6	245	246	247	Short
Common Yellowthroat	261	242	247	284	Short
Connecticut Warbler	6	268	272	276	Long
MacGillivray's Warbler	3	271	271	271	Short
Magnolia Warbler	28	244	268	271	Short
Mourning Warbler	3	242	242	244	Long
Nashville Warbler	29	242	269	290	Short
Northern Parula	30	242	269	274	Short
Northern Waterthrush	45	242	244	270	Long
Palm Warbler	52	242	269	284	Short
Wilson's Warbler	48	242	245	269	Long
Yellow Warbler	40	244	244	269	Long
Yellow-breasted Chat	11	269	269	274	Long
Emberizidae					
Chipping Sparrow	6	246	268.5	291	Long
Dark-eyed Junco	16	268	284	290	Short
Field Sparrow	6	292	292	292	Short
Lincoln's Sparrow	21	268	270	291	Long
Savannah Sparrow	23	242	270	271	Long
Song Sparrow	73	245	250	292	Short
Swamp Sparrow	42	245	272	284	Long
White-throated Sparrow	68	243	270	284	Short

warblers and sparrows as migrating either less or farther than the median migration distance for the family (Table 10.1). The final generalized linear model took the same form as the top-performing model, but included only captures of warblers and sparrows. We were unable to explore the effect of relative migratory distance on patch use for vireos and flycatchers due to a low number of captures among species in the short-distance category (flycatchers: n = 17, vireos: n = 29, Table 10.1).

RESULTS

Over 429 patch-capture days across 24 habitat patches and four sites, we captured 1,678 birds of 34 species in four different families: flycatchers (F. Tyrannidae), vireos (F. Vireonidae), warblers (F. Parulidae), and sparrows (F. Emberizidae, Table 10.1). Daily patch capture totals varied widely through the season by family (Figure 10.1).

Relative Patch Characteristics Vary in Space and Time

Static habitat variation among patches was greater than variation in the mean dynamic variable values among patches (Table 10.2). We transformed both principal components by adding one to make the mean one instead of zero and compared coefficients of variation for all patches. Height-PC ranged from 2.02 times higher (fruit abundance) to 12.8 times higher (shrub leaf greenness) than the dynamic variables (mean = 6.4 times higher). Similarly, cover-PC was more variable among patches than the dynamic variables, ranging from 1.90 to 12.0 times higher (mean = 6.0 times higher). Variation of the dynamic patch variables, however, was as high within patches versus among patches, indicating that patches did not change together in parallel through the season (Table 10.2). Mean within-patch variation ranged from 0.26 (tree leaf greenness) to 1.73

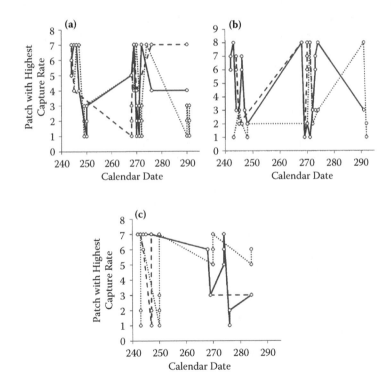

Figure 10.1. Patch capture rate (corrected for effort) variability through time for three migratory stopover sites: (a) Metinic Island, (b) Petit Manan Point, and (c) Mount Desert Island. Each point indicates the numbered patch (the number of patches ranges from eight to nine depending on the site, and number designation within each site is irrespective of patch characteristics) that achieved the highest daily capture rate (corrected for effort) at that site. Vertical lines indicate days when multiple patches achieved the highest daily capture rate, and horizontal lines indicate periods where one patch maintained the highest capture rate through time. The solid lines connect points indicating patches with the highest capture rate for warblers, the dotted lines for sparrows, and the dashed lines for vireos.

TABLE 10.2

Coefficients of variation for patch habitat characteristics habitat variable.

	Among patches	Within patches
Ripe fruit availability	0.959	0.856
Invertebrate biomass	0.959	0.576
Tree leaf greenness	0.794	0.171
Shrub leaf greenness	0.151	0.261
Soil moisture	0.154	0.202
Height-PC	1.938	
Cover-PC	1.819	

(ripe fruit abundance) times that of among-patch variation (mean = 0.85). Shrub leaf greenness and soil moisture exhibited more within-patch than among-patch variation, while the remaining variables exhibited more among-patch than within-patch variation.

Our top-ranked model predicting family-specific variation in apparent habitat patch use included only the dynamic variables of structure and food ($w_i = 0.95$), controlling for capture effort, stopover site differences, and family-specific migration phenology (Table 10.3). This model strongly outperformed the next best model, which also included the static habitat structure ($\Delta QICu = 5.9$, $w_i = 0.05$) and the null model ($\Delta QICu = 997.6$, $w_i < 0.0001$).

Within the top model we found significant effects of fruit availability ($\chi^2 = 15.4$, P = 0.004), invertebrate biomass ($\chi^2 = 11.3$, P = 0.02), shrub leaf greenness ($\chi^2 = 10.6$, P = 0.03), tree leaf greenness ($\chi^2 = 15.7$, P = 0.003), and soil moisture ($\chi^2 = 19.6$, P = 0.0006). These patterns, however, varied by taxonomic family (Figure 10.2). Fruit abundance increased capture rates similarly for warblers ($\beta \pm SE = 0.42 \pm 0.09$, Z = 4.5, P < 0.0001), sparrows (0.37 ± 0.12, Z = 3.2, P = 0.001), and vireos (0.30 ± 0.12, Z = 2.5, P = 0.01), but was not related to captures of flycatchers (Z = −0.29, P = 0.77). Invertebrate biomass had a negative effect on capture rates of flycatchers ($\beta \pm SE = -0.35 \pm 0.12$, Z = −3.0, P = 0.003) and vireos (−0.28 ± 0.10, Z = −3.0, P = 0.003), but no effect on warblers (Z = −0.6, P = 0.58) or sparrows (Z = 1.3, P = 0.20). Shrub leaf greenness was negatively related to captures of sparrow ($\beta \pm SE = -1.33 \pm 0.49$, Z = −2.7, P = 0.007) and flycatchers (−1.41 ± 0.64, Z = −2.2, P = 0.03), but showed nonsignificant trends with captures of warblers (Z = −1.8, P = 0.07) and vireos (Z = 1.9, P = 0.06). Tree leaf greenness was positively related to captures of sparrows ($\beta \pm SE = 0.55 \pm 0.16$, Z = 3.4, P = 0.0006) and vireos (1.05 ± 0.28, Z = 3.7, P = 0.0002), but had no effect on captures of warblers (Z = 1.8, P = 0.07) or flycatchers (Z = 1.0, P = 0.34). Captures of sparrows ($\beta \pm SE = 2.40 \pm 0.48$, Z = 5.0, P < 0.0001), warblers (1.30 ± 0.29, Z = 4.5, P < 0.0001), and vireos ($\beta \pm SE = 1.63 \pm 0.69$, Z = 2.4, P = 0.02) were all

TABLE 10.3

Model selection based on QICu for eight models predicting apparent habitat patch use as a function of different combinations of dynamic food availability,[a] static habitat structure,[b] and dynamic habitat structure[c].

Model	QICu	$\Delta QICu$	w_i
Dynamic structure + dynamic food	848.8	0.0	0.95
Dynamic structure + static structure + dynamic food	854.7	5.9	0.05
Dynamic + static structure	872.6	23.8	<0.0001
Static structure + dynamic food	913.4	64.6	<0.0001
Dynamic food	914.1	65.4	<0.0001
Dynamic structure	968.3	119.5	<0.0001
Static structure	974.3	125.5	<0.0001
Null model	997.6	148.8	<0.0001

[a] Fruit abundance and invertebrate biomass.

[b] Vegetation height, vegetative cover, and stem density.

[c] Leaf greenness and soil moisture.

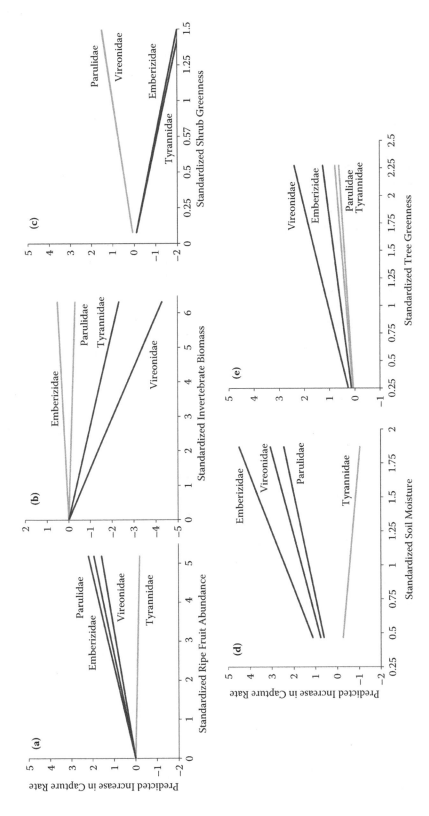

Figure 10.2. The relationships between the predicted values of daily bird captures for each of four taxonomic families and (a) ripe fruit abundance, (b) invertebrate biomass, (c) shrub leaf greenness, (d) soil moisture, and (e) tree leaf greenness within each sampled patch. Black lines are statistically significant, and gray lines are nonsignificant at $\alpha > 0.05$. All variable and predicted values were standardized for direct comparison by dividing by their means and slopes.

strongly positively related to soil moisture, but there was no relationship for this variable with captures of flycatchers ($Z = -0.9$, $P = 0.35$). Controlling for these dynamic resource effects, there were no significant differences in site use ($\chi^2 = 6.5$, $P = 0.09$).

Dynamic Habitat Importance Varies by Migration Distance

The preceding patterns varied within sparrows and warblers as a function of migration distance. Habitat use of long- and short-distance migrants was predicted by dynamic habitat attributes. Habitat use by long-distance migratory warblers (11 species) was explained only by variation in invertebrate biomass ($\beta \pm SE = -0.18 \pm 0.08$, $Z = 2.3$, $P = 0.02$). Habitat use by short-distance migratory warblers (eight species) was explained by fruit abundance ($\beta \pm SE = 0.46 \pm 0.09$, $Z = 5.3$, $P < 0.0001$), shrub leaf greenness (-1.30 ± 0.50, $Z = -2.7$, $P = 0.008$), and soil moisture (1.84 ± 0.33, $Z = 6.3$, $P < 0.0001$), but not by invertebrate biomass ($Z = -0.3$, $P = 0.76$).

Among sparrows, all patterns of habitat use were similar for both short- (four species) and long- (four species) distance migrants. Both groups preferred to use patches with higher fruit abundance (long-distance migrants: $\beta \pm SE = 0.48 \pm 0.15$, $Z = 3.3$, $P = 0.001$; short-distance migrants: 0.30 ± 0.10, $Z = 3.0$, $P = 0.003$) and patches with wetter soils (long-distance: 3.13 ± 0.76, $Z = 4.1$, $P < 0.0001$; short-distance: 2.20 ± 0.35, $Z = 6.3$, $P < 0.0001$). Similarly, both groups preferred to use patches with more shrub defoliation (long-distance: $\beta \pm SE = 1.38 \pm 0.60$, $Z = -2.3$, $P = 0.02$; short-distance: -1.30 ± 0.50, $Z = -2.6$, $P = 0.009$) but less tree defoliation (long-distance: 0.39 ± 0.19, $Z = 2.1$, $P = 0.04$; short-distance: 0.91 ± 0.18, $Z = 5.1$, $P < 0.0001$). Neither group, however, altered patch use as a function of insect biomass (long-distance: $Z = 0.7$, $P = 0.48$; short-distance: $Z = 1.5$, $P = 0.14$).

DISCUSSION

Apparent habitat patch use during autumnal migratory stopover was best explained by the variation in local resources. The model with these dynamic predictors outperformed models that included either static patch differences alone (such as woody vegetation structure and height)

or a combination of both static and dynamic habitat characteristics.

All four common families of North American songbirds were sensitive to dynamic resource changes in fruit abundance and invertebrate biomass (although direction of relationships varied), as well as soil moisture and leaf cover, and shifted their use to different patches throughout the migratory cycle at each stopover site (Figure 10.1). Thus, songbirds within each family with similar or identical ecological requirements ranked the quality of available migratory stopover patches differently as autumn progressed. Patch quality ranking, assessed by capture rates, would have remained stable within a given stopover site if high-quality patches consistently outproduced low-quality patches, or if dietary niches were broad enough to allow songbirds to meet their energetic requirements in different habitats. We can reject both of these alternatives. At a larger scale, our results suggest that the costs and benefits of a given stopover site may be temporally dynamic and that songbirds should maximize their fitness by timing their passage to match peak periods of stopover site quality. If the local phenology of fruiting and invertebrate abundance outpaces the plasticity of migrants to find adequate patches within a site, mismatches are possible that will alter the costs of migration, the relative fitness of individuals, and local population dynamics.

The Relative Taxonomic Sensitivity to Dynamic Resources

Habitat use by all four taxonomic families was predicted significantly by both food and nonfood resources, although the strength of relationships within each family varied. Not surprisingly, the importance of different food resources varied by dietary niche. Fruit abundance was positively related to apparent patch use by all but the most specialist invertivore family (flycatchers). Invertebrate biomass, on the other hand, was inversely related to apparent patch use for flycatchers and more generalist invertivores (vireos and long-distance migratory warblers), but had no effect on stopover dynamics of two groups with diets of seeds or larger fruits (short-distance migratory warblers and sparrows). Nonfood resources were also successful in predicting apparent patch use across all taxonomic families—but, again, the patterns for each family were variable (Figure 10.2).

Nonfood resources exerted a larger effect on apparent patch use than food resources for all taxonomic families. Changes in capture rate for a given change in soil moisture and shrub or tree greenness were two to three times higher on average than a similar change in either of the food resources. Shrub leaf greenness and soil moisture were unusual as the only two variables where within-patch variation was greater than among-patch variation. Thus, even among the dynamic variables, the two variables that exhibited the greatest within-patch variation exhibited the strongest effect on bird patch use. It is clear, therefore, that the efficacy of these variables is not due simply to their ability to differentiate among patches within stopover sites.

In post hoc phylogenetic contrasts, habitat patch use was predicted by different dynamic variables for the long- versus short-distance species of migratory warblers. Patch use for long-distance migrants was related strictly to invertebrate biomass, which is not surprising since these species tend to rely on more invertebrate prey. Patch use for short-distance migrants was predicted by fruit abundance and nonfood resources. There were no apparent differences among sparrows that were comparable to the warblers, and both sparrow migratory groups behaved similarly to the short-distance warbler migrants. The range of migration distances was smaller among sparrows relative to warblers (median migration distance is 3.1 times farther for warblers), and there may not have been enough of a difference between long- and short-distance migrants to alter overall stopover strategy of sparrows. Our observation that habitat patch use in the shortest distance warbler migrants was similar to sparrows suggests that habitat patch use may be more similar among species with more similar migratory distances. More families would need investigation to support the generality of this conclusion. Patch use was predicted by dynamic habitat resources for all of the migratory groups, but we found no support for the increased sensitivity of any group over another, as has been shown among breeding birds in Europe (Both and Visser 2001, Moussus et al. 2011).

Overall, the temporal dynamics of both food and nonfood resources predicted apparent patch use, regardless of variation in the breadth of the dietary niche or migratory distance. All of the taxa we examined shifted apparent habitat use among patches throughout the migration season, thereby exhibiting behavioral plasticity to changing ecological condition within each site. Plasticity suggests that species benefit from phenological matching and will thus likely exhibit mismatches if temporal variability outpaces their capacity for change.

Alternative Explanations

It is unclear whether any of the specific associations we report here are truly causal or whether temporal variation in these food resources is merely correlated with an unmeasured dynamic resource. Given the negative relationship between invertebrate biomass and apparent patch use by flycatchers and warblers (long-distance migrants only), we suspect the latter explanation. Our invertebrate collection methods were biased against flying insects and may reflect community trade-offs among different guilds of invertebrates, although many other such linkages are possible. The unexpected patterns are unlikely to be due to association among our predictor variables, given their weak relationships to each other (range in $r^2 = 0.002 - 0.13$; mean $= 0.06$).

Tree leaf greenness was positively associated with sparrow patch use, and conifer forests received the highest greenness scores later in the season when migratory sparrows were passing through our region. A strong association between sparrows and evergreen tree habitats could explain the pattern. The same cannot be said for shrub leaf greenness scores, however, as our study sites possessed no evergreen shrub species. Rare evergreen saplings were included in this layer, but the shrub layer of every patch was heavily dominated by deciduous shrubs.

Despite a lack of evidence that the measured dynamic resources were the habitat characteristics under selection by migrants, we maintain that either they are under selection or they represent an unmeasured dynamic resource. One alternative, however, is that dynamic resources performed better because migration stopover behaviors are themselves dynamic for numerous reasons unrelated to stopover site phenology, including weather, habitat quality on previous stopover sites, point of origin, and breeding season dynamics. We think bias is unlikely for two reasons:

1. All of our dynamic variables were averaged over 8-day periods. While capture rates of individual taxa can vary dramatically from day to day at a given site, our averaged dynamic habitat variables are unlikely to track the daily dynamics of migrant arrival.
2. We only used bird capture data during periods without significant precipitation because our insect collection methods precluded otherwise. Our results do not include periods with environmental conditions that cause migratory "fall-outs," when stopover site use can be over an order of magnitude higher than normal, and captures on a single day could bias an 8-day average.

Our scale of inquiry among families could be obscuring considerable variation in species- or population-level dynamics. Averaging should be more likely to decrease the importance of dynamic resources because within-family niche variation should allow different species to use a given patch for a longer period of time than a single species. We would thus predict that a species-level investigation would have shown greater sensitivity to dynamic resources than what we have shown here with taxonomic families. Two of our study families (Tyrannidae and Vireonidae) were heavily weighted by the capture rates of a single species or species complex and may perform similarly to results for individual species.

Taxonomic families would not obscure the importance of dynamic resources, however, if species within a family possessed distinct niches and migrated through a site at distinct times. What we have interpreted as differential patch use in the face of dynamic resources could simply be different niches within a family being replaced by each other over time. Given the large overlap in passage times of species within each family, however, we do not think this explanation is likely (Table 10.1). More detailed exploration into species-specific niches relative to passage time would be required to test this explanation.

The general patterns reported here, while replicated across four stopover sites within four taxonomic families, represent phenological variation within a single year. The specific relationships may vary with climatic conditions across years. Our primary conclusion—that songbirds match habitat use to temporally variable resources and therefore could experience phenological mismatch—remains even if the matching we demonstrate here occurs only in some years. The long-standing questions are related to biological relevancy: whether future phenological changes in this region will result in a higher probability of mismatch and what degree of fitness consequences such mismatches might cause.

Incorporating Timing into Habitat Selection Theory

The current paradigm for habitat selection in migratory birds describes a hierarchical process where individuals must first select flyway, then region, then habitat patch (*sensu* Johnson 1980, Hutto 1985), but this approach is focused on spatial and not temporal scales. Timing of migration in birds has been well studied, but the objectives and conclusions of most studies have been largely removed from our understanding of habitat selection. We hope that our results here will suggest that the field needs a more formal inclusion of timing into our understanding of habitat selection in migratory animals.

Timing decisions certainly have the potential to influence habitat selection at almost any spatial scale. For instance, departure time (or even more broadly, departure season) can determine the broad route of migration (Cooke 1910, Myers et al. 1990, Delmore et al. 2012) or patterns of stopover timing (O'Reilly and Wingfield 1995). Our work here suggests that timing may also be important at the spatial scale of a stopover site. More empirical research is needed to understand how timing modifies the traditional hierarchical habitat selection model. It is widely recognized that habitat selection can influence the speed and timing of migration (Lehikoinen and Sparks 2010), but our work suggests that the timing of migration also may affect which local habitats are selected.

CONCLUSIONS

Across the four most common songbird families that migrate along the Gulf of Maine during the autumn, patterns of habitat patch use within stopover sites varied through time. The importance of particular patches to a group of birds varied through the migration season, and patch use was predicted better by within-season changes in invertebrate and fruit availability, soil moisture, and leaf greenness than by habitat characteristics that varied only spatially and not temporally. Our results suggest that songbirds may be sensitive to changes in the resource phenology of stopover sites in ways that are similar to recently discovered sensitivities to breeding resources. Future research should explicitly test for the population and evolutionary implications of mismatch between migratory timing and stopover resource phenology. Further, the within-season variability in stopover site quality should be considered when making conservation priorities. Successful conservation actions may need to protect patches or networks of patches that maintain sufficiently high habitat quality throughout the migratory season.

ACKNOWLEDGMENTS

Data collection at Metinic Island was overseen and supported by Rebecca Holberton. Thanks to Sara Williams and the Maine Coastal Islands National Wildlife Refuge for access to, financial support of, and logistical assistance with work at Metinic Island and for sharing data collected by refuge staff at Petit Manan Point. Thanks to Bruce Connery, Bik Wheeler, and Acadia National Park for similar collaboration and support of work on Mount Desert Island and to John Anderson, the College of the Atlantic, and the Nature Conservancy for work on Great Duck Island. Last, thanks to Abe Miller-Rushing for helpful comments on an initial draft of this chapter. The project was funded by grants from the US Fish and Wildlife Service, the National Park Service, the National Institute of Food and Agriculture, and the University of Maine.

LITERATURE CITED

Alerstam, T., and A. Hedenström. 1998. The development of bird migration theory. Journal of Avian Biology 29:343–369.

Baker, A. J., P. M. González, T. Piersma, L. J. Niles, I. de L. S. do Nascimento, P. W. Atkinson, N. A. Clark, C. D. T. Minton, M. K. Peck, and G. Aarts. 2004. Rapid population decline in Red Knots: fitness consequences of decreased refueling rates and late arrival in Delaware Bay. Proceedings of the Royal Society of London B 271:875–882.

Balbontín, J., A. P. Møller, I. G. Hermosell, A. Marzal, M. Reviriego, and F. de Lope. 2009. Individual responses in spring arrival date to ecological conditions during winter and migration in a migratory bird. Journal of Animal Ecology 78:981–989.

Berthold, P. 1975. Migration: control and metabolic physiology. Pp. 77–128 in D. S. Farner and J. R. King (editors), Avian biology. Academic Press, New York, NY.

Blem, C. R. 1980. The energetics of migration. Pp. 59–113 in S. A. Gauthreaux (editor), Animal migration, orientation, and navigation. Academic Press, New York, NY.

Borgmann, K. L., S. F. Pearson, D. J. Levey, and C. H. Greenberg. 2004. Wintering Yellow-rumped Warblers (Dendroica coronata) track manipulated abundance of Myrica cerifera fruits. Auk 121:74–87.

Both, C. 2010a. Flexibility of timing of avian migration to climate change masked by environmental constraints en route. Current Biology 20:243–248.

Both, C. 2010b. Food availability, mistiming, and climatic change. Pp. 129–147 in A. P. Møller, W. Fiedler, and P. Berthold (editors), Effects of climate change on birds. Oxford University Press, New York, NY.

Both, C., S. Bouwhuis, C. M. Lessells, and M. E. Visser. 2006. Climate change and population declines in a long-distance migratory bird. Nature 441:81–83.

Both, C., and M. E. Visser. 2001. Adjustment to climate change is constrained by arrival date in a long-distance migrant bird. Nature 411:296–298.

Both, C., and M. E. Visser. 2005. The effect of climate change on the correlation between avian life-history traits. Global Change Biology 11:1606–1613.

Castro, G., and J. P. Myers. 1993. Shorebird predation on eggs of horseshoe crabs during spring stopover on Delaware Bay. Auk 110:927–930.

Cooke, W. W. 1910. Distribution and migration of North American shorebirds. Pp. 100 in US Dept. of Agriculture, Biological Survey Bulletin 35.

Delmore, K. E., J. W. Fox, and D. E. Irwin. 2012. Dramatic intraspecific differences in migratory routes, stopover sites and wintering areas revealed using light-level geolocators. Proceedings of the Royal Society of London B 279:4582–4589.

Drent, R. H. 2006. The timing of birds' breeding seasons: the Perrins hypothesis revisited especially for migrants. Ardea 94:305–322.

Erickson, L., K. McGowan, H. Powell, M. Savoca, and C. Sedgwick. [online]. 2011. All about birds: bird guide. <http://www.allaboutbirds.com/guide>

Hardin, J. W., and J. M. Hilbe. 2007. Generalized linear models and extensions. Stata Press, College Station, TX.

Hüppop, O., and K. Hüppop. 2003. North Atlantic oscillation and timing of spring migration in birds. Proceedings of the Royal Society of London B 270:233–240.

Hutto, R. L. 1985. Habitat selection by nonbreeding, migratory land birds. Pp. 455–476 in M. L. Cody (editor), Habitat selection in birds. Academic Press, New York, NY.

Johnson, D. H. 1980. The comparison of usage and availability measurements for evaluating resource preference. Ecology 61:65–71.

Lack, D. L. 1950. The breeding seasons of European birds. Ibis 92:288–316.

Lack, D. L. 1968. Ecological adaptations for breeding in birds. Methuen, London, UK.

Lank, D. B., R. W. Butler, J. Ireland, and R. C. Ydenberg. 2003. Effects of predation danger on migration strategies of sandpipers. Oikos 103:303–319.

Lehikoinen, E., and T. H. Sparks. 2010. Changes in migration. Pp. 89–112 in A. P. Møller, W. Fiedler, and P. Berthold (editors), Effects of climate change on birds. Oxford University Press, New York, NY.

Marra, P. P., C. M. Francis, R. S. Mulvihill, and F. R. Moore. 2005. The influence of climate on the timing and rate of spring bird migration. Oecologia 142:307–315.

McKinney, A. M., P. J. CaraDonna, D. W. Inouye, B. Barr, C. D. Bertelsen, and N. M. Waser. 2012. Asynchronous changes in phenology of migrating Broad-tailed Hummingbirds and their early-season nectar resources. Ecology 93:1987–1993.

McLaughlin, J. F., J. J. Hellmann, C. L. Boggs, and P. R. Ehrlich. 2002. Climate change hastens population extinctions. Proceedings of the National Academy of Sciences of the USA 99:6070–6074.

Møller, A. P. 1994. Phenotype-dependent arrival time and its consequences in a migratory bird. Behavioral Ecology and Sociobiology 35:115–122.

Møller, A. P., D. Rubolini, and E. Lehikoinen. 2008. Populations of migratory bird species that did not show a phenological response to climate change are declining. Proceedings of the National Academy of Sciences of the USA 105:16195–16200.

Moussus, J.-P., J. Clavel, F. Jiguet, and R. Julliard. 2011. Which are the phenologically flexible species? A case study with common passerine birds. Oikos 120:991–998.

Myers, J. P., A. M. Sallaberry, E. Ortiz, G. Castro, L. G. Gordon, J. L. Maron, C. T. Schick, E. Tabilo, P. Antas, and T. Below. 1990. Migration routes of New World Sanderlings (Calidris alba). Auk 107:172–180.

Newton, I. 2007. Weather-related mass-mortality events in migrants. Ibis 149:453–467.

O'Reilly, K. M., and J. C. Wingfield. 1995. Spring and autumn migration in Arctic shorebirds: same distance, different strategies. American Zoologist 35:222–233.

Pan, W. 2001. Akaike's information criterion in generalized estimating equations. Biometrics 57:120–125.

Parrish, J. 1997. Patterns of frugivory and energetic condition in Nearctic landbirds during autumn migration. Condor 99:681–697.

Place, A. R., and E. W. Stiles. 1992. Living off the wax of the land: bayberries and Yellow-rumped Warblers. Auk 109:334–345.

Rubolini, D., A. Møller, K. Rainio, and E. Lehikoinen. 2007. Intraspecific consistency and geographic variability in temporal trends of spring migration phenology among European bird species. Climate Research 35:135–146.

Sanz, J. J., J. Potti, J. Moreno, S. Merino, and O. Frias. 2003. Climate change and fitness components of a migratory bird breeding in the Mediterranean region. Global Change Biology 9:461–472.

Sillett, T. S., and R. T. Holmes. 2002. Variation in survivorship of a migratory songbird throughout its annual cycle. Journal of Animal Ecology 71:296–308.

Smith, A. D., and S. R. McWilliams. 2009. Protocol for the rapid assessment of fruit abundance on New England Wildlife Refuges. University of Rhode Island, Kingston, RI.

Sokolov, L. V., M. Y. Markovets, A. P. Shapoval, and Y. G. Morozov. 1998. Long-term trends in the timing of spring migration of passerines on the Courish Spit of the Baltic Sea. Avian Ecology and Behaviour 1:1–21.

Streby, H. M., S. M. Peterson, and D. E. Andersen. 2011. Invertebrate availability and vegetation characteristics explain use of nonnesting cover types by mature-forest songbirds during the postfledging period. Journal of Field Ornithology 82:406–414.

Suomala, R. W., S. R. Morris, K. J. Babbitt, and T. D. Lee. 2010. Migrant songbird species distribution and habitat use during stopover on two islands in the Gulf of Maine. Wilson Journal of Ornithology 122:725–737.

Tietz, J. R., and M. D. Johnson. 2007. Stopover ecology and habitat selection of juvenile Swainson's Thrushes during fall migration along the northern California coast. Condor 109:795–807.

Tsipoura, N., and J. Burger. 1999. Shorebird diet during spring migration stopover on Delaware Bay. Condor 101:635–644.

Verboven, N., J. M. Tinbergen, and S. Verhulst. 2001. Food, reproductive success and multiple breeding in the Great Tit, Parus major. Ardea 89:387–406.

CHAPTER ELEVEN

Annual Variation in Autumn Migration Phenology and Energetic Condition at a Stopover Site in the Western United States[*]

Robert A. Miller, Jay D. Carlisle, Neil Paprocki, Gregory S. Kaltenecker, and Julie A. Heath

Abstract. Climate change is having a dramatic effect on many migratory species. Changes in climate may lead to changes in food availability or other proximate cues that affect migratory behavior. We used 13 years (2000–2012) of data on songbird banding and raptor migration counts and captures during autumn migration in the intermountain West to evaluate whether regional temperature or precipitation or hemispheric climate indices predicted autumn migratory timing and energetic condition. We examined overall trends and evaluated the effects of diet and migratory distance on phenology and conditional responses. For the 13-year study period, no temperature, precipitation, or climate index trends were evident. There was no change in migratory timing for all species combined, but trends were apparent when evaluated by diet and migratory distance. The magnitude of these changes varies by diet and by migratory distance, but not as predicted by previous research of autumn timing in other parts of the globe. Long-distance migrants tended to migrate later in autumn, whereas short-distance migrants exhibited no change in timing. Annual variation in timing was predicted by regional temperature and precipitation and by hemispheric climate indices, and the predicted effects differed by diet and migratory distance. Granivores responded to the broadest set of climate indices, whereas avivores responded to the least. Frugivores responded with the greatest magnitude to annual variation in climate. We did not measure food availability but in most cases the predictive effect of climate on migratory timing of birds was consistent with predicted effects on food. Frugivorous birds migrated earlier in warmer years when fruit quality and quantity were expected to be lower. Energetic condition measurements supported the food hypotheses in some, but not all cases. The different responses of species to annual variation in climate suggest that different species integrate difference cues in their decision to migrate.

Key Words: Accipiter cooperii, A. striatus, Empidonax oberholseri, Idaho, *Junco hyemalis, Oporornis tolmiei, Pipilo maculatus, Spizella passerina,* timing, *Zonotrichia leucophrys.*

[*] Miller, R. A., J. D. Carlisle, N. Paprocki, G. S. Kaltenecker, and J. A. Heath. 2015. Annual variation in autumn migration phenology and energetic condition at a stopover site in the western United States. Pp. 177–191 in E. M. Wood and J. L. Kellermann (editors), Phenological synchrony and bird migration: changing climate and seasonal resources in North America. Studies in Avian Biology (no. 47), CRC Press, Boca Raton, FL.

Climate change has influenced biotic communities and those impacts may be amplified in the coming decades. Plant and animal distributions (Parmesan and Yohe 2003, La Sorte and Thompson 2007, Huntley et al. 2008) and phenologies are changing (Cotton 2003, Gordo 2007, Nufio et al. 2010, Anderson et al. 2012). Changes in distributions and phenology may have complex and dramatic impacts on food webs when responses to climatic shifts differ between species in different trophic levels. For example, some butterflies and moths have been shown to migrate earlier as spring warming advances, whereas timing is unchanged for their predator, the Pied Flycatcher (*Ficedula hypoleuca*; Altermatt 2010). A mismatch has been the hypothesized cause of population declines observed in the Pied Flycatcher (Both and Visser 2001, Both et al. 2006).

Annual variation in climate adds additional complexity, making it more difficult to identify responses of organisms in complex food webs to these changes. Prey abundance often varies with daily weather patterns as well as annual fluctuations. Blancher and Robertson (1987) found that flying insect abundance varied with daily temperature and date, while annual differences were explained by the previous year's precipitation. Bell (1985) found that arthropod abundance declined in periods of drought. An increase in a previous year's precipitation and earlier spring warming can advance all phases of plant phenology including fruiting and seeding (Inouye 2008, Lambert et al. 2010). An advancement may cause plants to flower earlier, exposing them to greater frost mortality, which can decrease food for frugivorous and granivorous birds (Inouye 2008). Warmer temperatures may decrease the number of flowering plants, number of seeds per unit area, and the availability of lightweight seeds that some species require, or increase seed production and germination rates (Meunier et al. 2007, Gao et al. 2012). Mismatch of timing has larger potential effects among individual species in habitats where prey fluctuates sharply than in habitats with a more constant prey supply such as forests versus marshes (Both et al. 2009, Zuckerberg et al. 2011).

For migratory species such as birds, the ability to respond to changing resources may depend on annual cycle constraints such as diet, migration distance, and weather. During spring migration, short-distance migrants have shown a greater response to change in climate, possibly because of a greater sensitivity to changing weather patterns (Butler 2003). Earlier spring migration and earlier breeding can lead to greater re-nesting or earlier autumn departures (Cotton 2003, Halupka et al. 2008, Lehikoinen et al. 2010), but advancement of spring migratory timing may be constrained in some species by a lack of physiological or behavioral plasticity (Both and Visser 2001, Dawson 2008).

The majority of avian climate studies have focused on timing of spring migration and the initiation of breeding. The effect of climate change on autumn migration has generally received less attention, with a few notable exceptions (Jenni and Kéry 2003, Van Buskirk et al. 2009, Filippi-Codaccioni et al. 2010, Rosenfield et al. 2011). Autumn migration represents a significant portion of the annual cycle of avian migratory species and plays a large role in their annual survival (Sillett and Holmes 2002). Evolutionary pressure on energetic condition can be strong, but few studies have evaluated how condition is affected by changes in climate or phenology (Swanson et al. 1999). In eastern North America, autumn migration has become earlier among long-distance Neotropical migrants, but is later for short-distance temperate migrants (Van Buskirk et al. 2009, Rosenfield et al. 2011). In Europe, autumn migration has also advanced among long-distance migrants, while no change or later departures were measured in short-distance migrants or bird species with a variable number of broods (Jenni and Kéry 2003, Filippi-Codaccioni et al. 2010).

We used 13 years of data (2000–2012) from two migratory monitoring stations to evaluate the correlation of climate change on the autumn migration of songbirds and raptor species that prey on songbirds in the western United States. We hypothesized that annual variation in climate would explain annual variation in timing and energetic condition as we expect weather to impact food responses directly; this is believed to be the ultimate driver of migratory behavior (Newton 2008). We hypothesized that changes over time in migration phenology would depend on migration distance, diet, or both. Specifically, we predicted that long-distance Neotropical migrants would migrate earlier or have no change, whereas short-distance temperate migrants would migrate later (Jenni and Kéry 2003, Van Buskirk et al. 2009, Rosenfield et al. 2011). Also, we predicted that climate effects on phenology would differ dependent upon diet as weather should influence food availability differently. Alternatively, migratory birds could show more plasticity in body condition than in phenology.

METHODS

Study Sites and Species

The study was conducted in southwest Idaho along the Boise Foothills, which comprise north–south trending peaks and hills in the Boise Mountains. The foothills form the southernmost extent of the central Idaho mountains. Specifically, our two study sites were Lucky Peak (1,845 m), located 12 km east of Boise, Ada County, Idaho (43° 36′ N, 116° 03′ W), and Boise Peak (1,992 m), located 14 km northeast of Boise, Boise County, Idaho (43° 42′ N, 116° 05′ W). We obtained songbird counts and measures at Lucky Peak, raptor counts at Lucky Peak, and raptor measures at both

Lucky Peak and Boise Peak. The study area is part of the intermountain corridor migratory flyway (Goodrich and Smith 2008) and is located at the boundary between the mostly forested mountains to the north and the shrub steppe to the south (Carlisle et al. 2004).

We started with an initial list of abundant birds, including 25 species of songbirds and two species of primary bird-eating raptors. We selected the top 17 species of songbirds and top two species of raptors based on capture volumes that did not show trends in total volume or the ratio of juveniles to adults during the duration of our study (Table 11.1). Estimates of phenology trends may be biased if the sampling of cohorts with different migratory timing changes over time (R. A. Miller, pers. obs.). We chose to focus on the most abundant species to maximize compatibility with our generalized data collection approach, to minimize the influence of outliers, and to enable the broadest general conclusions possible (minimum

TABLE 11.1

Seventeen songbird species and two raptor species, in taxonomic order, used for analysis of diet, migratory distance, and climate on passage date, and energetic condition (mean ± SE) of migratory landbirds in the Boise Foothills, Idaho, 2000–2012.

Species	Diet[a]	Distance	Mean passage	Mean condition
Sharp-shinned Hawk (*Accipiter striatus*)	Birds	Short	266.2 ± 0.10	2.04 ± 0.00
Cooper's Hawk (*Accipiter cooperii*)	Birds	Short	262.6 ± 0.10	2.67 ± 0.01
Hammond's Flycatcher (*Empidonax hammondii*)	Insects	Long	237.2 ± 0.55	3.37 ± 0.01
Dusky Flycatcher (*Empidonax oberholseri*)	Insects	Long	222.9 ± 0.24	3.94 ± 0.01
Warbling Vireo (*Vireo gilvus*)	Insects	Long	227.2 ± 0.37	4.24 ± 0.01
Mountain Chickadee (*Poecile gambeli*)	Insects	Short	243.8 ± 0.98	3.75 ± 0.01
Red-breasted Nuthatch (*Sitta canadensis*)	Insects	Short	250.6 ± 0.66	3.58 ± 0.01
Ruby-crowned Kinglet (*Regulus calendula*)	Insects	Short	266.1 ± 0.15	3.34 ± 0.00
Hermit Thrush (*Catharus guttatus*)	Insects	Short	259.6 ± 0.87	3.54 ± 0.02
American Robin (*Turdus migratorius*)	Fruit	Short	259.9 ± 0.98	3.60 ± 0.02
Townsend's Warbler (*Setophaga townsendi*)	Insects	Long	243.7 ± 0.48	3.69 ± 0.01
Wilson's Warbler (*Cardellina pusilla*)	Insects	Long	241.2 ± 0.43	4.51 ± 0.01
Spotted Towhee (*Pipilo maculatus*)	Seeds	Short	237.4 ± 0.41	6.56 ± 0.01
Chipping Sparrow (*Spizella passerina*)	Seeds	Short	230.4 ± 0.48	3.68 ± 0.01
White-crowned Sparrow (*Zonotrichia leucophrys gambelii*)	Seeds	Short	261.3 ± 0.11	5.79 ± 0.01
Oregon Junco (*Junco hyemalis, oreganus* group)	Seeds	Short	271.7 ± 0.20	4.15 ± 0.01
Western Tanager (*Piranga ludoviciana*)	Fruit	Long	237.7 ± 0.23	4.00 ± 0.01
Black-headed Grosbeak (*Pheucticus melanocephalus*)	Fruit	Long	225.1 ± 0.56	4.85 ± 0.02
Pine Siskin (*Spinus pinus*)	Seeds	Short	228.4 ± 0.57	3.40 ± 0.03

[a] Diet guild information sourced from Sedgwick 1993, 1994, Chilton et al. 1995, Greenlaw 1996, Dawson 1997, Middleton 1998, Wright et al. 1998, Ammon and Gilbert 1999, Ghalambor and Martin 1999, Hudon 1999, Mccakkum et al. 1999, Sallabanks and James 1999, Bildstein and Meyer 2000, Gardali and Ballard 2000, Nolan et al. 2002, Curtis et al. 2006, Swanson et al. 2008, Ortega and Hill 2010, and Dellinger et al. 2012.

capture rates > 30 birds per year per species). We acknowledge that climate is likely influencing rare species as well, but a more focused study approach would be warranted for those situations. We categorized the migratory distance for each species as long distance if their winter range was clearly separated from their breeding range and the bulk of their winter range was south of the US–Mexican border; otherwise, we classified them as short-distance migrants (Table 11.1). We classified the diet for each species as avivore, insectivore, granivore, or frugivore based upon their primary diet during breeding and early autumn migration stages. Nearly all nonraptorial species of birds consume insects to some degree, but we assigned species to other diets if plant materials made up a significant portion of their diet during breeding or autumn migration. For example, Western Tanagers (Piranga ludoviciana) eat insects but we classified the species as a frugivore because the diet is primarily fruits during autumn. In contrast, the Ruby-crowned Kinglet (Regulus calendula) was classified as an insectivore because its diet is almost entirely insects and insect products (Hudon 1999, Swanson et al. 2008).

Songbird and Raptor Survey Methods

We captured songbirds at Lucky Peak using 12 m × 2.5 m × 36 mm mesh mist nets in mountain shrubland habitat (Carlisle et al. 2004). The standard operation consisted of 10 nets operated daily for 5 hours beginning at sunrise from 16 July to 15 October, except in the case of high winds or continuous precipitation. We trapped migrating raptors at both sites using a variety of traps (dho-gaza, bow net, and mist net) and avian lures (Bloom et al. 2007). Captured birds were identified to species, age, and sex (Pyle 1997, 2008). We recorded date of capture, wing chord length, mass, and other morphological characteristics of each bird. For the purpose of this analysis, birds were not counted on their second or subsequent captures within or among years (Miller et al. 2011). We used Julian date to represent each bird's passage date. We divided mass by wing chord length cubed to calculate a size-corrected mass as an index to energetic condition (Swanson et al. 1999, DeLong and Gessaman 2001). We multiplied this value by 100,000 to avoid influences of rounding and to make the calculated measurements easier to report (Winker et al. 1992). Within a given

species, we considered birds with a higher size-corrected mass to be in better energetic condition relative to birds with lower size-corrected mass.

We conducted raptor migration counts at Lucky Peak using standardized methods (Hoffman and Smith 2003). Migrating raptors were counted daily from 25 August through 31 October by a minimum of two trained observers. Counts were curtailed only during periods of prolonged precipitation. Counts began at 12:00 MST during August and 10:00 MST for the remainder of the season and continued throughout the day until raptor flights ceased, usually between 17:00 and 19:00 MST. Best efforts were made to ensure that only migrating raptors were counted (Kaltenecker et al. 2012).

Climate Data

We obtained monthly temperature and precipitation data during the study period from the Global Historical Climatology Network (GHCN) Daily, version 2 (US Department of Commerce 2012). The data are provided as monthly means for temperature and monthly totals for precipitation, and they have been subjected to a suite of quality assurance. We chose climate data from stations spread across the Northern Rockies Bird Conservation Region (BCR, US NABCI Committee 2000, Figure 11.1). We restricted our area of consideration to portions of the Northern Rockies BCR north of our monitoring station and west of the continental divide to best represent the breeding areas of the birds migrating through southwestern Idaho. We further restricted the data to that gathered from weather stations greater than 50 km apart and with complete data sets over our 13-year study period, resulting in the use of data from 17 weather stations (Figure 11.1). We averaged the data from the 17 stations to produce a monthly index for temperature and precipitation across the region. Our intent was to generate broad weather averages across the region where our sampled birds breed, and we made no attempt to further correct for latitudinal or elevation effects.

We obtained monthly data for two atmospheric pressure indices from the National Center for Atmospheric Research. The North Pacific index (NPI) is an area-weighted sea level pressure measurement from the North Pacific intended to measure variations in atmospheric circulation

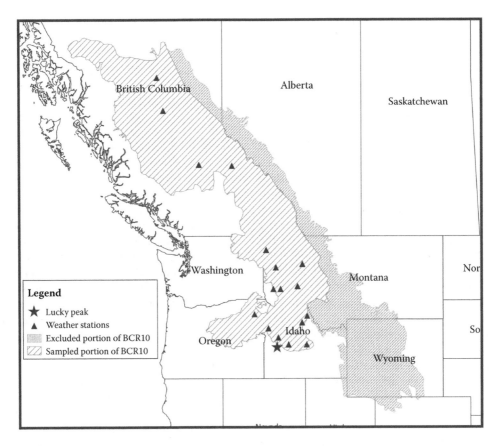

Figure 11.1. Map of Bird Conservation Region 10 in the northern Rocky Mountains, separated by the continental divide with eastern portion excluded from consideration. The map includes locations of 17 weather stations in the western portion of the BCR that had consistently reliable data over the period of this study and the location of our "Lucky Peak" monitoring station. The "Boise Peak" monitoring station was situated 11 km NNW of Lucky Peak (not pictured).

(Trenberth and Hurrell 1994, Hurrell and National Center for Atmospheric Research Staff 2013). The multivariate El Niño/southern oscillation index (MEI) is a standardized combination of six common El Niño measures focusing on the broader Pacific region (Wolter and Timlin 1993, National Center for Atmospheric Research Staff 2013). We included the hemispheric-scale indices to represent more general and larger scale climate influences not captured by regional temperature and precipitation indices and because they have been shown to be influential on the timing of avian migration (Nott et al. 2002, Van Buskirk et al. 2009). Higher values for each index are expected to be correlated with warmer temperatures, drier winters, and extreme weather events in the northern–western United States.

For each climate/weather index we created a 3-month index (July–September), a 6-month index (April–September), a 9-month index (January–September), and a 12-month index (previous October–September). The indices align well with peak periods of avian migration in Idaho and with standardized "water-year" measurements reported for precipitation. We created the indices for temperature, NPI, and MEI by averaging the monthly values across 3, 6, 9, and 12 months. We created the index for precipitation by summing across 3, 6, 9, and 12 months. Climate indices and year were scaled and centered to assist in model convergence. Centering was performed by subtracting the mean of the index over the length of the study period from each value. Scaling was performed by dividing each value by the standard deviation of the index.

Statistical Analyses

We checked for trends in each climate index at each time scale using linear models with year as the predictor and the climate index as the response. We used an alpha level of 0.05 to measure significance of these trends. For analyses of migratory timing and energetic condition we followed the same analytical procedure. We used generalized linear mixed modeling and maximum likelihood with a Gaussian distribution for all analyses (Zuur et al. 2009). In each case, we included species and year (scaled and centered) as random effects. For each climate index (temperature, precipitation, NPI, and MEI), we first chose the time scale that best explained each response variable by comparing Akaike information criterion (AIC) values for each model of each index and time scale (3 months, 6 months, 9 months, and 12 months; Burnham and Anderson 2002). There were no correlations between pairs of climate indices.

We created a global model including the best time scale for each of the four climate indices, year, diet, migratory distance, the interactions between diet and year and diet and each climate index, and the interactions between migratory distance and year and migratory distance and each climate index as fixed effects, with year and species as random effects to predict passage date or energetic condition of each bird. Including year as a fixed effect enabled us to test for an overall trend in migratory timing. Interaction terms were included to test whether the effects of year or climate on migratory timing or energetic condition were dependent on diet or migratory distance. We compared all subsets of the global model using AIC to choose the top model (Burnham and Anderson 2002). Lower ranked models were considered parsimonious if they were ΔAIC ≤ 2 of the top model, contained informative parameters, and were not a subset of the higher ranked model (Burnham and Anderson 2002, Arnold 2010). Coefficients and 85% confidence intervals are presented before back-transformation. We report predicted effect size for each variable whose 85% confidence interval of the coefficient failed to overlap zero after back-transformation (Arnold 2010).

We conducted all statistical analyses in Program R (version 2.15.2, R Foundation for Statistical Computing, Vienna, Austria). We used functions on the lme4 package for mixed-model analyses (version 1.0-4, D. Bates, M. Maechler, B. Bolker, and S. Walker). We calculated standard errors with the function described in the package psych

(version 1.2.12, W. Revelle). All means are presented with ±SE.

RESULTS

We captured a total of 48,602 individuals of 17 songbird species over a 13-year study period (2000–2012). Mean passage date for songbird species was 9 September, but ranged among species from 11 August for Dusky Flycatchers (*Empidonax oberholseri*) to 28 September for Oregon Juncos (*Junco hyemalis, oreganus* group; Table 11.1). Mean energetic condition for songbirds was 4.15 ± 0.005 g/mm^3. We counted 25,096 individuals of migrating raptors. Mean passage date for raptors was 22 September, but ranged from 19 September for Cooper's Hawks (*Accipiter cooperii*) to 23 September for Sharp-shinned Hawks (*A. striatus*; Table 11.1). We captured a total of 9,795 individuals of two raptor species. Mean energetic condition for the raptors was 2.21 ± 0.004 g/mm^3.

There were no significant trends in any of the climate indices at any time scale over the duration of our study (Table 11.2). In predicting passage date, the 3-month time scale was chosen for temperature, 9-month for precipitation, 9-month for NPI, and 3-month for MEI (Table 11.2). In predicting energetic condition, the 9-month time scale was chosen for temperature, 12-month for precipitation, 12-month for NPI, and 12-month for MEI (Table 11.2).

The top model predicting migratory passage date included 3-month temperature, 9-month precipitation, 9-month NPI, and 3-month MEI, year, diet, migratory distance, the interactions between diet and year and diet and each climate index, and the interactions between migratory distance and year and migratory distance and each climate variable (Table 11.3). There was no overall trend in migratory timing among all species over the study period (Figure. 11.2). Frugivores exhibited the largest trend in timing, shifting 1.48 d/y later over the study period. Insectivores shifted 0.27 d/y earlier, granivores shifted 0.38 d/y later, and avivores had no shift in timing (Figure 11.2). Short-distance migrants showed no trend in autumn migratory timing, whereas long-distance migrants trended later at our study site (0.53 d/y; Figure 11.2).

Each of the climate indices retained in the top model exhibited effects on the migratory timing of birds at Lucky Peak (Table 11.4). For each 0.1°C

TABLE 11.2
Mean values, standard error, and results of statistical tests evaluating trends in four climate indices measured at four time scales across the Northern Rockies Bird Conservation Region in western North America from 2000 to 2012.

Index	3 Months	6 Months	9 Months	12 Months
Temperature	$15.5 \pm 0.19°C^a$	$12.2 \pm 0.16°C$	$7.2 \pm 0.18°C^b$	$5.4 \pm 0.15°C$
	$F_{1,11} = 0.12,$ $P = 0.74$	$F_{1,11} = 0.53,$ $P = 0.48$	$F_{1,11} = 0.10,$ $P = 0.76$	$F_{1,11} = 0.23,$ $P = 0.64$
Precipitation	96.4 ± 8.7 mm	282.5 ± 12.3 mm	484.5 ± 14.4 mm^a	710.4 ± 13.4 mm^b
	$F_{1,11} = 4.54,$ $P = 0.06$	$F_{1,11} = 0.002,$ $P = 0.96$	$F_{1,11} = 0.05,$ $P = 0.82$	$F_{1,11} = 0.18,$ $P = 0.68$
NPI	1015.8 ± 0.19	1015.5 ± 0.18	1013.3 ± 0.26^a	1012.6 ± 0.27^b
	$F_{1,11} = 0.001,$ $P = 0.97$	$F_{1,11} = 0.002,$ $P = 0.97$	$F_{1,11} = 0.72,$ $P = 0.41$	$F_{1,11} = 2.46,$ $P = 0.15$
MEI	0.08 ± 0.20^a	0.10 ± 0.13	-0.02 ± 0.13	-0.06 ± 0.15^b
	$F_{1,11} = 0.76,$ $P = 0.40$	$F_{1,11} = 0.60,$ $P = 0.45$	$F_{1,11} = 0.62,$ $P = 0.45$	$F_{1,11} = 0.64,$ $P = 0.44$

NOTES: Trends measured with linear regression with an alpha value of 0.05 but no trends were present in any climate variable measured at any time scale.

[a] Time scale chosen as best for predicting migratory passage date.
[b] Time scale chosen as best for predicting mean energetic condition.

TABLE 11.3
Top model, closest model, and "null" model from AIC model selection of various climate indices predicting Julian passage date of migratory birds past Lucky Peak, Idaho.

Model	K	AIC	ΔAIC	w_i	Cum. w_i	LL
npi9 + mei3 + Temp3 + Precip9 + diet:npi9 + diet:mei3 + diet:Temp3 + diet:Precip9 + distance:npi9 + distance:mei3 + distance:Temp3 + distance:Precip9 + diet:year + distance:year + diet + distance + year	33	612910.4	0.00	1	1	−306422.2
npi9 + mei3 + Temp3 + Precip9 + diet:npi9 + diet:mei3 + diet:Temp3 + diet:Precip9 + diet:year + distance:year + diet + distance + year	29	612925.4	15.01	0	1	−306433.7
NULL (distance:year + diet:year + distance + diet + year)	13	613333.6	423.26	0	1	−306653.8

NOTES: Each model includes additional random effects for year and species. Models with AIC weights less than 0.01 are not shown. npi9 = 9-month North Pacific index; mei3 = 3-month multivariate El Niño/southern oscillation index; Temp3 = 3-month temperature index from northwestern Rocky Mountains bird conservation region; Precip9 = 9-month precipitation index from northwestern Rocky Mountains bird conservation region.

increase in the 3-month temperature index there was a predicted shift in timing of 0.56 d earlier in frugivores, 0.19 d earlier in granivores, and 0.21 d earlier in short-distance migrants (Figure 11.3). For each 0.1-mm increase in the 9-month precipitation index there was a predicted shift in timing of 0.003 d later for insectivores, 0.002 d later for avivores, 0.0007 d earlier for granivores and 0.002 d later for long-distance migrants (Figure 11.3). For each unit increase in the 9-month NPI there was a predicted shift in timing of 0.50 d earlier for avivores, 0.96 d later for granivores, 0.19 d earlier for

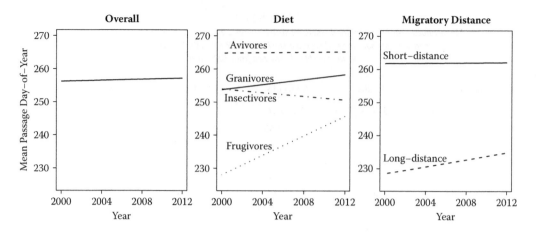

Figure 11.2. Mean passage dates during a 13-year study at Lucky Peak in southwest Idaho, 2000–2012. We present all sampled birds combined, all sampled birds separated by diet, and all sampled birds separated by migratory distance. No change was detected for the overall population, avivores, or short-distance migrants (confidence interval overlaps zero).

TABLE 11.4

Subset of model coefficients interactions and 85% confidence intervals from top model-predicting Julian passage date of migratory birds past Lucky Peak, Idaho.

Predictor variable	Avivore	Frugivore	Granivore	Insectivore	Short distance	Long distance
3-Month temperature	−0.01 (−0.32, 0.29)	**−3.92** (**−4.52, −3.33**)	**−1.36** (**−1.70, −1.02**)	0.08 (−0.20, 0.35)	**−0.45** (**−0.84, −0.06**)	0.30 (−0.07, 0.67)
9-Month precipitation	**−0.72** (**−0.99, −0.45**)	0.33 (−0.21, 0.87)	**−1.95** (**−2.25, −1.64**)	**1.59** (**1.34, 1.84**)	0.31 (−0.03, 0.65)	**1.43** (**1.12, 1.76**)
9-Month North Pacific index	**−0.64** (**−0.94, −0.34**)	−0.05 (−0.58, 0.48)	**0.74** (**0.42, 1.05**)	0.18 (−0.11, 0.46)	**−0.75** (**−1.14, −0.36**)	**0.55** (**0.17, 0.93**)
3-Month multivariate El Niño/ southern oscillation index	−0.02 (−0.35, 0.31)	**3.00** (**2.34, 3.66**)	**1.96** (**1.59, 2.34**)	**−0.65** (**−0.97, −0.34**)	**0.87** (**0.43, 1.31**)	**−1.09** (**−1.51, −0.66**)

NOTES: All predictor variables were scaled and centered. Bold text indicates coefficients whose 85% confidence intervals did not overlap zero.

short-distance migrants, and 0.59 d later for long-distance migrants (Figure 11.3). Last, for each unit increase in the 3-month MEI there was a predicted shift in timing of 3.22 d later for frugivores, 1.80 d later for granivores, 0.90 d earlier for insectivores, 1.40 d later for short-distance migrants, and 0.20 d later for long-distance migrants (Figure 11.3).

The top model-predicting energetic condition included 9-month temperature, 12-month precipitation, 12-month NPI, and 12-month MEI, year, diet, migratory distance, and the interactions between diet and year and diet and each climate

index, and the interaction of migratory distance and year (Table 11.5). In general, energetic condition has improved over the study period at a rate of 0.014 g/mm³/y (Figure 11.4). However, differing trends became apparent when analyzed by diet and migratory distance (Figure 11.4). Granivores exhibited the largest trend in energetic condition, increasing at a rate of 0.038 g/mm³/y over the study period. Insectivores increased at a rate of 0.01 g/mm³/y, frugivores declined at a rate of 0.01 g/mm³/y, and avivores exhibited no trend in energetic condition. Short-distance migrants

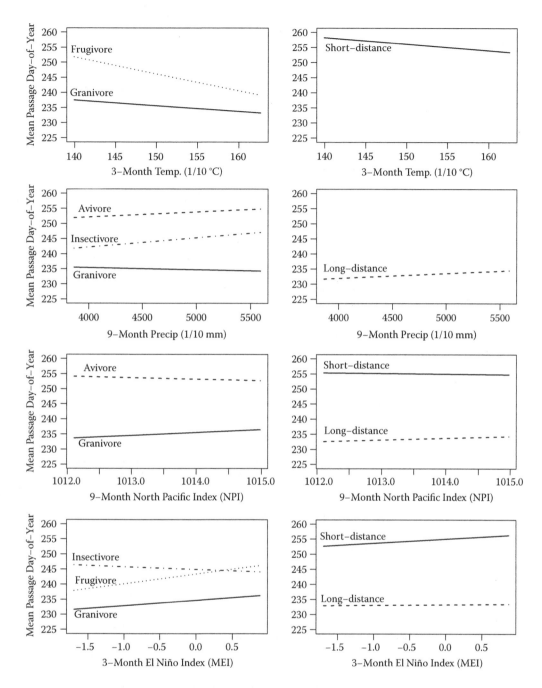

Figure 11.3. Predicted effect sizes of each of the four climate variables represented in the top ranked model for mean passage date of migratory birds by Lucky Peak in southwest Idaho, 2000–2012. Effect sizes were calculated separately by primary diet and then by migratory distance. For each sample, all other covariates in the top model are held at their mean values. Trend lines are shown for groups where confidence interval of the coefficient did not overlap zero.

increased at a rate of 0.013 g/mm³/y, whereas long-distance migrants increased at a rate of 0.01 g/mm³/y (Figure 11.4).

Each of the climate indices retained in the top model exhibited effects on the energetic condition of birds at Lucky Peak and Boise Peak (Table 11.6). For each 0.1°C increase in the 9-month temperature index there was a predicted shift in energetic condition of 0.005 g/mm³ decrease for granivores, a 0.004 g/mm³ decrease for frugivores,

TABLE 11.5

Top model, closest model, and "null" model from AIC model selection of various climate indices predicting energetic condition of migratory birds past Lucky Peak, Idaho.

Model	K	AIC	ΔAIC	w_i	Cum. w_i	LL
npi12 + mei12 + Temp9 + Precip12 + diet:npi12 + diet:mei12 + diet:Temp9 + diet:Precip12 + diet:year + distance:year + diet + distance + year	29	56476.63	0.00	0.63	0.63	−28209.30
npi12 + mei12 + Temp9 + diet:npi9 + diet:mei3 + diet:Temp3 + diet:year + distance:year + diet + distance + year	25	56477.95	1.33	0.32	0.95	−28213.96
npi12 + mei12 + Temp9 + diet:npi12 + diet:mei12 + diet:Temp9 + distance:npi12 + distance:mei12 + distance:Temp9 + diet:year + distance:year + diet + distance + year	28	56482.83	6.21	0.03	0.98	−28213.40
npi12 + mei12 + Temp9 + Precip12 + diet:npi12 + diet:mei12 + diet:Temp9 + diet:Precip12 + distance:npi12 + distance:mei12 + distance:Temp9 + distance:Precip12 + diet:year + distance:year + diet + distance + year	33	56483.33	6.70	0.02	1.00	−28208.64
NULL (distance:year + diet:year + distance + diet + year)	13	56558.53	81.90	0	1	−28266.26

NOTES: Each model includes additional random effects for year and species. Models with AIC weights less than 0.01 are not shown. npi9 = 9-month North Pacific index; mei3 = 3-month multivariate El Niño/southern oscillation index; Temp3 = 3-month temperature index from northwestern Rocky Mountains bird conservation region; Precip9 = 9-month precipitation index from northwestern Rocky Mountains bird conservation region.

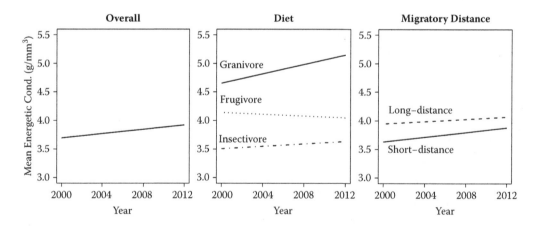

Figure 11.4. Mean energetic condition of birds during a 13-year study at Lucky Peak in southwest Idaho, 2000–2012. Condition is shown for all sampled birds combined, all sampled birds separated by diet, and all sampled birds separated by migratory distance. Overall, energetic condition tended to be higher but no trend was present for avivores.

and a 0.001 g/mm³ decrease for insectivores. For each 0.1 mm increase in the 9-month precipitation index, there was a predicted shift in energetic condition of 3.6×10^{-5} g/mm³ decrease for frugivores and a 4.4×10^{-7} g/mm³ decrease for insectivores. For each unit increase in the 12-month NPI there was a predicted increase in energetic condition of 0.04 g/mm³ for frugivores and 0.02 g/mm³ for granivores. Last, for each unit increase in the 12-month MEI there was a predicted shift in energetic condition of 0.06 g/mm³ increase for granivores and 0.01 g/mm³ decrease for insectivores.

TABLE 11.6
Subset of model coefficient interactions and 85% confidence intervals from top model predicting energetic condition of migratory birds past Lucky Peak and Boise Peak, Idaho.

Predictor variable	Avivore	Frugivore	Granivore	Insectivore
9-Month temperature	0.0002	**−0.0023**	**−0.0340**	**−0.0079**
	(−0.0119, 0.0124)	**(−0.0366, −0.0100)**	**(−0.0428, −0.0252)**	**(−0.0160, 0.0001)**
12-Month precipitation	0.0121	**−0.0293**	−0.0084	**−0.0124**
	(−0.0008, 0.0250)	**(−0.0441, −0.0146)**	(−0.0176, 0.0009)	**(−0.0211, −0.0036)**
12-Month North Pacific index	−0.0054	**0.0479**	**0.0239**	0.0074
	(−0.0180, 0.0072)	**(0.0362, 0.0595)**	**(0.0150, 0.0327)**	(−0.0009, 0.0157)
12-Month multivariate El Niño/southern oscillation index	0.0110	−0.0059	**0.0169**	**−0.0184**
	(−0.0034, 0.0255)	(−0.0216, 0.0099)	**(0.0065, 0.0272)**	**(−0.0282, −0.0085)**

NOTES: All predictor variables were scaled and centered. Bold text indicates coefficients whose 85% confidence intervals did not overlap zero.

DISCUSSION

Global climate change could have dramatic impacts on bird species, particularly migratory species with life histories that require use of multiple geographically dispersed habitat types, and are often dependent upon synchronized availability of prey. To better understand the connection between climate and species life histories, long-term avian monitoring projects are important (Porzig et al. 2011). We used 13 years of data for 19 species of birds to explore trends in migratory timing, energetic condition, and the effects of year-to-year variation in climate on these trends. While climate change has occurred around the globe and in the western United States, the Northern Rockies Bird Conservation Region has not experienced any significant trends in climate over the duration of our study.

Contrary to our initial hypothesis, and to other studies of migratory timing based on migration distance (Jenni and Kéry 2003, Van Buskirk et al. 2009, Rosenfield et al. 2011), we found no trend in the timing of autumn migration for short-distance migrants, but long-distance migrants tended to migrate later. We detected no significant trends in climate over the duration of our 13-year study, and we might expect little change in migratory timing of the short-distance migrants. However, long-distance migrants are exposed to a greater diversity of environments where greater change might be evident, which could explain their response. The magnitude and direction of trends in timing vary by the diet of the individuals.

The avivore raptors showed no trend in timing, whereas the frugivores shifted more than 15 days later over the duration of our study. As there were no overall trends present in climate over the duration of the study, the trends in timing by diet must be influenced by other factors not measured or climatic factors operating on different scales than we evaluated.

We hypothesized that annual climate variation would explain annual variation in timing and energetic condition as we expect regional climate directly to impact food resources, which is believed to be the ultimate driver of migratory behavior (Newton 2008). Our results showed relationships between annual variation in climate and annual variation in both timing and energetic condition. Furthermore, the strength and direction of these relationships varied by the diet of migrant birds. Generally, higher average within-year temperatures were correlated with birds migrating earlier—specifically granivores, frugivores, and short-distance migrants—with a corresponding decrease in energetic condition of these species in warmer years. Changes in condition could be the results of less availability and palatability of seeds and fruits. Insectivores showed no correlation between annual temperatures and timing, but did show a decrease in energetic condition in years with warmer average temperatures. The result is counter to our expectations as ectothermic arthropods are expected to be more active and at higher abundances in warmer years (Tulp and Schekkerman 2008). Increased precipitation was correlated with many species migrating later, with the exception of the granivores, which migrated earlier in years of greater precipitation. Increased precipitation has been shown to influence arthropod abundance and fruit abundance positively, which in turn would be expected to influence the insectivores and frugivores to migrate later. However, the earlier migration of granivores in wet years was unexpected because summer precipitation has been shown to increase seed production (Dunning and Brown 1982). Also unexpected was a decrease in body condition of insectivores and frugivores in wetter years.

We included the hemispheric climate models to capture general climate influences not captured by temperature and precipitation. Retention in the top model indicated that hemispheric climatic conditions had some explanatory power for patterns of bird migration. The NPI in general had the least predictive power of the two measures, which may explain why this index is not used more broadly in ecological studies. The MEI exhibited larger influences, particularly on frugivores and short-distance migrants; however, the direction of the influence contradicts our expectations based upon the results of the temperature index. However, during our study, there were no strong El Niño events.

We further hypothesized that factors influencing songbird migration would similarly influence raptors that feed on songbirds. Raptors exhibited no trend in migratory timing and no trend in condition. Raptor responses to annual climate variation were negligible. Rosenfield et al. (2011) documented a delayed migration of Sharp-shinned Hawks in the midwestern United States, but we did not detect a significant shift in the timing of passage for Sharp-shinned Hawks or Cooper's Hawks. The difference may be the result of the longer duration of their study (35 years) than our study (13 years), which can improve the ability to detect gradual trends. Additionally, generalist raptors may have more flexibility to shift among various prey sources while maintaining their migratory timing and their average energetic condition, especially if different guilds of prey are responding in opposite directions and thus collectively maintaining a relatively constant food source throughout the autumn migration. However, as shifts within their prey populations continue, a mismatch in timing between their migration and that of their prey could eventually exist, requiring a response.

Some species shifted timing while holding energetic condition constant, while others held timing constant while shifting condition. Our results suggest that different species may have different abilities to adapt to annual variation, possibly integrating day length, fat stores, and other environmental factors in their decision to migrate (Sandberg and Moore 1996, Helm et al. 2009). Individuals not reaching a minimum threshold of body condition may perish before reaching our study site.

Our study focused on correlations between a limited number of climate factors and migratory behaviors of a diverse set of birds. Our results confirmed a number of hypotheses but contradicted others. Clearly, the timing and condition of migrants are subject to many ecological forces and cannot be fully explained by a few climate indices. The findings of this study could be enhanced by the direct measurement of food availability on the breeding grounds and en route to our monitoring station.

In conclusion, timing of autumn migration is changing for a number of avian species migrating through the western United States. The magnitude of these changes varies by diet and by migratory distance, but is not as predicted by previous research. The annual variation in timing for a given class of migrants is predicted by regional temperature and precipitation, and by hemispheric climate indices. We did not measure food availability, but the predictive effect of climate on migratory timing of birds is mostly consistent with its presumed effect on food availability. Energetic condition measurements supported the food hypotheses in some, but not all cases. The different responses of species to annual variation in climate suggest that different species integrate different cues in their decision to migrate.

ACKNOWLEDGMENTS

We would like to thank all of the individual and organizational supporters of the Intermountain Bird Observatory including Boise State University, the Boise State University Raptor Research Center, the Idaho Department of Fish and Game, the Southwestern Idaho Birder's Association, and the Golden Eagle Audubon Society. In addition we thank all of the volunteers and crew members who have worked tirelessly for more than 13 years gathering valuable data on bird migration.

LITERATURE CITED

Altermatt, F. 2010. Tell me what you eat and I'll tell you when you fly: diet can predict phenological changes in response to climate change. Ecology Letters 13:1475–1484.

Ammon, E. M., and W. M. Gilbert. [online]. 1999. Wilson's Warbler (*Cardellina pusilla*). In A. Poole (editor), The birds of North America. Cornell Lab of Ornithology, Ithaca, NY. <http://bna.birds.cornell.edu/bna/species/478>

Anderson, J. T., D. W. Inouye, A. M. McKinney, R. I. Colautti, and T. Mitchell-Olds. 2012. Phenotypic plasticity and adaptive evolution contribute to advancing flowering phenology in response to climate change. Proceedings of the Royal Society of London B 279:3843–3852.

Arnold, T. W. 2010. Uninformative parameters and model selection using Akaike's information criterion. Journal of Wildlife Management 74:1175–1178.

Bell, H. L. 1985. Seasonal variation and the effects of drought on the abundance of arthropods in savanna woodland on the northern tablelands of New South Wales. Australian Journal of Ecology 10:207–221.

Bildstein, K. L., and K. Meyer. [online]. 2000. Sharp-shinned Hawk (*Accipiter striatus*). In A. Poole (editor), The birds of North America. Cornell Lab of Ornithology, Ithaca, NY. <http://bna.birds.cornell.edu/bna/species/482>

Blancher, P. J., and R. J. Robertson. 1987. Effect of food supply on the breeding biology of Western Kingbirds. Ecology 68:723–732.

Bloom, P. H., W. S. Clark, and J. W. Kidd. 2007. Capture techniques. Pp. 193–220 in D. M. Bird and K. L. Bildstein (editors), Raptor research and management techniques. Hancock House, Blaine, WA.

Both, C., S. Bouwhuis, C. M. Lessells, and M. E. Visser. 2006. Climate change and population declines in a long-distance migratory bird. Nature 441:81–83.

Both, C., C. A. M. Van Turnhout, R. G. Bijlsma, H. Siepel, A. J. Van Strien, and R. P. B. Foppen. 2009. Avian population consequences of climate change are most severe for long-distance migrants in seasonal habitats. Proceedings of the Royal Society of London B 277:1259–1266.

Both, C., and M. E. Visser. 2001. Adjustment to climate change is constrained by arrival date in a long-distance migrant bird. Nature 411:296–298.

Burnham, K., and D. R. Anderson. 2002. Model selection and multi-model inference: a practical information-theoretic approach. Springer-Verlag, New York, NY.

Butler, C. J. 2003. The disproportionate effect of global warming on the arrival dates of short-distance migratory birds in North America. Ibis 145:484–495.

Carlisle, J. D., S. L. Stock, G. S. Kaltenecker, and D. L. Swanson. 2004. Habitat associations, relative abundance, and species richness of autumn landbird migrants in southwestern Idaho. Condor 106:549–566.

Chilton, G., M. C. Baker, C. D. Barrentine, and M. A. Cunningham. [online]. 1995. White-crowned Sparrow (*Zonotrichia leucophrys*). In A. Poole (editor), The birds of North America. Cornell Lab of Ornithology, Ithaca, NY. <http://bna.birds.cornell.edu/bna/species/183>

Cotton, P. A. 2003. Avian migration phenology and global climate change. Proceedings of the National Academy of Sciences of the USA 100:12219–12222.

Curtis, O. E., R. N. Rosenfield, and J. Bielefeldt. [online]. 2006. Cooper's Hawk (*Accipiter cooperii*). In A. Poole (editor), The birds of North America. Cornell Lab of Ornithology, Ithaca, NY. <http://bna.birds.cornell.edu/bna/species/075>

Dawson, A. 2008. Control of the annual cycle in birds: endocrine constraints and plasticity in response to ecological variability. Philosophical Transactions of the Royal Society B 363:1621–1633.

Dawson, W. R. [online]. 1997. Pine Siskin (*Spinus pinus*). In A. Poole (editor), The birds of North America. Cornell Lab of Ornithology, Ithaca, NY. <http://bna.birds.cornell.edu/bna/species/280>

Dellinger, R., P. B. Wood, P. W. Jones, and T. M. Donovan. [online]. 2012. Hermit Thrush (*Catharus guttatus*). In A. Poole (editor), The birds of North America. Cornell Lab of Ornithology, Ithaca, NY. <http://bna.birds.cornell.edu/bna/species/261>

DeLong, J. P., and J. A. Gessaman. 2001. A comparison of noninvasive techniques for estimating total body fat in Sharp-shinned and Cooper's Hawks. Journal of Field Ornithology 72:349–364.

Dunning, J. B., and J. H. Brown. 1982. Summer rainfall and winter sparrow densities: test of the food limitation hypothesis. Auk 99:123–129.

Filippi-Codaccioni, O., J. P. Moussus, J. P. Urcun, and F. Jiguet. 2010. Advanced departure dates in long-distance migratory raptors. Journal of Ornithology 151:687–694.

Gao, S., J. Wang, Z. Zhang, G. Dong, and J. Guo. 2012. Seed production, mass, germinability, and subsequent seedling growth responses to parental warming environment in *Leymus chinensis*. Crop and Pasture Science 63:87–94.

Gardali, T., and G. Ballard. [online]. 2000. Warbling Vireo (*Vireo gilvus*). In A. Poole (editor), The birds of North America. Cornell Lab of Ornithology, Ithaca, NY. <http://bna.birds.cornell.edu/bna/species/551>

Ghalambor, C. K., and T. E. Martin. [online]. 1999. Red-breasted Nuthatch (*Sitta canadensis*). In A. Poole (editor), The birds of North America. Cornell Lab of Ornithology, Ithaca, NY. <http://bna.birds.cornell.edu/bna/species/459>

Goodrich, L. J., and J. P. Smith. 2008. Raptor migration in North America. Pp. 37–150 in K. L. Bildstein, J. P. Smith, E. R. Inzuna, and R. R. Veit (editors), State of North America's birds of prey. Nuttall Ornithological Club, Cambridge, MA, and American Ornithologists' Union, Washington, DC.

Gordo, O. 2007. Why are bird migration dates shifting? A review of weather and climate effects on avian migratory phenology. Climate Research 35:37–58.

Greenlaw, J. S. [online]. 1996. Spotted Towhee (*Pipilo maculatus*). In A. Poole (editor), The birds of North America. Cornell Lab of Ornithology, Ithaca, NY. <http://bna.birds.cornell.edu/bna/species/263>

Halupka, L., A. Dyrcz, and M. Borowiec. 2008. Climate change affects breeding of Reed Warblers, *Acrocephalus scirpaceus*. Journal of Avian Biology 39:95–100.

Helm, B., I. Schwabl, and E. Gwinner. 2009. Circannual basis of geographically distinct bird schedules. Journal of Experimental Biology 212:1259–1269.

Hoffman, S. W., and J. P. Smith. 2003. Population trends of migratory raptors in western North America, 1977–2001. Condor 105:397–419.

Hudon, J. [online]. 1999. Western Tanager (*Piranga ludoviciana*). In A. Poole (editor), The birds of North America. Cornell Lab of Ornithology, Ithaca, NY. <http://bna.birds.cornell.edu/bna/species/432>

Huntley, B., Y. C. Collingham, S. G. Willis, and R. E. Green. 2008. Potential impacts of climatic change on European breeding birds. PLoS One 3:e1439.

Hurrell, J., and National Center for Atmospheric Research Staff. [online]. 2013. The climate data guide: north pacific (NP) index by Trenberth and Hurrell; monthly and winter. <https://climatedataguide.ucar.edu/climate-data/north-pacific-np-index-trenberth-and-hurrell-monthly-and-winter> (16 September 2013).

Inouye, D. W. 2008. Effects of climate change on phenology, frost damage, and floral abundance of montane wildflowers. Ecology 89:353–362.

Jenni, L., and M. Kéry. 2003. Timing of autumn bird migration under climate change: advances in long-distance migrants, delays in short-distance migrants. Proceedings of the Royal Society of London B 270:1467–1471.

Kaltenecker, G. S., J. D. Carlisle, J. Pollock, G. Rozhon, J. Butch, and M. J. Bechard. 2012. 2011 annual report Idaho Bird Observatory fall migration monitoring of raptors and songbirds. Boise Ridge, Idaho. Idaho Bird Observatory, Boise, ID.

Lambert, A. M., A. J. Miller-Rushing, and D. W. Inouye. 2010. Changes in snowmelt date and summer precipitation affect the flowering phenology of *Erythronium grandiflorum* (glacier lily; Liliaceae). American Journal of Botany 97:1431–1437.

La Sorte, F. A., and F. R. Thompson. 2007. Poleward shifts in winter ranges of North American birds. Ecology 88:1803–1812.

Lehikoinen, A., P. Saurola, P. Byholm, A. Lindén, and J. Valkama. 2010. Life history events of the Eurasian Sparrowhawk *Accipiter nisus* in a changing climate. Journal of Avian Biology 41:627–636.

McCallum, D. A., R. Grundel, and D. L. Dahlsten. [online]. 1999. Mountain Chickadee (*Poecile gambeli*). In A. Poole (editor), The birds of North America. Cornell Lab of Ornithology, Ithaca, NY. <http://bna.birds.cornell.edu/bna/species/453>

Meunier, C., L. Sirois, and Y. Bégin. 2007. Climate and *Picea mariana* seed maturation relationships: a multi-scale perspective. Ecological Monographs 77:361–376.

Middleton, A. L. [online]. 1998. Chipping Sparrow (*Spizella passerina*). In A. Poole (editor), The birds of North America. Cornell Lab of Ornithology, Ithaca, NY. <http://bna.birds.cornell.edu/bna/species/334>

Miller, R. A., J. D. Carlisle, and G. S. Kaltenecker. 2011. Effects of regional cold fronts and localized weather phenomena on autumn migration of raptors and landbirds in southwest Idaho. Condor 113:274–283.

National Center for Atmospheric Research Staff. [online]. 2013. The climate data guide: multivariate ENSO index. <https://climatedataguide.ucar.edu/climate-data/multivariate-enso-index> (16 September 2013).

Newton, I. 2008. The migration ecology of birds. Academic Press, Boston, MA.

Nolan, V., Jr., E. D. Ketterson, D. A. Cristol, C. M. Rogers, E. D. Clotfelter, R. C. Titus, S. J. Schoech, and E. Snajdr. [online]. 2002. Dark-eyed Junco (*Junco hyemalis*). In A. Poole (editor), The birds of North America. Cornell Lab of Ornithology, Ithaca, NY. <http://bna.birds.cornell.edu/bna/species/716>

Nott, M. P., D. F. Desante, R. B. Siegel, and P. Pyle. 2002. Influences of the El Niño/southern oscillation and the North Atlantic oscillation on avian productivity in forests of the Pacific Northwest of North America. Global Ecology and Biogeography 11:333–342.

Nufio, C. R., C. R. McGuire, M. D. Bowers, and R. P. Guralnick. 2010. Grasshopper community response to climatic change: variation along an elevational gradient. PLoS One 5:e12977.

Ortega, C., and G. E. Hill. [online]. 2010. Black-headed Grosbeak (*Pheucticus melanocephalus*). In A. Poole (editor), The birds of North America. Cornell Lab of Ornithology, Ithaca, NY. <http://bna.birds.cornell.edu/bna/species/143>

Parmesan, C., and G. Yohe. 2003. A globally coherent fingerprint of climate change impacts across natural systems. Nature 421:37–42.

Porzig, E. L., K. E. Dybala, T. Gardali, G. Ballard, G. R. Geupel, and J. A. Wiens. 2011. Forty-five years and counting: reflections from the Palomarin Field Station on the contribution of long-term monitoring and recommendations for the future. Condor 113:713–723.

Pyle, P. 1997. Identification guide to North American birds, part 1. Slate Creek Press, Bolinas, CA.

Pyle, P. 2008. Identification guide to North American birds, part 2. Slate Creek Press, Bolinas, CA.

Rosenfield, R. N., D. Lamers, D. L. Evans, M. Evans, and J. A. Cava. 2011. Shift to later timing by autumnal migrating Sharp-shinned Hawks. Wilson Journal of Ornithology 123:154–158.

Sallabanks, R., and F. C. James. [online]. 1999. American Robin (*Turdus migratorius*). In A. Poole (editor), The birds of North America. Cornell Lab of Ornithology, Ithaca, NY. <http://bna.birds.cornell.edu/bna/species/462>

Sandberg, R., and F. R. Moore. 1996. Migratory orientation of Red-eyed Vireos, *Vireo olivaceus*, in relation to energetic condition and ecological context. Behavioral Ecology and Sociobiology 39:1–10.

Sedgwick, J. A. [online]. 1993. Dusky Flycatcher (*Empidonax oberholseri*). In A. Poole (editor), The birds of North America. Cornell Lab of Ornithology, Ithaca, NY. <http://bna.birds.cornell.edu/bna/species/078>

Sedgwick, J. A. [online]. 1994. Hammond's Flycatcher (*Empidonax hammondii*). In A. Poole (editor), The birds of North America. Cornell Lab of Ornithology, Ithaca, NY. <http://bna.birds.cornell.edu/bna/species/109>

Sillett, T. S., and R. T. Holmes. 2002. Variation in survivorship of a migratory songbird throughout its annual cycle. Journal of Animal Ecology 71:296–308.

Swanson, D. L., J. L. Ingold, and G. E. Wallace. [online]. 2008. Ruby-crowned Kinglet (*Regulus calendula*). In A. Poole (editor), The birds of North America. Cornell Lab of Ornithology, Ithaca, NY. <http://bna.birds.cornell.edu/bna/species/119>

Swanson, D. L., E. T. Liknes, and K. L. Dean. 1999. Differences in migratory timing and energetic condition among sex/age classes in migrant Ruby-crowned Kinglets. Wilson Bulletin 111:61–69.

Trenberth, K. E., and J. W. Hurrell. 1994. Decadal atmosphere–ocean variations in the Pacific. Climate Dynamics 9:303–319.

Tulp, I., and H. Schekkerman. 2008. Has prey availability for Arctic birds advanced with climate change? Hindcasting the abundance of tundra arthropods using weather and seasonal variation. Arctic 61:48–60.

US Department of Commerce. [online]. 2012. Monthly summaries of global historical climatology network (GHCN) daily, version 2. <http://www.climate.gov> (10 December 2013).

US NABCI Committee. [online]. 2000. North American bird conservation initiative bird conservation region descriptions. US NABCI Committee, Arlington, VA. <http://www.nabci-us.org/aboutnabci/bcrdescrip.pdf>

Van Buskirk, J., R. S. Mulvihill, and R. C. Leberman. 2009. Variable shifts in spring and autumn migration phenology in North American songbirds associated with climate change. Global Change Biology 15:760–771.

Winker, K., D. W. Warner, and A. R. Weisbrod. 1992. Daily mass gains among woodland migrants at an inland stopover site. Auk 109:853–862.

Wolter, K., and M. S. Timlin. 1993. Monitoring ENSO in COADS with a seasonally adjusted principal component index in Proceedings of the 17th Climate Diagnostics Workshop 52–57.

Wright, A. L., G. D. Hayward, S. M. Matsuoka, and P. H. Hayward. [online]. 1998. Townsend's Warbler (*Setophaga townsendi*). In A. Poole (editor), The birds of North America. Cornell Lab of Ornithology, Ithaca, NY. <http://bna.birds.cornell.edu/bna/species/333>

Zuckerberg, B., D. N. Bonter, W. M. Hochachka, W. D. Koenig, A. T. DeGaetano, and J. L. Dickinson. 2011. Climatic constraints on wintering bird distributions are modified by urbanization and weather. Journal of Animal Ecology 80:403–413.

Zuur, A. F., E. N. Ieno, N. J. Walker, A. A. Saveliev, and G. M. Smith. 2009. Mixed effects models and extensions in ecology with R. Springer, New York, NY.

CHAPTER TWELVE

Autumn Migration of North American Landbirds[*]

Elizabeth R. Ellwood, Amanda Gallinat, Richard B. Primack, and Trevor L. Lloyd-Evans

Abstract. Recent research has shown that bird species are responding to changes in temperature and that spring migrations are shifting earlier for many species. Much less research has examined if and how fall migrations are changing, particularly in the United States. Here, we present an analysis of fall migration for 37 species of landbirds captured over a 44-year period at Manomet Center for Conservation Sciences in Manomet, Massachusetts. In the years 1969–2012, migration cohorts of 28 species significantly declined in abundance and two increased. For 14 species, the mean dates of fall passage at Manomet changed—nine migrated later and five earlier. Over this time span, the mean temperature of the region during the months of August and September has warmed. Fall passage dates for 14 species were correlated with temperature—13 species tended to migrate later with warmer temperatures and one species migrated earlier. We also included fall plant phenology in our analyses—four bird species tended to migrate later when leaf color change is delayed. Linear mixed-effects analysis indicates that species that winter in the tropics migrate earlier in the fall than birds that winter in other areas, and single-brooded species migrate earlier in the fall than double-brooded birds. Patterns of fall migration at Manomet show relatively little correlation with comparable data reported from the banding station at Powdermill Nature Reserve in Rector, Pennsylvania, suggesting spatial variation in the responses of fall migration to a warming climate. It is evident that fall migration dates vary considerably among species and that species are responding in different ways to the changing climate.

Key Words: climate change, fall, landbirds, Manomet, Massachusetts, migration, phenology.

Phenological observations have become increasingly central to our understanding of the effects of climate change on various taxa (Walther et al. 2002, Parmesan 2006). Recent years have broken records for heat, drought, and extreme weather events throughout the United States, and many plants and animals have responded with altered phenology (Jochner et al. 2011, Wilson 2012, Ellwood et al. 2013). The majority of this research has focused on spring phenology events. A spring bias is due in large part to the burst of springtime activity that follows

[*] Ellwood, E. R., A. Gallinat, R. B. Primack, and T. L. Lloyd-Evans. 2015. Autumn migration of North American landbirds. Pp. 193–205 in E. M. Wood and J. L. Kellermann (editors), Phenological synchrony and bird migration: changing climate and seasonal resources in North America. Studies in Avian Biology (no. 47), CRC Press, Boca Raton, FL.

cold winter weather and to the relative prevalence of historical records that exist as a result. Likewise, in observational studies it is often easier to record the presence of an organism, such as detections of a newly arrived migratory bird, as opposed to the absence of an organism or when a bird has departed on migration. Many plants and insects are now active earlier in the spring in response to warmer spring temperatures (Walther et al. 2002, Menzel et al. 2006, Parmesan 2006). Many migratory songbird species have adjusted to ecological change and arrive earlier in the spring as well (Lehikoinen et al. 2004, Gordo 2007, Van Buskirk et al. 2009). The timing of spring migrations for a smaller number of bird species has not changed or been delayed, possibly due to migratory cues that are inflexible, such as photoperiod, or are decoupled from temperate breeding areas (Both and Visser 2001, Gordo et al. 2005, Weidinger and Král 2007).

A lack of fall phenology records and investigations, especially for migratory birds, has left a gap in our understanding of the effects of climate change on phenology (Møller et al. 2010, Lehikoinen and Jaatinen 2012). Recent studies of fall migration of birds southbound from North America and Europe to their respective wintering grounds in South America and Africa have given mixed results. Most field studies have been conducted in Europe and have found that the timing of fall migration is spatially and temporally variable, and strategies differ among species (Sparks and Braslavská 2001, Cotton 2003, Jenni and Kéry 2003, Gordo and Sanz 2006, Knape et al. 2009, Hüppop and Hüppop 2011, Tøttrup et al. 2012). The few studies conducted in North America confirm these findings (MacMynowski and Root 2007, Van Buskirk et al. 2009, Smith and Paton 2011). However, species-specific pressures appear to be driving certain long-distance migrants toward earlier fall migration and short-distance migrants to later migration (Jenni and Kéry 2003, Thorup et al. 2007). Jenni and Kéry (2003) found that reproductive strategy influenced trends toward earlier or later fall migrations in Switzerland. Multiple-brooded species may have a second or third clutch following an early spring, which could delay fall migration (Møller 2007), whereas single-brooded species may simply depart after their brood has matured, essentially shifting the annual life cycle earlier in years with warmer springs (Sokolov et al. 2000, Gordo 2007). Van Buskirk et al. (2009), on the other hand, found that the number of clutches did not explain the variation among species in Pennsylvania. The authors did find evidence that migrating species are either expanding or contracting their breeding seasons, depending on the species.

The difficulty in identifying patterns may be due in part to regional variation associated with different migration routes. Migratory cohorts traveling along coastal flyways, including birds using Manomet as a stopover point, often contain a high proportion of hatch-year birds (Brooks 2008, Finnegan 2008), so the effects of climate change on migration in these coastal pathways may strongly affect recruitment of certain species. Changes in the length of the breeding season, time to molt and put on fat before departure, and changes in synchrony with fruit maturation, leaf senescence, or other plant resources may have a positive or negative impact on cohorts during migration. Bird migration times might also be affected by the timing of leaf senescence, which represents the end of the growing season, and the time of fruit maturation and the onset of insect diapause, two key food sources.

In the present study, we focused on fall migration of landbirds through the Manomet Center for Conservation Sciences in Massachusetts, where researchers have been consistently banding birds since the fall of 1969. This long-term data set from a single location provides us with long-term data to investigate the following questions:

1. How is the timing of fall migration changing over time?

2. How are dates of fall migration correlated with climate variables or life history traits of different bird species?

3. Is fall migration of birds synchronous with leaf senescence?

In addition, the proximity of Manomet to Powdermill Nature Reserve, as well as overlap in species composition, allows us to compare phenology trends to investigate the regional extent of responses to climate change.

METHODS

Banding Data

All banding data are from Manomet Center for Conservation Sciences, Manomet, Massachusetts

(41° 50′ N, 70° 30′ W), where staff and volunteers maintain 45–50 mist nets over the 7-ha coastal forested property. The data analyzed here represent initial capture dates, excluding recaptures of individuals, of 37 species of birds during fall passage between 15 August and 15 November, for the years 1969–2012 (44 years), for a total of 139,546 individual birds captured.

In some cases, captures of birds were due to natal dispersal or food-related irruptions of resident species not normally regarded as migratory, but these records were relatively rare events (e.g., Black-capped Chickadee and Tufted Titmouse). Nevertheless, southward movements can still involve large numbers of individuals, such as 28,535 Black-capped Chickadees captured during autumn over 44 years. Thirty-three of our study species were captured in every year, and the remaining four species were captured in at least 39 of the 44 years. The number of captures in a given year also serves as a proxy for migration cohort size and was used here to approximate the degree to which cohorts are increasing or decreasing in size over time (Lloyd-Evans and Atwood 2004).

Captures are highly variable from year to year, so to calculate the percentage of change in migration cohort sizes, we calculated mean number of captures for the first 5 years of data and compared that to the mean of the last 5 years. Mist net locations, quantity, and the number of hours each net was deployed were essentially unchanged for all 44 years. Preliminary analysis revealed that effort was consistent and as such was not included in further analysis. Captured birds were collected from mist nets and brought to the banding lab, where they were individually banded and identified according to the latest nomenclature (American Ornithologists' Union 1998, 2012; Chesser et al. 2012).

Climate Data

Monthly mean temperatures for the months before and during fall migration were acquired from the five weather stations closest to Manomet and based on available data at the National Climatic Data Center of the National Oceanic and Atmospheric Administration (www.ncdc. noaa.gov). Nearby stations were located in the Massachusetts towns of Brockton, East Wareham, Plymouth, Rochester, and Taunton and monthly temperatures were averaged across all stations.

Compiling data from several stations provided an average temperature for a small region instead of a single site and ensured that anomalous values from a single station would not bias the data set.

In addition to temperature, we included a 3-month running mean of the monthly North Atlantic oscillation (NAO) index from June through October as an additional weather variable. The NAO index measures atmospheric pressure differences in the North Atlantic Ocean; these pressure differences affect weather in the northeastern United States, with positive values generally associated with warmer temperatures. Also, the phase of the NAO affects the location and severity of hurricanes on the eastern coast of the United States (Elsner 2003). Data for the NAO index were taken from the Climate Prediction Center of the National Weather Service (www. cpc.ncep.noaa.gov/data/teledoc/nao.shtml).

Tree Phenology Data

To test the synchrony of fall bird migrations and fall plant phenology, we obtained data of leaf color and leaf fall from Harvard Forest, Petersham, Massachusetts, for the years 1991–2010 (O'Keefe 2000). Petersham is ~150 km northwest of Manomet, but provides a broad-scale index of regional plant phenology. The length of this time series and quality of the data make it the best available information for investigating questions of tree phenology over time in this region. The percentages of leaf color and leaf drop of two to five marked individuals of common woody species were recorded once or twice weekly. For our analysis, we calculated the mean dates for which leaf color and leaf fall each reached 90%. Birds are likely to see a lower proportion of leaf color and leaf fall on migration, but the categorical nature of the tree phenology data limited the levels of leaf color and leaf drop that could be used. The values were useful for differentiating between "early" and "late" years of fall tree phenology.

Statistical Analysis

All statistical analyses were conducted with R statistical software (www.r-project.org). We used mean fall migration date of initial captures of individually banded birds to avoid biases inherent in first or last dates due to changes in size of migratory cohorts over the 44 years (Miller-Rushing et al. 2008a). Migration dates are often normally distributed within a year, which allowed us to use mean date as a reasonable estimate of bird stopover date. We used linear regression to determine the combination of months for which mean temperature was a significant explanatory variable for stopover dates of individual species. Our approach limited the number of candidate variables for final model selection and restricted the combinations of models that were fully tested. Results of the analysis show that the greatest number of species (14 species) responded significantly to mean August + September temperature, as determined by a significance level of $P < 0.05$. We repeated modeling for NAO where June, July, and August were significant for the greatest number of species (two species with $P < 0.05$, six species with $P < 0.10$). As such, further analyses included only

temperatures and NAO for these months. The relationship between stopover date versus year, leaf color, leaf drop, and changes in migratory cohort size over time were also determined using linear regressions.

We performed linear mixed-effects analyses to determine the relationship between mean bird stopover dates and the explanatory factors described later with the lme4 package of R (Bates et al. 2012). Fixed effects in the multivariate analysis included mean August and September temperature, June–August NAO, leaf color, leaf drop, broodedness (Ehrlich et al. 1988), wintering region, wintering habitat (Rappole et al. 1983)—as well as whether a species is a farmland breeder (Poole and Gill 2002), winters in the tropics (Ehrlich et al. 1988), or is sexually dimorphic (Sibley 2000; Table 12.1). Values for these factors

TABLE 12.1
Study species and life-history traits of migratory landbirds captured during autumn migration at Manomet, Massachusetts, 1969–2012.

Common name	Scientific name	No. years	n	Mean dep. date	Brood	Farm	Tropics	Sex. dim.	Winter hab.	Winter region
Downy Woodpecker	*Picoides pubescens*	44	789	269	1	No	No	Yes	Woods	NA
Eastern Phoebe	*Sayornis phoebe*	44	654	262	2	Yes	No	No	Dry	NA
Blue-headed Vireo	*Vireo solitarius*	44	432	280	1	No	Yes	No	Woods	CA
Red-eyed Vireo	*Vireo olivaceus*	44	3,417	265	1	No	Yes	No	Woods	SA
Blue Jay	*Cyanocitta cristata*	44	2,260	270	1	No	No	No	All	NA
Black-capped Chickadee	*Poecile atricapillus*	44	28,535	281	1	No	No	No	All	NA
Tufted Titmouse	*Baeolophus bicolor*	44	5,256	284	1	No	No	No	All	NA
Brown Creeper	*Certhia americana*	44	1,624	283	1	No	No	No	Woods	NA
Carolina Wren	*Thryothorus ludovicianus*	39	544	249	2	Yes	No	No	All	NA
Golden-crowned Kinglet	*Regulus satrapa*	44	3,007	290	2	No	No	Yes	Woods	NA
Ruby-crowned Kinglet	*Regulus calendula*	44	1,834	287	1	No	Yes	Yes	Woods	CA
Veery	*Catharus fuscescens*	44	532	256	1	No	Yes	No	Woods	SA
Swainson's Thrush	*Catharus ustulatus*	44	1,196	270	1	No	Yes	No	Woods	SA
Hermit Thrush	*Catharus guttatus*	44	1,844	294	2	No	No	No	Woods	NA
American Robin	*Turdus migratorius*	44	7,091	270	2	Yes	No	Yes	All	NA
Gray Catbird	*Dumetella carolinensis*	44	21,484	251	2	No	Yes	No	Dry	CA

TABLE 12.1 (continued)
Study species and life-history traits of migratory landbirds captured during autumn migration at Manomet, Massachusetts, 1969–2012.

Common name	Scientific name	No. years	n	Mean dep. date	Brood	Farm	Tropics	Sex. dim.	Winter hab.	Winter region
Ovenbird	*Seiurus aurocapilla*	44	643	252	1	No	Yes	No	Woods	CA
Northern Waterthrush	*Parkesia noveboracensis*	44	900	250	1	No	Yes	No	Wet	CA
Black-and-white Warbler	*Mniotilta varia*	44	1,085	250	1	No	Yes	Yes	Woods	CA
Nashville Warbler	*Oreothlypis ruficapilla*	44	640	268	1	No	Yes	Yes	Woods	CA
Common Yellowthroat	*Geothlypis trichas*	44	2,044	256	2	Yes	Yes	Yes	Wet	CA
American Redstart	*Setophaga ruticilla*	44	3,787	254	1	No	Yes	Yes	Woods	CA
Magnolia Warbler	*Setophaga magnolia*	44	793	262	1	No	Yes	Yes	Woods	CA
Blackpoll Warbler	*Setophaga striata*	44	7,768	272	1	No	Yes	Yes	Woods	SA
Black-throated Blue Warbler	*Setophaga caerulescens*	42	649	269	2	No	Yes	Yes	Woods	CA
Myrtle Warbler	*Setophaga c. coronata*	44	20,204	293	2	No	No	Yes	All	NA
Black-throated Green Warbler	*Setophaga virens*	44	338	267	1	No	Yes	Yes	Woods	CA
Wilson's Warbler	*Cardellina pusilla*	44	855	256	1	No	Yes	Yes	Woods	CA
Yellow-breasted Chat	*Icteria virens*	44	1,099	262	2	No	Yes	Yes	Dry	CA
Eastern Towhee	*Pipilo erythrophthalmus*	44	868	273	2	No	No	Yes	Dry	NA
Song Sparrow	*Melospiza melodia*	44	2,959	272	3	Yes	No	No	Dry	NA
Swamp Sparrow	*Melospiza georgiana*	44	1,106	286	2	No	No	No	Wet	NA
White-throated Sparrow	*Zonotrichia albicollis*	44	8,298	285	1	No	No	No	All	NA
Slate-colored Junco	*Junco h. hyemalis*	44	2,205	293	2	No	No	Yes	Dry	NA
Northern Cardinal	*Cardinalis cardinalis*	44	1,646	278	2	Yes	No	Yes	All	NA
Baltimore Oriole	*Icterus galbula*	41	734	241	1	Yes	Yes	Yes	Woods	CA
American Goldfinch	*Spinus tristis*	42	426	292	1	Yes	No	Yes	Dry	NA

NOTES: Number of years refers to how many years this bird was captured in autumn over the 44-year span and n is the total number of captures over the length of the study. Mean dep. date is the mean day of year (as days past 31 December) that this species was captured in autumn at Manomet across all years. Brood is the typical number of broods that a species has in a breeding season. Farm, tropics, and sex. dim. are dichotomous variables that refer to whether a species is a farmland breeder, winters in the tropics, or exhibits sexual dichromatism. Winter hab. is the preferred habitat of species on their wintering grounds and includes the variables: "dry," which includes farmland, grassland, and other dry habitats; "woods," which includes forested areas; "wet" for wetlands; and "all" for species that can be found in a variety of terrestrial habitats. Winter region is the broadly defined region where species overwinter: NA for North America, CA for Central America, and SA for South America.

were based on typical occurrences for each species and the more conservative estimate was used if factors were variable among different regions. We conservatively assigned wintering region to each species as North, Central (including Caribbean Islands), or South America, which also served as a crude surrogate of migration distance. For wintering habitat, we designated species as wintering in one of four general habitats: woods/forest, farmland/grassland/dry habitats, wetlands, or all terrestrial habitats. The four habitats are consistent with previous publications using the same data set (Miller-Rushing et al. 2008b). Species was included as a random variable. Linear mixed-effects modeling allows for data to be input as a panel, enabling analysis of data over time, as well as with respect to the chosen variables. Akaike's information criterion (AIC) was used to assess model strength, and 22 combinations of variables were systematically tested and dropped or switched until the model with the lowest AIC was identified.

Last, we compared timing of migration at Manomet and at Powdermill Nature Reserve in Rector, Pennsylvania, by regressing slopes of change over time in species' passage dates at each location (Van Buskirk et al. 2009). The analysis included both significant and nonsignificant responses because significance values were not available for the published results of Van Buskirk et al. (2009).

RESULTS

Multivariate Results

The full multivariate model included all possible variables and traits. The strongest model, based on AIC, was the following:

Fall migration date = temperature + tropics + wintering habitat + leaf color + broodedness + captures × species

This model, therefore, is the most robust and represents the best model given all combinations of biotic and abiotic predictor variables (Table 12.2).

In this model, the relationship between fall migration of all species and temperature was positive, such that birds depart 0.43 days later for each 1°C increase in temperature (SE = 0.42, t = 1.01). Species that winter in the tropics migrate ~ 21 days before their north temperate

TABLE 12.2

Model selection for stopover dates of migratory landbirds as a function of different environmental factors and life-history traits.

Stopover date ~	AIC	ΔAIC
a. Temperature + tropics + wintering habitat + leaf color + broodedness + captures × species	5109	0
b. Temperature + wintering region + leaf color + broodedness + captures	5122	13
c. Temperature + NAO index + wintering region + leaf color + broodedness + captures × species	5123	14
d. Temperature + NAO index + wintering region + leaf color + captures × species	5126	17
e. Temperature + tropics + leaf color + broodedness + captures × species	5127	18
f. Temperature + wintering habitat + leaf color + leaf drop + broodedness + captures × species	5128	19

counterparts. Habitat generalists that winter in "all" terrestrial habitats are the latest group of fall migrants, with species in each of the other habitats—wet, dry, and woods—migrating earlier by 4.24, 4.44, and 10.31 days, respectively. Positive relationships also exist between migration date and the date of leaf color change ($\beta = 0.11$, SE = 0.07, t = 1.51) and between migration date and broodedness ($\beta = 1.02$, SE = 5.09, t = 0.20). In other words, migration occurs earlier when leaves are changing color earlier, and in species that are more likely to have only one brood. None of the categorical explanatory variables such as wintering habitat or broodedness interacted with temperature (Tables 12.3 and 12.4).

Species Responses

The mean date of fall capture at Manomet for all species and over all years was 2 October. Baltimore Orioles (*Icterus galbula*) tended to be the first species to pass through Manomet with a mean fall migration date of 29 August (SE = 0.48, n = 734). Hermit Thrushes (*Catharus guttatus*) were generally the last to pass through, with a mean fall migration date of 21 October (SE = 0.25, n = 1844). Linear regression analyses demonstrated that 14 of the 37 species had stopover dates that have changed significantly over time (Table 12.5).

TABLE 12.3

Coefficients for the minimum AIC model explaining variation in stopover date versus temperature + tropics + wintering habitat + leaf color + broodedness + captures × species.

Variable	Estimate	SE	t value
Mean August–September temp.	0.43	0.42	1.01
Leaf color	0.11	0.07	1.51
Broodedness	0.74	5.02	0.15
Tropics	−21.46	5.90	−3.63
Winter habitat			
All: wet	4.24	12.72	0.39
All: dry	4.44	7.51	0.59
All: woods	10.31	7.08	1.51

NOTES: Positive coefficients indicate that the relationship between stopover date and the given variable was positive. For example, the relationship between stopover date of all species and temperature is positive, such that birds depart 0.43 days later for each 1°C increase.

TABLE 12.4

Variance, standard deviation, and correlation among groups.

Group	Variance	Std. Dev.	Corr.
Species	165.52	12.87	
Captures	0.0004	0.02	0.095
Residual	50.45	7.10	

Nine species are now recorded with significantly later movements during fall migration, with Blackpoll Warblers (*Setophaga striata*) exhibiting the greatest delay of ~0.3 days per year (SE = 0.04, n = 7768, P < 0.001). Stopover dates of five species were significantly earlier: Black-capped Chickadees (*Poecile atricapillus*, n = 28,535), Downy Woodpeckers (*Picoides pubescens*, n = 789), Eastern Phoebes (*Sayornis phoebe*, n = 654), Gray Catbirds (*Dumetella carolinensis*, n = 21,484), and Tufted Titmice (*Baeolophus bicolor*, n = 5,256). Downy Woodpeckers had the greatest adjustment, moving 0.5 days earlier per year (SE = 0.10, P < 0.001). Twenty-three species had stopover dates that did not change significantly: 15 of these species are trending toward slightly later stopover dates, and seven toward earlier dates; one species, Swamp Sparrow (*Melospiza georgiana*), has not changed its fall stopover date (n = 1,106).

Mean August and September temperature has significantly warmed during this period at a rate of approximately 0.04°C per year (P < 0.001). Mean August and September temperature is a significant factor in the passage dates of 14 species. Thirteen species move later in warm years, with the Slate-colored Junco (*Junco h. hyemalis*) exhibiting the greatest response to temperature, migrating 3.6 days later per 1°C increase in temperature (SE = 1.14, P = 0.003). Gray Catbirds are the only 1 of 14 species to migrate earlier in warmer years, and they do so at a rate of 1.1 days earlier per 1°C increase (SE = 0.49, P = 0.04). Species with later mean migration dates also appear to be the most responsive to temperature (Figure 12.1).

The NAO index was a significant factor for timing of migration in three bird species. When the NAO index is high, Black-throated Green Warblers (*Setophaga virens*, SE = 0.91, n = 338, P = 0.003) and Tufted Titmice travel later (SE = 0.75, n = 5,256, P = 0.009), possibly because the jet stream is weaker and temperatures are above average. In contrast, Black-and-white Warblers (*Mniotilta varia*) migrate earlier when the NAO index is high (SE = 0.80, n = 1,085, P = 0.01).

Captures of 28 species have declined significantly over the 44 years of this study, reflecting a decrease in passage cohort sizes of these species. Black-capped Chickadees have declined the most at 28 fewer captures per year (P = 0.006), and declines in Myrtle Warblers (*Setophaga c. coronata*) are also substantial, with 22 fewer captures per year (P < 0.001). Annual captures of Black-capped Chickadees are now ~7% of numbers at the start of the banding program: Captures averaged 1,392 ± 808SE per year in 1969–1973 versus 102 ± 37 captures per year in 2008–2012. Myrtle Warbler captures have decreased to 17% of historic rates: 979 ± 160SE versus 172 ± 31 captures per year. Carolina Wrens (*Thryothorus ludovicianus*) and Northern Cardinals (*Cardinalis cardinalis*) are the only two species whose captures have increased significantly, both by about one capture per year (±6 and 14, respectively, both P < 0.001). Only one Carolina Wren was captured in the first 5 years of banding, but they now average 25 captures per year. Northern Cardinal captures are almost eight times greater now than they were in the past with 8 versus 62 captures per year.

Long-term changes in plant phenology over 1991–2010 were not significant for dates when at least 90% of leaves changed color (P = 0.20) or there was at least 90% leaf drop (P = 0.20). However, the relationship between leaf color

TABLE 12.5

Linear regression of captures per year versus day of season, and mean capture date versus
annual variation in environmental factors of migratory landbirds captured
during autumn migration at Manomet, Massachusetts, 1969–2012.

Common name	Captures/year	Year	Temp: Aug., Sept.	NAO: June, July, Aug.	Leaf color	Leaf drop
Downy Woodpecker	−0.53**	−0.53***	−2.76	2.57	0.38	−0.32
Eastern Phoebe	−0.01	−0.23*	−0.01	1.78	0.21	−0.83
Blue-headed Vireo	0.03	0.11	2.88**	−0.30	0.46*	0.45
Red-eyed Vireo	−2.01***	0.14**	1.90**	−0.21	0.12	0.18
Blue Jay	−1.61*	−0.02	1.54	0.86	0.59	0.75
Black-capped Chickadee	−28.10**	−0.31*	−1.28	0.57	0.55	1.30
Tufted Titmouse	1.66	−0.21**	−0.57	2.06**	0.35	0.24
Brown Creeper	−1.53***	0.07	1.56*	0.53	0.34	−0.17
Carolina Wren	0.71***	0.05	0.11	−0.06	−0.10	−0.25
Golden-crowned Kinglet	−2.49***	0.09	0.94	−0.41	0.53*	0.56*
Ruby-crowned Kinglet	−1.12***	0.08	2.00**	−0.04	0.27	0.32
Veery	−0.44**	0.11	0.42	−0.10	0.05	−0.02
Swainson's Thrush	−1.14***	0.23***	2.31*	−0.73	−0.05	0.07
Hermit Thrush	−0.44*	0.15**	2.16**	−0.56	0.12	0.14
American Robin	−5.58***	0.17	2.97*	0.21	0.29	0.93*
Gray Catbird	−4.32**	−0.16***	−1.05*	0.69	0.19	−0.07
Ovenbird	−0.37***	−0.05	0.57	1.28	0.14	0.56
Northern Waterthrush	−0.82***	0.02	1.13	0.48	0.15	0.41
Black-and-white Warbler	−1.08***	0.12	−1.09	−2.09*	−0.16	0.03
Nashville Warbler	−0.26*	0.16	0.94	−0.11	0.24	−0.13
Common Yellowthroat	−1.83***	0.21***	1.85*	−0.51	0.05	−0.15
American Redstart	−4.19***	0.11*	1.66**	−0.22	0.30*	0.25
Magnolia Warbler	−0.31**	0.13	−0.25	0.24	−0.55	0.09
Blackpoll Warbler	−10.05***	0.27***	3.25***	0.20	0.57*	0.48
Black-throated Blue Warbler	−0.02	0.09	1.66	1.00	−0.09	−0.42
Myrtle Warbler	−22.05***	0.10	2.77**	1.14	0.33	0.34
Black-throated Green Warbler	−0.15**	0.07	2.41	3.08**	−0.54	0.12
Wilson's Warbler	−0.66***	0.16*	1.36	0.73	0.05	0.09
Yellow-breasted Chat	−0.91***	−0.05	1.41	0.71	0.55	0.23
Eastern Towhee	−1.25***	0.16	1.96	1.54	0.07	−0.06
Song Sparrow	−1.72***	0.10	1.54	1.10	0.06	−0.06
Swamp Sparrow	−0.43*	0.00	0.77	0.21	0.12	−0.16
White-throated Sparrow	−1.89	0.16*	2.32*	0.06	0.13	0.17
Slate-colored Junco	−2.70***	0.18*	3.58**	0.69	0.51	0.17
Northern Cardinal	1.44***	0.03	0.96	−0.99	−0.22	−0.06

TABLE 12.5 (continued)
TABLE 12.5 (continued)
Linear regression of captures per year versus day of season, and mean capture date versus annual variation in environmental factors of migratory landbirds captured during autumn migration at Manomet, Massachusetts, 1969–2012.

Common name	Captures/year	Year	Temp: Aug., Sept.	NAO: June, July, Aug.	Leaf color	Leaf drop
Baltimore Oriole	−0.65**	−0.03	−1.15	−1.27	0.04	−1.07
American Goldfinch	0.03	0.16	−0.71	−3.43	−0.56	0.55

NOTES: Linear regression of captures per year versus time represent long-term changes in population cohort size over the 44-year time span; positive values indicate population increases, whereas negative values are declines. The remaining five columns display the relationship between mean capture date and each explanatory factor. Positive values indicate that stopover dates have become later in recent years, with increasing temperature, with increasing NAO index values, with later leaf color change or later leaf drop.

Level of significance: * = P < 0.05, ** = P < 0.01, *** = P < 0.001.

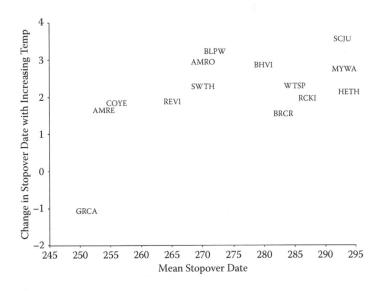

Figure 12.1. The relationship between mean stopover dates and changes in stopover date with August + September temperature for species demonstrating significant changes in stopover with temperature. The x-axis represents the mean stopover date of the species over all years, and the y-axis represents the slope of stopover date and temperature, days °C⁻¹. Species abbreviations are standardized codes: BHVI (Blue-headed Vireo), REVI (Red-eyed Vireo), BRCR (Brown Creeper), RCKI (Ruby-crowned Kinglet), SWTH (Swainson's Thrush), HETH (Hermit Thrush), AMRO (American Robin), GRCA (Gray Catbird), COYE (Common Yellowthroat), AMRE (American Redstart), BLPW (Blackpoll Warbler), MYWA (Myrtle Warbler), WTSP (White-throated Sparrow), and SCJU (Slate-colored Junco).

and August–September temperatures was significant ($r^2 = 0.37$, P = 0.006), as is the relationship between leaf color and leaf drop ($r^2 = 0.20$, P = 0.045). Leaf drop was not significantly correlated with temperature (P = 0.17). When these variables of leaf color and drop were regressed with fall migration dates, four species displayed a significant positive correlation with leaf color change and two species with leaf drop; these birds migrate later when leaf color and drop occur later. Fall migration dates for Golden-crowned Kinglets (*Regulus satrapa*) were correlated positively with both leaf color (0.5 days/°C, P = 0.02) and leaf drop (0.6 days/°C, P = 0.04).

Dates of fall migration for different species did not vary based on broodedness (P = 0.28), wintering habitat (P = 0.47), whether they are farmland breeders (P = 0.30), or sexual dichromatism (P = 0.99). Species that winter in the tropics migrated an average of 18 days earlier than north temperate species (P < 0.001). Similarly, there is a significant difference in the migration dates of species based on wintering ground (P < 0.001). Species that winter in Central America migrate earliest (mean departure date = 17 September), followed by species that winter in South America (23 September), and species that remain in North America were last to migrate in fall (6 October).

Analysis of data for species captured at the banding station at Powdermill Nature Reserve (Van Buskirk et al. 2009) shows generally similar patterns to our findings at Manomet. Some species changed over time and in relation to temperature, and long-distance migrants migrated earlier in the fall than short-distance migrants. We regressed slopes of change in stopover dates over time for the same set of study species at Powdermill (date of 50% capture, Van Buskirk et al. 2009) versus slopes for mean stopover at Manomet, but responses were not related (r² = 0.03, df = 26, P = 0.39). The analysis included nonsignificant responses for Powdermill, but it is clear that species exhibit spatial variability in the timing of their fall migrations and their responses to changes in temperature. The direction and degree of change in stopover date for the same species between these two locations were not the same.

DISCUSSION

Analysis of 44 years of banding data in Manomet reveals that about one-quarter of species are migrating later in the fall now than they did in the past. An additional five species (14%) are migrating earlier in the fall than they did previously. Comparisons between sites indicate that our results are not necessarily predictive of changes elsewhere, particularly farther inland and at other locations.

Manomet has warmed significantly in recent decades; 14 species have responded to temperature changes with altered fall passage dates, nine of which are species whose fall migration dates have changed over time. Other species traits, such as wintering region, broodedness, and wintering habitat, affected the sequence of migration dates, with some groups tending to migrate earlier or later, but none of the factors affected migratory responses to warming temperatures.

Our multivariate model indicated that birds tend to migrate later in years when leaves senesced later in the season. Few studies have incorporated fall plant phenology into research on timing of bird migration. Our result could reflect the tendency of birds to migrate and leaves to change color later in warmer autumns. Alternatively, birds may utilize the visual cue of leaf color change or associated cues, such as declines in fruit and insect abundance, along with temperature and other cues as a signal to migrate in the fall. Many of the trees in this region begin changing color in mid-September, a couple of weeks before the peak of migration at Manomet, and leaf coloration could provide a cue for some species, particularly those that migrate late in the season. Furthermore, if bird migration is correlated with leaf senescence, it could provide a way to estimate fall migration through future research utilizing remote sensing imagery.

Delays in autumn passage could be due to delays in departure from the breeding grounds, decreased speed of migration, increased time spent at stopover locations, or a combination of factors (Gordo 2007). Any of these factors could result in changes in synchrony between migratory birds and their autumn food sources. However, the implications of our results for food synchrony are unclear. Information is limited regarding specific food preferences of particular birds, fall phenology of key foods, and whether food supply is a limiting factor during fall migration. We know surprisingly little about temporal variation in fruit, seed, and insect abundance or their nutritional value. In some cases, fruits are maturing earlier in response to warming temperatures (Menzel and Dose 2005). However, some insects appear to delay diapause in warm falls (Altermatt 2010, Stoeckli et al. 2012). Given the variability we found in changes in fall migration phenology, it is possible that temporal mismatches are occurring in fall, as they may be in the spring (Both et al. 2006), but the patterns are difficult to identify. Seasonal phenology of food resources used in autumn migration is an area ripe for future research.

Our results are consistent with the expectation that some species may spend more time at breeding grounds in warmer years. Migrants of 21 bird species are captured at Manomet in the spring and fall (Miller-Rushing et al. 2008b), in sufficient numbers to examine changes in the length of time spent at breeding grounds. A comparison of spring and fall mean migration dates showed that only 1 of 21 bird species, Gray Catbirds, have significantly changed migration dates in both spring and fall.

Gray Catbirds are migrating through Manomet earlier in both seasons now than in the past. In terms of response to temperature, migration dates of only four species were correlated with temperature in the months preceding both spring and fall migration (Table 12.6). Gray Catbirds migrate earlier in both warm springs and falls. American Redstarts (*Setophaga ruticilla*), Hermit Thrushes, and White-throated Sparrows (*Zonotrichia albicollis*) respond significantly to spring and fall temperature, migrating earlier in the spring and later in the fall when the temperatures are warmer before migration; yet, the migration times of these species have not significantly changed over time. In warm years, the three species may be spending more time at breeding grounds, but the response has not been strong enough to be significant. The results must be interpreted cautiously because the spring and fall migration cohorts at Manomet are not necessarily from the same breeding populations, and the fall migration cohort contains a high proportion of first-year birds that are absent during spring migration (Brooks 2008, Finnegan 2008).

Delays in autumn passage may be indicative of changes in departures from the breeding grounds, but passage dates are driven by local temperatures for more species than were linked to the large-scale patterns of NAO. Links to local temperatures suggest that speed, stopover duration, and other events driven by local dynamics may play a more prominent role in passage than departure dates do. Future studies tracking the departures of individual breeding birds from their summer habitat to their wintering grounds could help to separate the effects of departure dates and local drivers on changing passage dates.

Last, we found that trends in fall migrations over time differed between cohorts of the same species at Manomet and Powdermill despite a relatively close geographical proximity. Our results could reflect differences in species' responses to climatic variability between the two sites. The differences between the two sites could also be influenced by differences in the breeding populations and ages of birds that pass through each location in the fall (Brooks 2008, Finnegan 2008). As a coastal site, Manomet has a higher proportion of first-year birds (71% of birds banded in 2007) than Powdermill (61% of birds banded in 2007; Brooks 2008, Finnegan 2008), and differences in age composition could influence migration timing at different sites and analytical comparisons between sites. A similar lack of correspondence for spring migration dates of birds was reported in a comparison of three sites in Massachusetts (Miller-Rushing et al. 2008c).

It is evident that the fall migration dates of birds captured at banding stations is changing over time and in response to a changing climate. To understand temporal changes in capture data from systematic bird banding stations, it is critical to determine where these birds spend their breeding season and where they overwinter, as well as the timing of departure from breeding areas and arrival at winter home ranges.

TABLE 12.6
Comparison of spring and fall migration at Manomet.

Species	Metric	Spring	Fall
Hermit Thrush	Temp	−1.69	2.16
Gray Catbird	Year	−0.12	−0.16
Gray Catbird	Temp	−1.09	−1.05
American Redstart	Temp	−1.13	1.66
White-throated Sparrow	Temp	−1.16	2.32

NOTES: Spring values are taken from Miller-Rushing et al. (2008b). The values for the metric "Year" are the slope of mean capture dates for the respective season and year, days year⁻¹, and for "Temp" are the slope of mean capture dates and mean temperature for the months preceding migration, days temp⁻¹. Only species with significant responses for both seasons are included here.

ACKNOWLEDGMENTS

We are deeply grateful to the dedicated staff and volunteers of the Manomet Center for Conservation Sciences for their tremendous banding efforts over the years. We are also thankful to C. McDonough MacKenzie, A. Miller-Rushing, and C. Polgar Perriello for helpful comments and suggestions.

LITERATURE CITED

Altermatt, F. 2010. Climatic warming increases voltinism in European butterflies and moths. Proceedings of the Royal Society of London B 277:1281–1287.

American Ornithologists' Union (A.O.U.). 1998. Check-list of North American birds. American Ornithologists' Union, Washington, DC.

American Ornithologists' Union (A.O.U.). [online]. 2012. Fifty-third supplement to the American Ornithologists' Union check-list of North American birds. Auk 129:573–588.

Bates, D., M. Maechler, and B. Bolker. [online]. 2012. lme4: linear mixed-effects models using S4 classes. R package version 0.999999-0: <http://CRAN.R-project.org/package=lme4>

Both, C., S. Bouwhuis, C. M. Lessells, and M. E. Visser. 2006. Climate change and population declines in a long-distance migratory bird. Nature 441:81–83.

Both, C., and M. E. Visser. 2001. Adjustment to climate change is constrained by arrival date in a long-distance migrant bird. Nature 411:296–298.

Brooks, E. W. 2008. Atlantic flyway review: region III (western ridge)—fall 2007 report. North American Bird Bander 33:192–202.

Chesser, R. T., R. C. Banks, F. K. Barker, C. Cicero, J. L. Dunn, A. W. Kratter, I. J. Lovette, P. C. Rasmussen, J. V. Remsen Jr., J. D. Rising, D. F. Stotz, and K. Winker. 2012. Fifty-third supplement to the American Ornithologists' Union check-list of North American birds. Auk 129:573–588.

Cotton, P. A. 2003. Avian migration phenology and global climate change. Proceedings of the National Academy of Sciences of the USA 100:12219–12222.

Ehrlich, P., D. S. Dobkin, and D. Wheye. 1988. Birder's handbook: a field guide to the natural history of North American birds. Simon & Schuster, New York, NY.

Ellwood, E. R., S. A. Temple, R. B. Primack, N. L. Bradley, and C. C. Davis. 2013. Record-breaking early flowering in the eastern United States. PLoS One 8:e53788.

Elsner, J. B. 2003. Tracking hurricanes. Bulletin of the American Meteorological Society 84:353–356.

Finnegan, S. 2008. Atlantic flyway review: region I (northeast)—fall 2007 report. North American Bird Bander 33:86–93.

Gordo, O. 2007. Why are bird migration dates shifting? A review of weather and climate effects on avian migratory phenology. Climate Research 35:37–58.

Gordo, O., L. Brotons, X. Ferrer, and P. Comas. 2005. Do changes in climate patterns in wintering areas affect the timing of the spring arrival of trans-Saharan migrant birds? Global Change Biology 11:12–21.

Gordo, O., and J. J. Sanz. 2006. Temporal trends in phenology of the honey bee Apis mellifera (L.) and the small white Pieris rapae (L.) in the Iberian Peninsula (1952–2004). Ecological Entomology 31:261–268.

Hüppop, O., and K. Hüppop. 2011. Bird migration on Helgoland: the yield from 100+ years of research. Journal of Ornithology 152:25–40.

Jenni, L., and M. Kéry. 2003. Timing of autumn bird migration under climate change: advances in long-distance migrants, delays in short-distance migrants. Proceedings of the Royal Society of London B 270:1467–1471.

Jochner, S. C., I. Beck, H. Behrendt, C. Traidl-Hoffmann, and A. Menzel. 2011. Effects of extreme spring temperatures on urban phenology and pollen production: a case study in Munich and Ingolstadt. Climate Research 49:101–112.

Knape, J., N. Jonzén, M. Sköld, and L. Sokolov. 2009. Multivariate state space modeling of bird migration count data. Pp. 59–79 in D. Thomson, E. Cooch, and M. Conroy (editors), Modeling demographic processes in marked populations. Springer, New York, NY.

Lehikoinen, A., and K. Jaatinen. 2012. Delayed autumn migration in northern European waterfowl. Journal of Ornithology 153:563–570.

Lehikoinen, E., T. H. Sparks, and M. Zalakevicius. 2004. Arrival and departure dates. Pp. 1–31 in A. P. Møller, W. Fielder, and P. Berthold (editors), Birds and climate change. Academic Press, London, UK.

Lloyd-Evans, T. L. and J. L. Atwood. 2004. 32 years of changes in passerine numbers during spring and fall migrations in coastal Massachusetts. Wilson Bulletin 116:1–16.

MacMynowski, D. P., and T. L. Root. 2007. Climate and the complexity of migratory phenology: sexes, migratory distance, and arrival distributions. International Journal of Biometeorology 51:361–373.

Menzel, A. and V. Dose. 2005. Detecting impacts of anthropogenic climate change on terrestrial ecosystems' phenology by Bayesian concepts. Geophysical Research Abstracts 7 (08397).

Menzel, A., T. H. Sparks, N. Estrella, E. Koch, A. Aasa, R. Ahas, K. Alm-Kubler, P. Bissolli, O. Braslavska, A. Briede, F. M. Chmielewski, Z. Crepinsek, Y.

Curnel, A. Dahl, C. Defila, A. Donnelly, Y. Filella, K. Jatcza, F. Mage, A. Mestre, O. Nordli, J. Penuelas, P. Pirinen, V. Remisova, H. Scheifinger, M. Striz, A. Susnik, A. J. H. Van Vliet, F. E. Wielgolaski, S. Zach, and A. Zust. 2006. European phenological response to climate change matches the warming pattern. Global Change Biology 12:1969–1976.

Miller-Rushing, A. J., D. W. Inouye, and R. B. Primack. 2008a. How well do first flowering dates measure plant responses to climate change? The effects of population size and sampling frequency. Journal of Ecology 96:1289–1296.

Miller-Rushing, A. J., T. L. Lloyd-Evans, R. B. Primack, and P. Satzinger. 2008b. Bird migration times, climate change, and changing population sizes. Global Change Biology 14:1959–1972.

Miller-Rushing, A. J., R. B. Primack, and R. Stymeist. 2008c. Interpreting variation in bird migration times as observed by volunteers. Auk 125:565–573.

Møller, A. P. 2007. Interval between clutches, fitness, and climate change. Behavioral Ecology 18:62–70.

Møller, A. P., W. Fiedler, and P. Berthold (editors). 2010. Effects of climate change on birds. Oxford University Press, Oxford, UK.

O'Keefe, J. 2000. Phenology of woody species at Harvard Forest since 1990. Harvard Forest Data Archive: HF003, Petersham, MA.

Parmesan, C. 2006. Ecological and evolutionary responses to recent climate change. Annual Review of Ecology Evolution and Systematics 37:637–669.

Poole, A., and F. Gill (editors). 2002. The birds of North America, Academy of Natural Sciences, Philadelphia, PA.

Rappole, J. H., E. S. Morton, T. E. Lovejoy, and J. Rous. 1983. Nearctic avian migrants in the Neotropics. US Fish and Wildlife Service, Washington, DC.

Sibley, D. A. 2000. The Sibley guide to birds. Alfred A. Knopf, New York, NY.

Smith, S. B., and P. W. C. Paton. 2011. Long-term shifts in autumn migration by songbirds at a coastal eastern North American stopover site. Wilson Journal of Ornithology 123:557–566.

Sokolov, L. V., V. D. Yefremov, M. Y. Markovets, A. P. Shapoval, and M. E. Shumakov. 2000. Monitoring of numbers in passage populations of passerines over 42 years (1958–1999) on the Courish Spit of the Baltic Sea. Avian Ecology and Behaviour 4:31–53.

Sparks, T. H., and O. Braslavská. 2001. The effects of temperature, altitude and latitude on the arrival and departure dates of the Swallow Hirundo rustica in the Slovak Republic. International Journal of Biometeorology 45:212–216.

Stoeckli, S., M. Hirschi, C. Spirig, P. Calanca, M. W. Rotach, and J. Samietz. 2012. Impact of climate change on voltinism and prospective diapause induction of a global pest insect—Cydia pomonella (L.). PLoS One 7:e35723.

Thorup, K., A. P. Tøttrup, and C. Rahbek. 2007. Patterns of phenological changes in migratory birds. Oecologia 151:697–703.

Tøttrup, A. P., R. H. G. Klaassen, R. Strandberg, K. Thorup, M. W. Kristensen, P. S. Jørgensen, J. Fox, V. Afanasyev, C. Rahbek, and T. Alerstam. 2012. The annual cycle of a trans-equatorial Eurasian-African passerine migrant: different spatio-temporal strategies for autumn and spring migration. Proceedings of the Royal Society of London B 279:1008–1016.

Van Buskirk, J., R. S. Mulvihill, and R. C. Leberman. 2009. Variable shifts in spring and autumn migration phenology in North American songbirds associated with climate change. Global Change Biology 15:760–771.

Walther, G.-R. R., E. Post, P. Convey, A. Menzel, C. Parmesan, T. J. C. Beebee, J.-M. Fromentin, O. Hoegh-Guldberg, and F. Bairlein. 2002. Ecological responses to recent climate change. Nature 416:389–395.

Weidinger, K., and M. Král. 2007. Climatic effects on arrival and laying dates in a long-distance migrant, the Collared Flycatcher Ficedula albicollis. Ibis 149:836–847.

Wilson, W. H. 2012. Spring arrivals of Maine migratory breeding birds: response to an extraordinarily warm spring. Northeastern Naturalist 19:691–697.

FIGURE A.1   (See color insert). Black-throated Green Warbler (*Setophaga virens*), Pheasant Branch, Wisconsin, 3 May 2011 (spring migration). (Photo credit: Eric M. Wood).

FIGURE A.2   (See color insert). Chestnut-sided Warbler (*Setophaga pensylvanica*), Pheasant Branch, Wisconsin, 17 May 2014 (spring migration). (Photo credit: Mike McDowell).

FIGURE A.3 **(See color insert).** Yellow Warbler (*Setophaga petechia*), Magee Marsh Wildlife Area, Ohio, 2 May 2013 (spring migration). (Photo credit: Kimberly Hall).

FIGURE A.4 **(See color insert).** Jennifer McCabe releases a Black-capped Chickadee (*Poecile atricapillus*) at the Seawall banding station in Acadia National Park, Maine, 2010. (Photo courtesy University of Maine).

FIGURE A.5   (See color insert). Ruby-crowned Kinglet (*Regulus calendula*), Pittsfield, Massachusetts, 2014 (fall migration). (Photo credit: Ian Davies).

FIGURE A.6   (See color insert). Blackpoll Warbler (*Setophaga striata*), Amherst, Massachusetts, 2014 (fall migration). (Photo credit: Ian Davies).

FIGURE A.7   (See color insert). Dark-eyed Junco (*Junco hyemalis*), Amherst, Massachusetts, 2014 (fall migration). (Photo credit: Ian Davies).

FIGURE A.8   (See color insert). Le Conte's Sparrow (*Ammodramus leconteii*), Pheasant Branch, Wisconsin, 27 October 2012 (fall migration). (Photo credit: Mike McDowell).

STUDIES IN AVIAN BIOLOGY
Series Editor: Brett K. Sandercock

34. *Beyond Mayfield: Measurements of Nest-Survival Data.* Jones, S. L., and G. R. Geupel, editors. 2007.

35. *Foraging Dynamics of Seabirds in the Eastern Tropical Pacific Ocean.* Spear, L. B., D. G. Ainley, and W. A. Walker. 2007.

36. *Status of the Red Knot (Calidris canutus rufa) in the Western Hemisphere.* Niles, L. J., H. P. Sitters, A. D. Dey, P. W. Atkinson, A. J. Baker, K. A. Bennett, R. Carmona, K. E. Clark, N. A. Clark, C. Espoz, P. M. González, B. A. Harrington, D. E. Hernández, K. S. Kalasz, R. G. Lathrop, R. N. Matus, C. D. T. Minton, R. I. G. Morrison, M. K. Peck, W. Pitts, R. A. Robinson, and I. L. Serrano. 2008.

37. *Birds of the US–Mexico Borderland: Distribution, Ecology, and Conservation.* Ruth, J. M., T. Brush, and D. J. Krueper, editors. 2008.

38. *Greater Sage-Grouse: Ecology and Conservation of a Landscape Species and Its Habitats.* Knick, S. T., and J. W. Connelly, editors. 2011.

39. *Ecology, Conservation, and Management of Grouse.* Sandercock, B. K., K. Martin, and G. Segelbacher, editors. 2011.

40. *Population Demography of Northern Spotted Owls.* Forsman, E. D., et al. 2011.

41. *Boreal Birds of North America: A Hemispheric View of Their Conservation Links and Significance.* Wells, J. V., editor. 2011.

42. *Emerging Avian Disease.* Paul, E., editor. 2012.

43. *Video Surveillance of Nesting Birds.* Ribic, C. A., F. R. Thompson III, and P. J. Pietz, editors. 2012.

44. *Arctic Shorebirds in North America: A Decade of Monitoring.* Bart, J. R., and V. H. Johnston, editors. 2012.

45. *Urban Bird Ecology and Conservation.* Lepczyk, C. A., and P. S. Warren, editors. 2012.

46. *Ecology and Conservation of North American Sea Ducks.* Savard, J.-P. L., D. V. Derksen, D. Esler, and J. M. Eadie, editors. 2014.

47. *Phenological Synchrony and Bird Migration: Changing Climate and Seasonal Resources in North America.* Wood, E. M., and J. L. Kellermann, editors. 2014.

T - #0402 - 101024 - C4 - 254/178/13 - PB - 9781138575783 - Gloss Lamination